Strategy and Politics

Strategy and Politics: An Introduction to Game Theory is designed to introduce students to the application of game theory for modeling political processes. This accessible text examines the phenomena that power political machineries—elections, legislative and committee processes, and international conflict—and answers fundamental questions about their nature and function in a clear, accessible manner.

Included at the end of each chapter is a set of exercises designed to allow students to practice the construction and analysis of political models. Although the text assumes only an elementary-level training in algebra, students who complete a course around this text will be equipped to read nearly all of the professional literature that makes use of game theoretic analysis.

Emerson M.S. Niou is Professor of Political Science at Duke University.

Peter C. Ordeshook is Professor of Political Science at California Institute of Technology.

Strategy and Politics

An Introduction to Game Theory

Emerson M.S. Niou
Duke University

Peter C. Ordeshook
California Institute of Technology

NEW YORK AND LONDON

First published 2015
by Routledge
711 Third Avenue, New York, NY 10017

and by Routledge
2 Park Square, Milton Park, Abingdon, Oxon, OX14 4RN

Routledge is an imprint of the Taylor & Francis Group, an informa business

Library of Congress Cataloging in Publication Data
Niou, Emerson M. S.
Strategy and politics : an introduction to game theory / Emerson Niou, Peter C. Ordeshook.
pages cm
1. Game theory. 2. Political science—Mathematical models. 3. Political science—Methodology I. Ordeshook, Peter C., 1942– II. Title.
JA72.5.N56 2015
320.01′5193—dc23 2014034874

ISBN: 978-1-138-01948-5 (hbk)
ISBN: 978-0-415-99542-9 (pbk)
ISBN: 978-1-315-73515-3 (ebk)

Typeset in Minion Pro
by Apex CoVantage, LLC

Printed and bound in the United States of America by
Edwards Brothers Malloy on sustainably sourced paper

Contents

1 Politics as a Game

1.1 Decision Versus Game Theoretic Decision Making

Over twenty five hundred years ago, the Chinese scholar Sun Tzu, in *The Art of War*, proposed a codification of the general strategic character of armed conflict and, in the process, offered practical advice for securing military victory. His advice is credited, for example, with having greatly influenced Mao Zedong's approach to conflict and the subtle tactics of revolution and the ways in which North Vietnam and the Viet Cong thwarted America's military advantages. The formulation of general strategic principles—whether applied to war, parlor games such as Go, or politics—has long fascinated scholars. And regardless of context, the study of strategic principles is of interest because it grapples with fundamental facts of human existence—first, people's fates are interdependent; second, this interdependence is characterized generally by conflicting goals; and, finally, as a consequence of the first two facts, conflicts such as war are not accidental but are the purposeful extension of a state's or an individual's motives and actions and must be studied in a rational way.

The Art of War is, insofar as we know, our first written record of the attempt to understand strategy and conflict in a coherent and general way. It is important, moreover, to recall that it was written at a time of prolonged conflict within an emerging China whereby the leaders of competing kingdoms possessed considerable experience not only in the explicit conduct of war, but also in diplomacy and strategic maneuver. As such, then, we should presume that it codifies the insights of an era skilled at strategy and tactics, including those of planning, deception and maneuver. This assumption, though, occasions a question: Although *The Art of War* was ostensibly written for the leader of a specific kingdom, what if all sides to a conflict have a copy of the book (or, equivalently, an advisor no less insightful than Sun Tzu)? How might our reading of Sun Tzu change if it is *common knowledge* that everyone studied *The Art of War* or its equivalent—where by "common knowledge" we mean that everyone knows that everyone has a copy of the book, everyone knows that everyone knows that everyone . . . and so on, ad infinitum. The assumption of common knowledge presumes that not only is each decision maker aware of the situation, but each is aware that the other is aware, each knows that the other knows, and so on and so forth, and after being told by Sun Tzu himself that the great trap to be avoided is underestimating the capabilities of one's opponents, it seems imperative that the implementation of his advice proceed with the presumption that common knowledge applies.

In the case of *The Art of War*, taking account of the possibility that both sides of a conflict have a copy of the book differentiates the social from the natural sciences. In physics or chemistry, including their practical applications, one does not assume that the scientist or engineer confronts a benevolent or malevolent nature that acts strategically to deliberately assist or thwart one's research or the application of natural laws as we understand them. Things might not function as designed, but only because our understanding or application of nature's laws is imperfect. In the social sciences, on the other hand, especially in the domain of politics, it is typically the case that individuals must choose and act under the assumption that others are choosing and acting in reaction to one's decisions or in anticipation of them, where those reactions can be either benevolent or malevolent.

Despite this fact, it is our experience that most readers of *The Art of War* implicitly or unconsciously (at least initially) take the view that the reader is the sole beneficiary of Sun Tzu's advice —that one's opponent is, much like nature, a "fixed target." This might have been a valid assumption in 225 BC China in the absence of the internet, printing presses and Xerox machines, but it is no longer valid given the worldwide distribution of the book, including having it as required reading in business schools and military war colleges. So a more sophisticated student of Sun Tzu's writings might suppose that one's opponents have read the book as well, and might then reasonably assume that their opponents' tactics and strategies conform to Sun Tzu's guidance. But suppose we take things a step further and try to interpret an opponent's actions not simply with the assumption that they've read the books with which we are familiar, but also that they know we are familiar with those books and that we are not only attempting to assess their tactics and strategies in light of the advice contained in *The Art of War*, but also that we are attempting to take into account the fact that they are attempting to take into account our familiarity with that advice.

If all of this sounds confusing, then referencing *The Art of War* as an introduction to this volume has served its purpose. Specifically, there are two general modes of decision making: *Decision Making Under Risk* and *Game Theoretic Decision Making*. In decision making under risk, one assumes, in effect, that although there may be inherent uncertainties associated with the consequences of one's actions due to chance events and the actions of others, probabilities can be associated with those events and actions and the right choice is the one that yields the greatest expected return, where that return can be expressed in monetary terms, as psychological satisfaction or whatever. In this world, one assumes that other decision makers, including those who might be opposed to your goals, have, in effect, a limited view and do not respond to the assumption of common knowledge. In other words, just as the engineer or natural scientist does not assume that nature has the capacity for logical thought, the notion of common knowledge plays little to no role in decision making under risk since here one sees opponents as non-strategic "fixed targets" whose likely actions can be guessed at on the basis of, say, past patterns of behavior, bureaucratic rigidities or simple stupidity.

In game theoretic decision making, in contrast, one assumes that one's opponents and other decision makers, in pursuit of their goals, take into account their knowledge of you, including the fact that you know that they know, etc. Other decision makers are no longer fixed targets. Now you must be concerned that, since they know you are aware of their past behaviors, they might try to confound your calculations by defecting in some way from whatever patterns their earlier decisions exhibit. And there is, moreover, an additional complication. Since in a game theoretic analysis we can also assume that they know you know their past history, they also know you know they might have an incentive to thwart your calculations by not changing past patterns of behavior at all. But, since you also know that they know that you know they might consider sticking to past patterns . . . and so on, ad infinitum once again.

Game theoretic decision making attempts to untangle such seemingly endless and convoluted thinking and in the process to define what it means to be rational in interactive decision contexts. This volume, then, attempts to lay out the rudiments of game theoretic analysis as it can be applied to situations we label "political." Our specific objective, however, is not to provide a text on the mathematics of game theory. There are any number of excellent books available for that purpose, and the subject itself can be as dense as any branch of mathematics. Rather, our goal is to show that a game theoretic approach to understanding individual action is an essential component not only of being skillful at war, but also of understanding the less violent aspect of politics. However, rather than try to argue this point here, let us consider a series of examples that perhaps more clearly illustrate the difference between decision theoretic and game theoretic reasoning.

The Atomic Bomb and Japan: On the morning of August 6, 1945, a single plane (preceded by two weather reconnaissance aircraft), the Enola Gay, flew to and dropped its bomb on the city of Hiroshima. Ignoring the debate over whether this act was warranted or unwarranted with respect to the goal of ending a war, the question that concerns us here is: Why only one plane, which so easily could have been intercepted? The answer is that America's strategic planners assumed that if the Enola Gay had been part of a fleet of bombers, the Japanese would have attempted to intercept the raid with its ground based fighters. A single plane, on the other hand, would be far less threatening and draw far less attention. That calculation turned out to be correct—based on earlier bomb raids over its cities, strategic planners correctly assessed Japan's approach to air defense, and when the two scout planes turned back to the Pacific, city sirens sounded the "all clear" on the ground. The logic behind sending a single plane, then, on its deadly mission seems straightforward. But then, three days later, another solitary plane, Bockscar, flew to Nagasaki and dropped America's second atomic bomb, and the question for us is: Did the same strategic calculation in choosing between a lone plane versus a plane imbedded in a fleet apply to Bockscar?

We do not know precisely what calculations were made in deciding to deliver the second bomb via another lone aircraft as opposed to a fleet. But certainly the calculation this time had to be different from the one that sent the Enola Gay on its way. In the case of the Enola Gay, America's strategic planners could reasonably assume that the Japanese had no idea as to the destructiveness of its cargo and, thus, no reason to fear it any more than any previous lone aircraft over Japan. The response of Japan's air defense command could be predicted with near certainty. But circumstances changed markedly once the Enola Gay delivered its payload. Now, presumably, there were those in Japan who knew the potential of a lone bomber, and the American decision to proceed as before had to be justified by a different calculation—one that took into account what the Japanese might now assume about lone bombers and how they might weigh that danger against the costs of scrambling interceptors against it. Might the Japanese assume that the United States wouldn't be bold enough to again send a single bomber to drop any additional atomic bombs and instead would now try to disguise any subsequent use of its atomic arsenal by imbedding the plane carrying it in a fleet of bombers? In other words, America's strategic planners now had to concern themselves with the possibility that Japan's approach to air defense had changed in a complex way dictated by its best guess as to America's guess about Japan's response to the first bomb.

The decision to send a single plane to Hiroshima, then, was decision theoretic: Japan's likely response to one plane versus many could be determined by its previous actions. All a strategic planner needs to do is to calculate the probability that Japan would try to intercept a single plane versus the likelihood that, if imbedded in a fleet, it would intercept the fleet and successfully shoot down the specific plane carrying a bomb. The decision to send only one plane to Nagasaki, in contrast, required an assessment of what Japan might have learned about the potential lethality of a single plane, whether Japan might assume the Americans would employ the same tactic a second time, how that tentative assessment might impact America's tactical calculations, and how in turn Japan should respond to what it thinks America's response would be to Japan's reassessment of things. The decision to use a single bomber versus a fleet over Nagasaki, then, was a game theoretic one.

The Boston Marathon 2013: For a more contemporary example of decision versus game theoretic choice, consider the FBI's move to release a department store security camera video of the two brothers who planted bombs at Boston's 2013 marathon. As portrayed by the media, that decision was intended to elicit the public's help in identifying the terrorists, and indeed the video was soon plastered across the internet's social media. From this perspective, then, the FBI's action appears to be a strictly decision theoretic move to increase the likelihood that their suspects would be recognized and identified. But suppose we give the FBI's personnel more credit in assessing motives. Suppose they anticipated the released video going viral on the internet and knew the suspects would soon realize that their identities could not be kept hidden. Thus, if they planned any additional terrorist acts, both men knew they had better act quickly with little opportunity to plan carefully. In other words, suppose the FBI intended to "smoke out" their Russian suspects and induce them to be less careful than they might otherwise be if they assumed their identities could remain hidden for a time. It might have been the case, of course, that the two brothers knew the FBI was trying to smoke them out, but as committed jihadists, what choice did they have? Thus, by anticipating the terrorists' response, the FBI can be said to have acted with a game theoretic understanding of things. And this is precisely what happened. A day or two after the bombing, at least one brother, seemingly oblivious to the fact that his identity would soon be known, was seen partying at the college he'd been attending. But following the video's release, the two brothers, with bomb parts still unassembled in their apartment, tipped their hand by hijacking a car that led to a shoot-out with the police wherein one brother was killed and the other injured and captured soon thereafter.

Voters and Interest Groups in Three Candidate Elections: It isn't always easy to decide how to vote in a three-candidate plurality rule (first-past-the-post) election. The problem here is the possibility of wasting one's vote by casting a ballot for a candidate, however strongly preferred, who stands no chance of winning. If the candidates' chances are unequal, it might be wise to vote for one's second preference. In making this decision, then, a voter might, after perhaps talking things over with family and friends, consult the polls to determine whether his or her preferred candidate is competitive. But now suppose our voter is not an ordinary citizen but heads some highly visible interest group—a labor union or citizen action committee—and that he or she must decide who that group should endorse. The endorsement decision is similar to that of an ordinary voter in that the relative competitiveness of the candidates should be taken into consideration, but it is different in that any decision should also take into account the likelihood that the endorsement will not only influence more than a mere handful of voters but also perhaps the actions of other interest groups. If their endorsement carries some weight and impacts the election's competitiveness, then presumably it will impact the calculations of others who might attempt to influence the election's outcome. Some of that influence might benefit the candidate in question if it leads other groups to endorse the same candidate. But it might also work against that group's interests if it results in any increase in the endorsements received by other candidates. Thus, your decision as a solitary voter, made under the assumption of a "fixed electorate" as reflected in the polls, is decision theoretic because your decision hardly affects anyone. But leaders of influential interest groups must not only concern themselves with their immediate impact on the electorate, but also with the responses of other interest groups. The decisions of each such group, then, should be made on the basis of a game theoretic analysis that attempts to take into account the reactions and counter-reactions of other groups and, ultimately, of the electorate as a whole.

The Electoral College and Bloc Voting: The U.S. Constitution leaves the door open to any number of schemes for translating individual votes into Electoral College votes and the ultimate determination of who wins a presidential election. Presently, nearly all Electoral College votes are determined by a winner-take-all system wherein whoever receives a plurality of popular votes in a state wins all of that state's electors. That, however, is not how it has always been. So suppose we step back in time to when individual states, as in the late 18th and early 19th centuries, employed a variety of schemes for allocating electoral votes among competing candidates, including selecting them proportionally or electing them by pre-defined districts. Suppose further that you are an advisor to some state legislature and are trying to convince them that they ought to use today's winner-take-all scheme (perhaps the state in question is Virginia, perhaps it is the election of 1800 and perhaps you are Thomas Jefferson). If, for whatever reason, you assume that no other state is likely to change its system for selecting electors, your argument is likely to be a simple one that focuses on the added weight and attention your state might enjoy by not splitting its vote among a multitude of candidates. It also seems an essential step to forestall (again if you are Jefferson) the reelection of your rival, John Adams, since otherwise some of Virginia's vote will go, if not to Adams, then perhaps to some third candidate. Suppose, on the other hand, that you think it's possible (as in fact happened), that one or more states will respond to Virginia's actions. No doubt, your calculations will differ since now you must concern yourself with guessing which states are likely to change their method of selecting electors and whom those changes will benefit. Thus, if you take the myopic view of supposing that your advice can treat the electoral schemes of all other states as fixed, your analysis is a decision theoretic one. But if, as actually occurred, a state such as Massachusetts responds by altering its method of choosing electors in order to aid its favorite son, John Adams, you'd best consider a game theoretic analysis that takes into account the likely responses of all other states.

Crime Control, Police Patrols and Crime Voting: Although we may not be experts, we all pretty much know the parameters with which a professional burglar or car thief must deal when they set out to ply their craft so as to minimize the likelihood of getting caught and convicted. Successfully implementing a crime may be difficult, especially if one is not a professional, but the decisions one makes seem straightforward and include such rules of thumb as "work fast, work at night, wear gloves, work quietly and discreetly." In this case, crime prevention requires an effective police force, a competent staff of prosecutors and perhaps an education process whereby ordinary law abiding citizens learn how not to make themselves a criminal's easy target. Now consider a different system as practiced in feudal Japanese villages. If the culprit of some crime could not be identified, the villagers themselves voted on who they thought was guilty, and anyone receiving more than some pre-established threshold of votes was summarily banished without compensation or trial. For example, then, in Fuse village (currently Chiba prefecture) in 1696, three bales of rice stored for tribute were stolen. After 10 days of searching, the thief could not be identified with any certainty. The village chief, section leaders, and 131 peasants thus agreed to hold an election to identify the thief. As a consequence, the two highest vote-getters were banished from the village and three others who received one or two votes each were sentenced to house arrest.

In predicting the actions of a potential criminal in the usual case, a decision theoretic analysis would most likely suffice. Using experience and common sense, we can suppose that all but the stupidest criminals can calculate the approximate likelihood of detection and apprehension under varied circumstances. This calculation, in combination with an assessment of the value of the crime in the event one is not apprehended, should suffice in providing a criminal with a good idea as to whether and/or where to strike. Correspondingly, those who have no intention of being criminals but who also do not wish to be victims can make the same calculations and take some simple measures to protect themselves. Similar calculations might apply to the example of Japan's crime voting system, but here things are more complicated. Not only must potential criminals be concerned with the likelihood of being discovered, but people generally must worry about what might happen to them if the culprit is not identified. A person might be subjected to banishment simply because their neighbors and acquaintances don't like them or seek retribution for some otherwise long forgotten slight. It seems only reasonable to suppose that hatreds and grudges were often reflected in the ballots. One might anticipate, then, that one's general social behavior is likely to entail a good deal of concern about how one is viewed by neighbors and acquaintances. Indeed, one can readily imagine society evolving to exhibit a great deal of deferential and overly courteous behaviors, including seeing those prone to commit crimes acting with extreme deference in their everyday lives. In other words, this somewhat strange judicial system will most likely induce a variety of strategic calculations of the sort "Do I appear too deferential? Am I deferential enough?" To understand what, if any, equilibrium of social behaviors is likely to emerge in this case requires something more than a simple decision theoretic analysis.

Anti-Missile Deployment: In the mid 1990s the United States set itself upon a course of convincing Poland and the Czech Republic that it was in their interest to allow the U.S. to install anti-ballistic missile (ABM) technology on their territory. The argument offered by American strategists seemed straightforward: There are those in the Middle East intent on developing and deploying offensive missiles that could target Europe—Poland and the

Czech Republic included. Armed with statistics on costs and the assessments of the likelihood that states such as Iran were pursuing the development of long range offensive systems, the argument for ABM seemed simple and incontrovertible. What that initial argument lacked, however, was an assessment of Russia's response, not only to a blunting of the capabilities of its client states, but also of its own missile system, since it seemed evident that a European-based ABM system could be directed at them as well as Iran. The Russians, unsurprisingly, were strongly opposed to the installation of any ABM system close to their borders, especially one controlled by its post-World War II foe, NATO. They thereafter initiated a contentious negotiation with whoever occupied the White House that included the threat to reignite the arms race if an ABM system were installed. Ultimately, the White House capitulated, ostensibly because it was attempting to secure agreements with Russia on other issues, while both Poland and the Czech Republic were left in the lurch after having committed to supporting American policy.

American policy here, then, illustrates the consequences of making foreign policy decisions by ignoring or by not being fully cognizant of the reactions of other relevant actors. We appreciate that a detailed historical analysis might tell us that one Presidential administration was fully cognizant of those reactions and preferred to ignore them while a subsequent administration was naïve in placing a different value on the threat of Russian retaliation and/or cooperation on other issues. Nevertheless, this example reveals how a decision theoretic approach can yield one policy while an analysis that makes even a minimal attempt at anticipating the responses of others might yield something wholly different.

Grading on a Curve: When administering a final exam, an instructor generally has two choices: to grade in absolute terms (i.e, an A requires a final grade of 90 to 100, a B requires 80 to 89, and so on) or on a curve wherein the class average grade is set at, say, B, even if the class, in the instructor's judgment, does poorly. Suppose you are a student in some class wherein everyone has, by some miracle, received an identical grade of B on the midterm exam (or where, perhaps, the final grade is determined solely by one's performance on the final). If the instructor grades on an absolute basis, how hard you study for the final will, we can assume, depend on the things that might serve as distractions, on how well you think you've mastered the subject and on your personal motivation to strive for an A versus the possibility of your final grade slipping to a C. Alternatively, suppose the instructor grades on a curve. If no one studies and everyone again does equally well on the final, you and everyone else will maintain a grade of B. But if a number of other students study and you do not, they will raise the curve and some mid-term Bs, including yours, will become Cs (or worse). Thus, how hard you choose to study for the final will depend not only on personal motivation and distractions, but also on how hard you think your classmates will study. But of course, how hard they study will depend on how hard they think others will study, including you.

Thus, while decision theoretic reasoning is most likely sufficient to predict the study habits of a student when the instructor grades on a fixed basis, a game theoretic analysis is required to account for behavior when grades are curved on a relative basis. In the case of an instructor who grades on an absolute basis, the study habits and performance of your classmates is irrelevant to the ultimate determination of your grade. But in the case of grading on a curve, not only is your final grade a function of the performance of your classmates, but the effort you put into studying for the final depends on your assessment of their actions, and by logical extension, your assessment of their assessment of your actions, and so on. This is precisely the sort of circumstance addressed by game theory.

Presidential Power: If we look at the formal, constitutionally prescribed powers of the presidency in the United States, we see a position with few powers that cannot be checked by other political actors. An American president plays no formal role in amending the Constitution, his veto over legislative acts can be over-ridden by the legislature, he cannot make formal appointments without the approval of the legislature, he cannot implement treaties with foreign powers without Senate approval, there is no constitutional provision that the legislature must consider any legislative proposal he might offer, he has no authority over state and local level offices, and he is now precluded from serving more than two terms of office. Yet, the assertion that an American president holds one of the most domestically powerful offices in the world would seem self-evident. This view, though, appears to fly in the face of the fact that presidents of countries elsewhere hold far greater formal constitutional powers, including the authority to veto regional laws and to appoint and discharge regional executives. The supposition, though, that granting a chief executive strong constitutional powers necessarily renders that office powerful commits the error of confusing decision theoretic with game theoretic reasoning. We might conclude that wide ranging constitutional authority renders that office powerful, but only if we impose a strong *ceteris paribus* condition on the responses of all other political actors. On the contrary, constitutionally strong powers might merely energize opponents to resist those powers while at the same time leading anyone who holds that office to rely solely on those powers and nothing else. Weak constitutional authority, on the other hand, might lessen the natural opposition of others while simultaneously inducing those who hold that office to develop less formal avenues of authority. In the case of the American presidency, for example, those weak powers have encouraged presidents to cultivate political parties and the non-constitutionally prescribed ways in which they can exert power through persuasion and the leadership of a party. Thus, to understand the implications of alternative political institutional designs requires a game theoretic treatment rather than a decision theoretic one—a treatment that examines how individual motives and choices influence each other as opposed to one that assumes the motives and choices of people are somehow fixed.

West Point Honor Code and Chinese Self-Reporting: The Honor Code as it is practiced in America's military academies, such as West Point, requires that, among other things, students report any observed instances of cheating. The code provides for consequences, moreover, in the event that someone observes cheating but fails to report it. Thus, this implementation of the code parallels a Chinese version that dates back to the Zhou dynasty (1088–221 BC), wherein a person failing to report a violation of the code is punished more harshly than if he or she had themselves committed the violation. In this scheme, we not only prosecute anyone who commits a crime, but we prosecute in a doubly harsh way anyone who had knowledge of the crime but fails to report it to the authorities. And to make this system even more interesting (and akin to "turning state's evidence"), suppose the perpetrator of a crime, after being caught, identifies others who knew of his illegal actions and, in so doing, either receives a more lenient sentence or none at all.

With the distinction between decision and game theoretic reasoning in mind, we can perhaps see more clearly the difference between an honor code system that prosecutes only a person who commits a violation versus one that also prosecutes a person who fails to report a violation. Aside from the agony you might experience if a code's violator were a close friend, in the first case deciding whether to report a violation might hinge on your assessment of the violation's severity. But in the second case, you also have to be concerned that the violator, in seeking to reduce his penalty, will report things on his own—in which case, if you fail to report, you will be punished . . . possibly even more heavily than if you had been the one who originally violated the code. In the first case, then, your choice is a decision theoretic one; in the second, it is game theoretic because you must anticipate the actions of another person who is, at the same time, attempting to assess the likelihood that you will turn him in.

It is also interesting here to compare the Japanese system of crime voting with China's self-reporting system. In the Japanese case, a person's probability of being ostracized by his neighbors as a criminal depends only in part on whether or not she is guilty of the crime under investigation. It is not unreasonable to suppose that a good many persons were "wrongly" convicted merely because those around them deemed some aspect of their personality distasteful or disreputable. In response, one can readily imagine a system of social norms arising whereby acting in accordance with those norms avoids having such descriptive words as "arrogant," "unfriendly," "intemperate," "mean," "boisterous" and "combative" appended to one's character. However, the more fully those norms take hold, the more difficult it is to sort people by their degree of conformity to them, in which case, signaling one's conformity may require overt and accentuated actions such as ritualized bowing as if one were being presented to a monarch. The important point here, however, is that the evolution of such norms and their ultimate manifestation must be viewed as the consequences of people's strategic interactions. If, for instance, one bows not enough, then that might be taken as a violation of the norm and a potential basis for people to vote for you in some criminal investigation. On the other hand, bowing too deeply might be viewed as a sign that one is indeed over-compensating for some prior criminal actions. There follows, then, a complex evolutionary process wherein people, across generations, learn and then codify "proper" methods of social greeting. In the case of Maoist China, in contrast, a different pattern of social behavior is likely to emerge: Since the innocent must be concerned that a criminal might attempt to implicate them when caught, the best approach is to isolate oneself from society to minimize the chance of being associated with anyone who might be accused of criminal activity. Thus, in both Japan and China people must make game theoretic decisions in assessing the reactions of their acquaintances to their everyday actions: How much deferential behavior is too much because it raises suspicions versus how much is too little and marks me as an ungracious and disliked member of the community? Or, do I dare make any friends at all since almost anyone might be a reader of pornography or of banned literature and likely to try to save his own skin by fingering me as an accomplice should they be discovered?

Fighting a War with Allies: It might seem that in confronting Japan in WWII, America and Britain simply had to ensure the effective coordination of their actions and the efficient allocation of their resources. If so, then whatever was to be decided could be decided by the generals (or admirals), with perhaps the assistance of a staff skilled in organizing each country's industrial capacity. Aside from various inter-service rivalries, a decision theoretic approach aided by such tools as operations research would appear to be adequate to the task of directing the actions of the two allies. Things, however, were a bit more complicated and only partially influenced by the shared goal of Japan's unconditional surrender. Britain (or at least Churchill) was also concerned about maintaining (or resurrecting) its empire and thus favored military actions and an allocation of resources that facilitated the recapture of Malaya and Singapore, moving the Japanese out of Burma (Myanmar), and maintaining its control of India. The United States (or at least Roosevelt), in contrast, was wholly unsympathetic with this goal, and simple logistics seemed to dictate focusing its resources on a Pacific campaign. It was well understood, of course, with Britain focused on the German threat to its homeland, that the main burden of the war against Japan would be borne by the United States. Nevertheless, cooperation was essential and to sustain it at an efficient level often required negotiation and anticipating the likely responses of one's ally. Churchill, of course, had to make certain Britain pursued a strategy that kept the U.S. committed to its "Germany first" policy and that it didn't pursue its Asian and Southeast Asian goals in a way that left the American

> public to view it as simply another imperialistic power. And as America's input into the overall war effort increased and then surpassed Britain's, Churchill sought a strategy whereby it would remain a great power after the war. The U.S., for its part, needed whatever assistance Britain could supply, especially in airlifting supplies to China, along with the unflagging commitment of the other Commonwealth countries of Australia and New Zealand. And it understood as well that the reconstruction of Asia after the war would benefit from Britain's input. Thus, Anglo-American relations during the war could not be modeled in simple decision theoretic terms but were more akin to the give and take that often describe legislative coalitions and the trading of votes across legislation—processes that cooperative game theory seeks to address.

Although the political content of some of the preceding examples is minimal, each suggests that if all of politics entailed simple decision theoretic reasoning, politics most likely would be utterly boring. But politics and the processes that characterize it entail, virtually by definition, the interactions of people wherein the consequences of their choices depend on what others do, and what everyone does depends on what everyone else does or is expected to do. Which candidate wins an election depends on the character and actions of his or her opponents; which bills pass a legislature depend on what vote trades individual legislators might make across even disparate legislation; what international alliances form depend, at least in part, on an assessment of what counter-coalitions are likely to form and the actions of states absent from those alliances. In other words, individual decisions we might label *political* do not arise in a vacuum and are rarely predicated on the assumption that only one decision maker's actions are relevant. Politics, then, is inherently game theoretic and understanding political processes either from the perspective of explaining what has happened or from that of predicting what will happen necessarily requires understanding how participants perceive (or misperceive) the game(s) they are playing. And to do that requires that we understand how to represent and analyze those games, and here our examples give us some idea as to the components of that representation. Specifically, a careful description of each of the above scenarios requires at least the following:

1. The identities of relevant decision makers;
2. the choices confronting decision makers, including the order in which decisions (choices) must be made;
3. a specification of outcomes and the linkage between choices and outcomes;
4. each decision maker's preferences over the set of possible outcomes; and
5. the perceptions of each decision maker about the components of the game that concern him or her.

In the case of grading on a curve, for instance, the relevant decision makers are the students, the choice confronting each is how much or how hard to study, the outcomes are final grades, the linkage between choices and outcomes is dictated by the instructor's grading scheme, and preferences are, presumably, "a higher grade is preferred to a lower grade and, *ceteris paribus*, less effort devoted to studying is preferred to more work studying." And since we are ostensibly speaking of students who are at least semi-conscious of their educational environment, we can assume that their perceptions of things correspond to our description of them.

1.2 Preferences, Risk and Utility

In expanding on the preceding list of the things that comprise a potential game theoretic representation, the easiest place to begin is with individual preferences. So suppose we start with an abstract list of outcomes $\boldsymbol{o} = (o_1, o_2, o_3, \ldots, o_n)$. In fact, to begin with the simplest quantifiable possibility, suppose the *o*'s correspond to different amounts of money, where o_1 denotes a greater amount than o_2, o_2 corresponds to a greater amount than o_3, and so on. It seems reasonable to suppose now that a person, *ceteris paribus*, will prefer more money to less so that o_1 is preferred to o_2, o_2 is preferred to o_3, etc. Moreover, given this preference, we can also say that o_i is preferred to o_j provided only that $j > i$. In this instance, then, a person's preferences are *complete* (i.e., between any two outcomes, o_i and o_j, o_i is preferred to o_j, o_j is preferred to o_i or indifference holds between them) and *transitive* (i.e., if the person prefers o_i to o_j and prefers o_j to o_k, then he or she prefers o_i to o_k).

To this point, nothing seems exceptional, and if there were nothing else to consider when abstractly describing preferences, the reader could legitimately claim we have introduced the idea of complete and transitive preferences merely to add some academic jargon to the discussion. Unfortunately (or fortunately, depending on one's perspective), things can quite readily become more complicated. Consider, for example, what is likely to happen if one of the authors of this book is taken to an art museum and asked to state a preference between successive pairs of paintings. Given our somewhat pedestrian understanding of art, when shown paintings #1 and #2, we might say we prefer #1 because it has more blue in it. Then when shown paintings #2 and #3, we might indicate a preference for #2 because, while the intent of each artist is unintelligible to our eyes, we find #2's frame more appealing. Finally, when asked to choose between paintings #1 and #3, we cannot preclude the possibility that we would state a preference for #3 because we have yet to be exposed to the current self-proclaimed purveyors of fashion and lack an appreciation, as art, for a painting of a blue soup can.

One could write this example off as aberrant and assert that our models and theories of politics can be limited to those situations where people know their preferences. Of course, excepting the tautological assertion that people are said to know their preferences only when those preferences match our assumptions, we are left with the question as to how and when we know what other people's preferences might be. Matters grow even more confusing, though, when we try to be anthropomorphic about things and attribute preferences or goals to groups, such as when we seek to explain a state's foreign policies while treating a state as a unified entity. Consider, for instance, the problems one encounters with attempting to assess Britain's goals prior to the outbreak of WWI. It seems easy to focus on its treaty commitments with France, its commitment to Belgian sovereignty and its longstanding policy of working against any one country becoming predominant on the continent in explaining its commitment of troops to the defense of France. But there were confounding matters. First, an equally salient issue for Britain at the time was that of home rule for Ireland and the conflict between Northern Ireland and the South. Policy makers in London could not discount the possibility that maintaining peace there would require whatever military resources it might otherwise allocate to the Continent. Second, there was a simmering diplomatic conflict with Russia over Iran. Britain was converting its navy from coal to oil, which required Iran's oil resources. But Russia was also attempting to extend its influence there, and, if one looked at its history with respect to the expansion of its territory, perhaps its sovereignty as well. So why join in an alliance, via France, with Russia against Germany? Indeed, Germany could be viewed as a counterweight to Russia in the rapidly decaying Ottoman Empire and, in particular, in helping forestall Russia's longstanding designs on Constantinople. It was anything but clear at the time, both to outside observers and to some within Britain's government, how these concerns would play out in dictating policy. At a minimum, attributing coherent transitive preferences to Britain then was fraught with difficulty.

We will, in fact, have other more theoretically exact reasons for questioning the advisability of attributing goals to groups, but setting such things aside for the moment, consider another problem with the preceding representation of preference, which concerns the possibility that outcomes arise only up to some probability. To see the problem here, suppose you are asked how much you are willing to pay to play the following "game" denoted The St. Petersburg Paradox, named after Daniel Bernoulli's presentation of the problem and his solution in 1738 in *The Commentaries of The Imperial Academy of Science of St. Petersburg* (although the problem was first stated by his cousin, Nicholas, in 1713): A fair coin will be tossed and if it comes up heads, you will be paid \$2, and the game ends. But if it comes up tails, the coin will be tossed again, and if it comes up heads on that second toss, you will be paid \$4, and the game will then end. But if it comes up tails twice in a row, the coin will be tossed yet a third time, and so on, until a heads finally appears, so that if a heads first appears on the nth toss, you will be paid 2^n dollars. Usually, now, when people are asked how much they are willing to pay to play this game, few give an answer in excess of \$20. Consider, though, the game's *expected* dollar value. The probability of earning only \$2 is ½ (the probability that a heads appears on the first toss); the probability of earning \$4 is ¼ (the probability of a tails on the first toss times the probability of a heads on the second); . . . the probability of earning $\$2^n$ is $1/2^n$ (the probability of $n - 1$ tails followed by a heads), and so on. Thus, the *expected dollar return* is

$$\$2(1/2) + \$4(1/4) + \ldots + \$2^n(1/2^n) + \ldots = \$1 + \$1 + \ldots + \$1 + \ldots = \infty$$

That is, the expected payoff from this game expressed in dollars is infinite—an infinite summation of 1's. We seriously doubt, however, that most people who initially said they wouldn't pay more than $20 to play this game would, after shown this calculation, increase their willingness to pay by more than a few dollars (if anything at all).

Now consider a second observation about human behavior: The vast majority of homeowners buy insurance that protects them against the possibility of their homes burning down or of someone tripping on their basement stairs and suing for bodily injury. We also know that the big prize from state lotteries commonly achieve a value of upwards of eight or nine digits and that hundreds of thousands, if not millions, of people buy lottery tickets in the hopes of winning that mega-prize. It seems safe to assume, then, that there are a considerable number of people who buy both insurance and lottery tickets. However, in one instance (buying insurance) a person is exhibiting *risk averse* behavior with respect to money, while in the second instance (buying a lottery ticket) that same person is exhibiting *risk acceptant* behavior. In the case of insurance, people are spending money to avoid risk; in the case of a state lottery, people are spending money in pursuit of risk. And in both cases, the expected return on their "investments" is negative because neither insurance companies nor state lotteries are in the business of losing money. More formally, suppose a person is presented with a lottery that affords them a probability p of receiving $\$X$ and $(1 - p)$ of $\$Y$, where $X < Y$ and where the expected dollar value of the lottery is $pX + (1 - p)Y = \$Z$. If given a choice, now, between $\$Z$ with certainty versus playing the lottery, a risk acceptant person prefers the lottery while a risk averse person prefers the certainty of $\$Z$. Thus, if given a 50–50 chance of winning $100 versus nothing, a risk acceptant person would choose the lottery to an offer of being given $50 with certainty whereas a risk averse person would take the $50.

It is important to note that nothing said here contradicts the reasonable assumption that people prefer more money to less or negates the assumptions of transitivity and completeness since our discussion merely introduces a new consideration into people's choices; their assessments of risk. In the case of the coin toss, it is surely true that $\$2^n$ is a considerable amount of money when n is large, and it doubtlessly remains true that $\$2^n$ is preferred to $\$2^{n-1}$. But it is also true that the number paired with $\$2^n$, the probability of winning that amount $(1/2^n)$, is quite small for large n—so small in fact that a person might reasonably choose to ignore the term entirely as a feasible possibility. Alternatively, while buying insurance and lottery tickets may also entail small probabilities and considerable sums of money, the choices here are qualitatively different. In the case of insurance, one is trading the certainty of an insurance premium for a guarantee against the threat of a disagreeable lifestyle-changing loss whereas in the case of the lottery ticket, one is trading the certainty of a small loss (the cost of the ticket) for a potentially wondrously lifestyle changing gain. And we should not be surprised that people will somehow treat risk differently, depending on whether we are speaking of significant gains versus significant losses.

What we require, then, is a way of representing preferences that parsimoniously summarizes people's attitudes toward risk. That thing is the concept of *utility*. So suppose instead of our previous coin toss calculation we instead, for the left hand side of the equation, write

$$U(\$2)(1/2) + U(\$4)(1/4) + U(\$8)(1/8) + \ldots + U(\$2^n)(1/2^n) + \ldots$$

with the assumption that $U(\$2) < U(\$4) < U(\$8) < \ldots < U(\$2^n) \ldots$ That is, suppose we define the function $U(x)$ such that it increases monotonically as x increases and require that $U(x) > U(y)$ if and only if the person prefers x to y. Then surely we have not violated the assumption that a person prefers more money to less. But we have instead substituted for that statement the requirement that "the greater the amount of money, the greater is that person's *utility*."

If one asks now about the form of the function $U(.)$ it is here that we gain our handle on representing preferences over choices that entail risk or uncertainty. First, notice that there is no reason to suppose that $U(\$)$ is a *linear* function of money—that the utility of a $1 increase in wealth is invariant with the amount of money a person currently has in their wallet. Indeed, speaking for ourselves, we can honestly say that the utility of, say, ten million dollars, given our current salaries, would surely outweigh the utility or pleasure we'd likely derive if we were already in possession of a hundred million dollars. At least for the authors of this volume, when speaking of substantial sums, money exhibits diminishing marginal value (and we are open to anyone who might wish to test this hypothesis). At the same time, differences in the value of various sums of money will vary depending on the range over which those differences will apply. In the case of insurance and lottery tickets, suppose the potential loss of one's home from a natural disaster or the amount we can be sued equals $\$X$, and that an insurance policy that protects us against such a possibility costs $\$Y << \$X$. Thus, if the perceived probability of incurring that loss is p, we are then choosing $U(-\$Y)$ over the lottery $pU(-\$X) + (1-p)U(0)$. At the same time, suppose we are one of those people who, when the potential winnings from a state run lottery reach, say, $Z, run out and immediately spend $W on lottery tickets. If the probability of winning the lottery is q, our actions reveal a preference $qU(\$Z) + (1-q)U(-\$W)$ over $U(0)$. Regardless of how small p and q might be and regardless of how large Y and W are, there is nothing in the definition of preferences or utility that renders these two preferences necessarily inconsistent. Indeed, we would be surprised to learn that the average person is anything but risk averse when confronting lotteries that entail large potential losses and risk acceptant when dealing with lotteries that open the door to large potential gains.

With respect to the St. Petersburg Paradox, to represent the idea that increasing amounts of money exhibit diminishing marginal value, suppose for purposes of a numerical example that $U(\$X) = X/(X+1)$. With this assumption, the value of tossing a fair coin becomes, in expected utility terms, the sum of the infinite series

$$(2/3)(1/2) + (4/5)(1/4) + (8/9)(1/8) \ldots$$

which sums to approximately 0.77. If we now set $X/(X + 1) = 0.77$, we find that $X \approx 3.35$. Thus, if a person's utility for money abided by the admittedly ad hoc function $X/(X + 1)$, he or she should be willing to pay no more than \$3.35 to play our coin toss game (as opposed to infinity).

As a side note, we emphasize that no one has ever seen a utility function (aside from those which academics postulate on paper). Utility is a contrived concept developed for the purpose of representing people's preferences over risky alternatives. Thus, they serve much the same function as did the concept of the electron in the 19th century. No one had ever seen or at the time hoped to see an electron, but positing its existence (and here credit is due to Benjamin Franklin) explained the observable phenomena associated with positive and negative charge. This isn't to say that someday we will not find a better and more theoretically satisfying way to deal with the complexities of individual choice. It may be that the concept of a utility function will have a half-life no greater than that of the ether, which scientists once thought necessary to explain the transmission of light.

So restating our assumptions about individual preferences, the requirement that preference is a complete relation is akin to supposing that between any two outcomes, o_1 and o_2, either $U(o_1) > U(o_2)$ or $U(o_1) < U(o_2)$ or $U(o_1) = U(o_2)$. Transitivity, in turn, requires that if $U(o_1) > U(o_2)$ and $U(o_2) > U(o_3)$, then $U(o_1) > U(o_3)$. In other words, we require that U act much like the natural number system. There is, though, one additional requirement. Suppose $\mathbf{p} = (p_1, 0, 1 - p_1)$ is a lottery that assigns o_1 the probability p_1, o_2 the probability 0, and o_3 the probability $1 - p_1$, suppose $\mathbf{q} = (0, 1, 0)$ is a "lottery" that assigns the probability 0 to both o_1 and o_3, and certainty to o_2, and suppose $U(o_1) > U(o_2) > U(o_3)$. Then a person is said to prefer $\mathbf{p}$ to $\mathbf{q}$ (or equivalently, $U(\mathbf{p}) > U(\mathbf{q})$) if and only if $p_1U(o_1) + (1 - p_1)U(o_3) > U(o_2)$. In other words, we assume that a person's utility function can be defined so that it not only represents a person's *ordinal* preferences over outcomes, but allows us to represent that person's preferences over risky prospects in terms of his or her preferences across the specific outcomes over which the risk is spread.

Before we elaborate on the concept of a utility function and some problems with it, let us first consider some examples to better appreciate the role risk plays in individual decision making:

Risk, Traffic Control and China's Media: People's attitudes toward risk can sometimes go a long way in explaining government policies or in understanding how governments might manipulate individual choice by manipulating risk. Consider, for example, China's policy with respect to its mass media. If one questions newspaper editors, columnists, and so on there, you will quickly learn that Beijing's policy seems at times mercurial—sometimes it is harsh and at other times lenient, with no apparent pattern to its forbearance of criticism. It is, of course, entirely possible that this ebb and flow merely reflects the shifting fortunes of individuals in authority within the People's Republic of China (PRC) hierarchy. But consider the possibility that a mercurial policy is wholly intentional and intended to keep publishers, commentators, newspaper editors and the like in line. Here the argument would be that with no clearly delineated and seemingly coherent policy, the PRC leadership is essentially making the likelihood of punishment a lottery—and if, as is likely, those publishers, etc., are risk averse with respect to their careers, they will adhere to a more docile and constrained agenda than if the regime established a hard and fast rule. Under a stable and well-defined rule, we can expect that publishers will "walk up to the line" as closely as possible and even, in a few cases, cross over it for short periods of time, knowing precisely when they are in compliance with the government's policy. But under an uncertain or unclear rule, individuals will make risk-avoidant choices and adhere more carefully to Beijing's ultimate (but imperfectly publicly stated) goal. To see what we mean here in a different context, consider normal behavior on a Los Angeles freeway unencumbered by gridlock (yes, that happens on occasion). With a posted speed limit of 65 mph and a general understanding that the police rarely ticket anyone driving within 10 mph of that limit, most traffic will move along at 75 mph and a few drivers will push the envelope a bit. Suppose, instead, that the state highway patrol adopts the publicly stated policy of choosing a number at random between 70 and 80 every day at midnight, and, without publicly revealing that number, tickets everyone who exceeds it on that day. Now we would expect the same average speed limit—75 mph—that was the de facto limit before, but the question is: How might the behavior of drivers change? If drivers are risk averse with respect to receiving speeding tickets and the time lost spent by the side of the road while the officer writes the ticket, our answer should be a decrease in average driving speeds to something below 75.

A Crime Control Proposal: In the attempt to insure that people—and convicted criminals, in particular—are not unduly penalized merely because of their race, ethnicity or economic status, state and local governments in the U.S. have, over time, instituted an admittedly varied system of sentencing guidelines for judges wherein two people convicted of a similar offense receive the same or approximately equivalent sentences. Such guidelines, then, like a posted speed limit, define one's sentence for, say, a first, second and third conviction of automobile theft or shoplifting. But suppose instead of penalties being defined in terms of fines or length of time incarcerated in a prison we instead formulate those guidelines as a probability—a probability of being put to death. Thus, when convicted of some minor offense (e.g., failing to stop at a stop sign) the assigned probability will be small (hopefully VERY VERY small). But when convicted of murder, that probability will be significant, perhaps even 0.99 or 1.0.

Following conviction, a lottery will be conducted in accordance with the assigned (sentencing) probabilities with the outcome of the lottery dictating whether that person will be immediately set free or put to death. We don't know about the readers of this volume, but we do know that in such a system, the authors herein would most definitely be very careful about stopping at every stop sign encountered when driving.

China, Taiwan, the United States and Strategic Deterrence Through Risk: The case of the United States' policy of strategic ambiguity toward the dispute between China and Taiwan serves as an additional illustration of the strategic manipulation of risk. China believes that Taiwan is but a renegade province, that the island's reunification with the Mainland is a domestic issue, and that force may legitimately be used to compel reunification. There is widespread agreement, however, that China at the present time prefers the status quo to entering into a military conflict with the United States over Taiwan. Taiwan, on the other hand, refuses to acknowledge the People's Republic of China as the legitimate representative government for all of China, and seeks increased international autonomy. It is also commonly believed that Taiwan ultimately wants to be de jure independent from the PRC regime, but prefers its de facto political independent status to fighting China without American assistance. Most strategic analysts agree that the U.S. prefers the status quo to all other feasible outcomes. The U.S., then, faces a standard dual deterrence dilemma: Announcing a policy of under-commitment to Taiwan raises the incentive for China to secure reunification by force whereas a policy that over-commits to Taiwan's defense risks emboldening Taiwan to move recklessly toward independence, thereby compelling China to upset the military status quo. Beginning with President Eisenhower in the early 1950s, the U.S. policy response has, therefore, been to be strategically ambiguous about the conditions under which it will defend Taiwan. Specifically, the policy of strategic ambiguity, which derives formally today from the Taiwan Relations Act, acknowledges that there is only one China, that Taiwan is part of China, that resolution of the Taiwan issue is a domestic matter, but at the same time regards any security threat to Taiwan as a "grave concern" to the U.S. This seemingly contradictory policy has the effect of signaling that the U.S. has a definite stake in the outcome of the conflict but prefers to abdicate any "first move" to China or Taiwan while leaving both sides uncertain as to its ultimate response to any change in the status quo. Uncertain about how the U.S. will respond, neither China nor Taiwan has chosen to take decisive provocative action, and as long as the U.S. enjoys an asymmetrical power advantage over both China and Taiwan, strategic ambiguity offers a better shot at maintaining things as they are than does strategic clarity.

The preceding examples demonstrate that the sources of risk need not derive from the things we don't know or cannot predict about "nature," such as the weather, but also include those risks deliberately contrived as an element of individual strategy. A good part of this volume, then, will consider the manipulation of risk as a strategy in human interactions. But before we do so, we need to confront the fact that when it comes to the analysis of risk and our treatment of preferences, neither life nor the study of politics is ever simple. To wit, consider the following three outcomes:

o_1 = \$5 million
o_2 = \$1 million
o_3 = \$0

Now we would like the reader to carefully consider these two lotteries over the outcomes:

$$\boldsymbol{p} = (0.10, 0.89, 0.01) \text{ versus } \boldsymbol{q} = (0, 1, 0)$$

After making the bold attempt at putting yourself in a situation where you might actually get to make such a choice, which would you choose? Done thinking? Now give some serious thought to the following two alternative lotteries:

$$\boldsymbol{p}' = (0.10, 0, 0.90) \text{ versus } \boldsymbol{q}' = (0, 0.11, 0.89)$$

It has been our experience now that when students (and most everyone else, including ourselves) are asked to choose between ***p*** and ***q***, a good share chooses ***q*** after reasoning that "a bird in the hand is worth two in the bush." Or, "a lot can be done with one million dollars, and think of the regret if ***p*** were chosen instead and I ended up with nothing." Now, when asked to choose between ***p'*** and ***q'***, a reasonable share of people who initially chose ***q*** would choose ***p'*** over ***q'***, with the rationalization that "there isn't much difference between the likelihood of getting five million with ***p'*** as opposed to one million with ***q'*** so why not go for the big bucks?"

We would hardly label these two choices—***q*** over ***p*** and ***p'*** over ***q'***—as irrational or illogical; they might be the ones we ourselves would make. The problem here, though, is that no utility function is consistent with them. By indicating a preference for ***q*** over ***p***, it must be that

$$0.10U(\$5 \text{ million}) + 0.89U(\$1 \text{ million}) + 0.01U(\text{nothing}) < U(\$1 \text{ million})$$

or equivalently,

$$0.10U(\$5 \text{ million}) + 0.01U(\text{nothing}) < 0.11U(\$1 \text{ million})$$

However, the choice of ***p'*** over ***q'*** requires,

$$0.10U(\$5 \text{ million}) + 0.90U(\text{nothing}) > 0.11U(\$1 \text{ million}) + 0.89U(\text{nothing})$$

or, after rearranging terms

$$0.10U(\$5 \text{ million}) + 0.01U(\text{nothing}) > 0.11U(\$1 \text{ million}),$$

which directly contradicts the implication of a choice of $\boldsymbol{q}$ over $\boldsymbol{p}$. What's going on here? There are, we suppose, any number of possible explanations for such seemingly inconsistent choices, but the one that especially appeals to us is that the 0.01 difference in the likelihood of getting nothing between $\boldsymbol{p}$ and $\boldsymbol{q}$ is not being evaluated in the same way as is the difference in these likelihoods between $\boldsymbol{p}$' and $\boldsymbol{q}$'. In the first case, moving from $\boldsymbol{q}$ to $\boldsymbol{p}$ renders something that is impossible (getting nothing) possible whereas, in the second case, moving from $\boldsymbol{q}$' to $\boldsymbol{p}$', something that is likely merely becomes a bit more likely. In other words, the 0.01 difference between the pairs of lotteries of coming away empty handed, while treated identically in an algebraic manipulation, has a different psychological impact in the two sets of decisions.

It would seem, then, that not only is the value we place on objects wholly subjective and dependent on context (e.g., how we value a million dollars depends on whether or not we are already rich), but the probabilities we associate with risky choices are subjective, as well, and dependent on context. Unsurprisingly, this fact is widely recognized by decision theorists and considerable effort has been given to seeing what generalizations can be devised about *subjective probability*—probabilities that do not necessarily adhere to the rules we impose on them in mathematics and statistics. In this volume, however, we will make little use of that research because it only complicates our attempt to lay out the fundamentals of game theory as applied to politics and because very little of that research has yet been applied to the study of politics. Thus, throughout this volume we will treat probabilities in much the same way a statistician might by assuming that they obey the laws of algebra, that they do not fall outside of the range [0, 1], and that when summed across all feasible outcomes for a particular problem, that sum equals 1.0. Once again, though, we realize that individual behavior will often violate this assumption and it is important that we keep this fact in mind before we draw too strong a conclusion from any analytical exercise.

Why Vote?: To perhaps better appreciate the role subjective probabilities might play in politics, consider the simple act of voting in mass elections. At least in a democracy there is perhaps no more fundamental act of citizenship than that of casting one's ballot for or against a candidate, a party or some proposition on a referendum. But suppose we ask why people vote. This might seem a question with a simple answer—people vote because they want to increase the likelihood their preferred outcome prevails. Presumably, however, voting is a costly act. Even if one ignores the costs of becoming informed about the alternatives on the ballot, it requires an allocation of time to simply get to the polling station and in important elections people have been known to stand in line for hours waiting for their turn to enter the voting booth. So, proceeding to some minimal formalism, let P be the probability that your favored candidate in a two-candidate contest wins if you do not vote, P' be that probability if you do vote, U be the value you associate with seeing that candidate victorious, U' be the value associated with that candidate losing, and C the cost of voting. Then ignoring any algebraic complexities occasioned by the possibility of making or breaking ties between the candidates, the expected utility of not voting and not incurring the cost C is

$$E(\mathrm{NV}) = PU + (1 - P)U'$$

while the expected utility of voting is

$$E(\mathrm{V}) = P'U + (1 - P')U' - C.$$

Presumably, then, a person should vote if and only if $E(\mathrm{V}) > E(\mathrm{NV})$, or equivalently,

$$(P' - P)(U' - U) - C > 0.$$

Admittedly, now, for people who intensely prefer a candidate, the difference $U' - U$ may be considerable. But consider $P' - P$, which is the probability of being pivotal in the election in terms of making or breaking ties. Such a probability might not be small if we are considering an election in some village with 100 or so voters. But what of a national election with millions of voters? Surely the probability of being pivotal then fades to insignificance. Indeed, to say that your favored candidate is more likely to win if you vote for him rather than abstain is akin to saying you are more likely to hit your head on the moon by standing on a chair. But if $(P' - P)$ is essentially zero, and if C is consequential, then few people should vote. Since this prediction is clearly at odds with the data, we must ask again why people take the time to cast ballots in mass elections.

There are, in fact, two alternative explanations for non-zero turnout in mass elections (aside from those countries that fine people for not voting). The first hypothesis is that voting gives people a sense of fulfilled citizen duty—a warm feeling in the tummy, you might say. That is, we might suppose that people derive utility from the mere act of voting regardless of what impact their vote has on the election outcome. Equivalently, we might suppose that failing to vote is costly. Anyone living with a 12-year-old daughter or granddaughter who, on the basis of what she has been taught in school, regards her parents or grandparents as beneath contempt if they do not vote understands this cost. An alternative hypothesis (which does not preclude the first explanation from applying) is to suppose that people, subjected with mass media reports of how close an election might be, subjectively over-estimate $(P' - P)$. In fact, it is possible that people partake of a rather strange form of backwards causality, reasoning that "there are millions of people like me, and if I decide not to vote, they most likely would reach the same decision. But if I decide to vote, they will as well because their thinking will be the same as mine. Thus, my decision isn't merely impacting one vote but millions." Such thinking, of course, inflates the probability that one's vote is pivotal, and far be it for us to say that such reasoning cannot describe the inner workings of the mind.

Some academics object to the idea that people vote because of a sense of citizen duty, arguing that such a supposition merely makes voting rational by assumption and thus tautological. However, it is no more tautological to suppose that people vote because they have been socialized to value the simple act of voting for its own sake any more than to say a person buys a red as opposed to blue car because he or she likes red. Similarly, to suppose that people partake of a seemingly perverse view of causality when voting might seem strange, but in modeling people we best be prepared to learn that the human brain can entertain or seemingly employ forms of logic that defy logic. Be that as it may, there is one final qualification we need to add to our presentation of the concept of utility.

To this point we've made the assumption, when speaking of money, that $U(\$X) > U(\$Y)$ if and only if $X > Y$. However, suppose to the description of outcomes we append the date at which the money is received. Specifically, suppose $\$X$ is "One hundred dollars a month from today" and $\$Y$ is "$50 today." In other words, even when speaking of a simple thing like money we suppose outcomes are multidimensional and their descriptions include not only the quantity of money but also when it is received. Here we know that people's preferences vary. Some will prefer receiving the $50 immediately whereas others will prefer to postpone things provided they are compensated by a larger amount. In other words, people's preferences are defined not only over monetary amounts but also over time. Such possibilities require a representation, and perhaps the simplest is to add a discount to the timing of an outcome, where that discount is calibrated by the units of time under consideration. For example, for $\$X$ received next month, we might write δX, where $0 < \delta < 1$ since presumably people will prefer $\$X$ today to $\$X$ next month. And for $\$X$ two months from now, we can doubly discount and write $\delta^2 X$, and for three months from now, $\delta^3 X$, and so on.

Time discounting applies to things other than monetary outcomes. For example, it is well known that the behavior of drug addicts is only imperfectly impacted by a knowledge of the long term medical consequences of their addiction. Attempting to cure addiction by educating the addict about the harmful medical consequences of their problem will almost certainly fail. This is because addicts, nearly by definition, have an overly strong preference for immediate self-gratification as opposed to the long-term benefits of abstention and recovery. Indeed, one might say that "getting hooked on drugs" is shorthand for saying that the drug itself alters a person's time discount. Time discounts can also be impacted by one's environment and the time discounts of others. Suppose you are contemplating an investment in a society rife with political corruption and where most persons, as a consequence (as is arguably the case, for example, in many of the states of the former Soviet Union), act with very short-term horizons. Those short horizons derive from the fact that, in a truly corrupt state where there is no line between the criminal and the government, the government today might encourage your investments but tomorrow, after being bribed by your competitors, act to confiscate everything. Confronted with such a state, most people would naturally prefer, when making any investment, to "take what they can and run." But this means people will have few incentives to abide by long-term contractual agreements, in which case, anyone entering that economy with a long-term planning horizon will be akin to a small fish in a pool of sharks.

1.3 Economics Versus Politics and Spatial Preferences

The notion of time discounting will bear substantive fruit later when, in addition to the matter of political corruption, we consider such things as how political constitutions survive or fail as well as how cooperation in any form emerges in a society. But before we proceed to modeling specific political processes or phenomena, we note that when attempting to theorize about politics or to construct a model of some particular political process it behooves us to use the weakest assumptions possible, if only to ensure the greatest generality of whatever insights we might establish. But while generality has a self-evident value, it is unfortunately also the case that the weaker our assumptions, the less substantively precise are our insights. Thus, theorizing about anything, be it physics, chemistry, biology, economics or politics, entails maintaining a balance between generality and substantive specificity. Political science, though, is a discipline that stands relatively high on the food chain of our knowledge of and theories about social processes. Indeed, one might even draw a parallel between various fields of engineering and design versus the more fundamental fields of physics, chemistry and mathematics. Political science is (or at least should be) an applied field that takes what we know from statistics, from decision theory, from psychology and from game theory (as well as from the other social sciences) and applies what is known to the social processes we label political, ostensibly with the goal of improving the lot of our species. Thus, while the political scientist is not required to be a game theorist per se who goes about proving mathematical theorems about this or that, he or she is expected to be able to say something about such things as constitutional design, coalition formation in legislatures and parliaments, the imperatives of various forms of democratic governance, the sources of international peace versus war and the operation of alternative electoral processes.

The engineer who wishes to design a more efficient gas turbine or faster aircraft illustrates the parallel in the physical sciences. While the engineer is not expected to advance fundamental laws of physics and thermodynamics, he must nevertheless make use of those laws (or at least not presume that a design can violate them) in a creative way, filling in the blanks of abstract constructs with specific measurements or assumptions while at the same time making approximations that allow for the formulation of a substantively (physically) meaningful design proposal. The same is true in economics wherein those attempting to gauge trends in interest rates or the impact of some regulatory edict on firm behavior know that the laws of supply and demand will constrain events. And just as those elementary economic principles begin with highly abstract formal representations of consumer preferences and firm objectives, the political scientist, when modeling political processes, must often begin with abstract representations of preference and uninterpreted functions that denote utility, supplying them later with specific substantive meaning.

The Grocery Store: To see what we mean by all of this, let us attempt to gain a better understanding of the differences between economics and political science (without presuming that these two disciplines are necessarily disjoint) with a somewhat fanciful scenario. Consider the simple act of purchasing groceries in a supermarket—but to make our life simple, suppose there are but two distinct commodities in that store, X and Y. Your decision, then, is to choose how many items of X to buy, denoted x, and how many of Y to buy, denoted y, where your decision is subject to a budget constraint, B. Thus, the most of X and Y you can purchase is $xp_X + yp_Y = B$, where p_X and p_Y are the per unit prices of X and Y respectively. Assuming that you prefer as much of X and of Y as possible (i.e., you don't confront a problem of storing either commodity and neither is perishable), we can assume you'll balance off your purchases of these two goods so as to maximize your overall utility.

This much, of course, is little more than the introductory chapter of Elementary Economics 101 and corresponds to the economist's classical representation of a trip to the grocery store. Now, however, imagine a somewhat modified scenario. Rather than visit the store whenever you feel the need to replenish your supply of X and Y, suppose you are assigned a specific day and time to go to the store and that you are also required to bring with you a certain amount of money. Upon arriving at the store you find that 100 other people have been assigned the same time as you to shop and have been told to bring the same amount of money with them. However, upon entering the store the door is locked behind all of you, you are all led into a back room and told, after your money has been collected, that what you purchase today will be determined by a majority vote among all 101 of you. More precisely, suppose two of you are chosen at random and labeled "candidates." Each candidate must then propose a package of X and Y whose cost equals the sum of money collected from you with the presumption that everyone's budget will be spent in an identical way. The 99 of you who are now designated "voters" must then vote for one of the two candidates, and the candidate receiving the most votes will be declared the winner. Each voter and both candidates will then be given a shopping bag that matches the proposal of that winner with the winning candidate receiving an additional side payment of some sort so that both candidates have an incentive to win (as opposed to merely proposing their ideal allocation as their "campaign platform").

This might seem a truly strange way of organizing grocery shopping, but it does illustrate some of the differences between economics and politics, which in this case is simply the difference between two ways of allocating the goods and services people value. Now, though, consider the full implications of this difference. In the more regular way of allocating groceries, each person is free to choose the combination of X and Y that best serves their tastes whereas, in the second, each person is, in effect, a prisoner of the tastes of a majority or of the two competing proposals offered by the candidates. In the economic realm, then, we might attempt to predict how many of X and Y will sell by a careful examination of individual consumer tastes with the understanding that ultimate demand will equal the sum of demands. In the more collective or political realm, on the other hand, ultimate demand will depend on learning what proposals the candidates are likely to make and how voters will vote when confronted by alternative proposals. In the economic realm we need only identify that specific combination of X and Y that maximizes a consumer's utility, given their budget constraint. In the political realm, in order to learn how they might vote between the proposals of the two competing candidates, we must also be concerned with what their preferences look like over combinations they might not choose were they dictator of their own budgets.

We warn that we should not draw too sharp a distinction between economics and politics, since often politics entails deciding how to organize our "shopping"—should, for example, the purchase of health care insurance be a private or public matter, should people be free to discriminate against certain classes or races when selling their own homes, and should even a long-established public retirement program be made a partially private affair with both public and private options? Surely, few would argue that the answers to these questions are straightforward or without controversy. The same is true with our grocery store example. Suppose X and Y correspond to beer and baby food, and suppose that a clear majority of the 101 people sharing the back room of the supermarket are mothers with babies. Then suppose that you are an unmarried male. I suspect you would then hold a strong preference for the usual way of buying groceries (unless you have a perverse taste for crushed peas and strained carrots). Conversely, your preferences for how grocery shopping might best be organized could change if mothers with babies constituted only a minority of those present. Absent a concern for mothers with hungry and crying babies, you might see this as an opportunity to have parents subsidize your consumption of beer.

Politics is often a choice of how to allocate goods and resources—what to relegate to the private sector and what to allocate by some collective process. But in making that decision it is important to understand how different institutional forms—different methods for making social decisions—perform. For example, in lieu of selecting two persons at random to play the role of candidates, suppose we simply let the 101 people in the room negotiate directly among themselves until a majority reach an agreement and terminate further discussion. Or suppose we divide them into three constituencies of 33, 34, and 34 people, let each of them in a manner of their own choosing select a representative who will then negotiate with the two representatives from the other constituencies until they reach an agreement? What difference, if any, will each of these schemes imply in terms of the agreements reached?

To answer such questions—to conduct a comparative analysis of institutional forms—requires a common underlying structure for modeling the alternatives and individual preferences over the outcomes with which they deal. Returning, then, to our two commodities X and Y, and for a specific (albeit arbitrary) analytic example, let us suppose that the utility a person associates with a combination of X and Y is given by

$$U(x, y) = [5 - 5/(x + 1)] + [4 - 4/(y + 1)]$$

As complex as this expression might seem, it has a simple interpretation: If $x = y = 0$, then $U(0, 0) = 0$, but as either x or y increase, the subtractions in the expression decrease at a decreasing rate. Thus, as x or y increase, $U(x, y)$ increases (but at a decreasing rate) and approaches the upper bound of 9 as the amount of both commodities approaches infinity. The two commodities, though, are not perfect substitutes. For example, $U(2,0) = 10/3$ whereas $U(0,2) = 8/3$. In other words, if you have two units of X, you would require more than two units of Y to be compensated for the loss of your holdings of X. The relationship between X and Y in a person's preferences can be portrayed, then, as in Figure 1.1. The horizontal axis denotes units of X while the vertical axis denotes units of Y. The curves in turn correspond to *indifference curves*—combinations of X and Y that yield the same value for $U(x, y)$ and where combinations on curves further from the origin are preferred to combinations that fall on curves closer to the origin. Figure 1.1 also portrays a person's decision when choosing some combination of X and Y subject to a budget constraint. Here we assume that the per unit cost of X exceeds that of Y, so if a person spends their entire budget on one commodity, they can buy more units of Y than of X. Finally, the indifference curve that is tangent to this line represents the highest level of utility our decision maker can achieve, given their budget, so that x^* and y^* are the combination of goods we can assume they will purchase if they are dictator over their purchases.

Figure 1.1 is common to any introductory economics text and its discussion of consumer behavior in markets. But now let us again shift back to our peculiar (collective) method of grocery shopping. Here a person can no longer ensure that (x^*, y^*) is chosen, since the outcome depends on the preferences of other voters and the packages proposed by the candidates. In this instance, any point along the budget constraint line is a possibility (recall our assumption that each person brought the same sum of money to the store). But notice that the shape of the indifference curves in Figure 1.1 tells us something about the nature of this person's preferences across that line. Specifically, if we label the point o^*, which corresponds to the combination (x^*, y^*), the person's *ideal point*, then as we move away from that point in either direction, we move to lower and lower indifference curves. That is, the further we move from o^*, the less our abstract person/voter likes it.

If we now lay out the budget constraint line horizontally, we can draw a *preference* or *utility curve* such as the one shown in Figure 1.2, which for obvious reasons we refer to as a *single peaked preference curve*. The horizontal axis

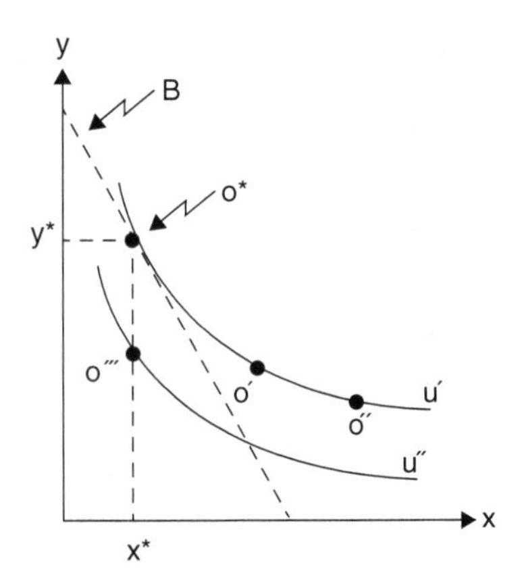

Figure 1.1 Consumer indifference curves with budget constraint

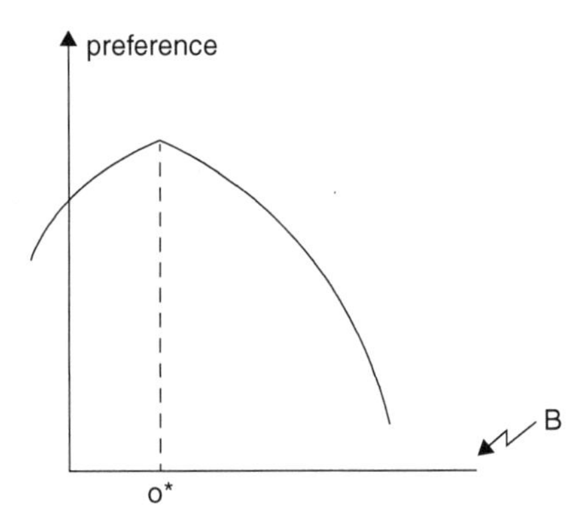

Figure 1.2 A single peaked preference

now corresponds to different allocations of the person's budget between X and Y, while the vertical axis denotes the person's preference or utility—which we know decreases as we move away from o^*, either to the left or right.

While Figure 1.2 might offer information about preferences that we need not concern ourselves with when discussing choices in a supermarket when the usual rules apply, it may be critical for determining how a person votes when those store purchases are made using some collective mechanism. Suppose, for example, that X = beer and Y = baby food. Then an unmarried male might have the preferences denoted by the rightmost curve in Figure 1.3 (not setting that person's ideal point at Y = 0 allows for the possibility that he might be curious as to what crushed peas or strained carrots taste like or that he feels some degree of sympathy toward mothers with babies). In contrast, the left-most curve with an ideal point at A might correspond to one of those women with babies who, nevertheless, is willing to allocate a small part of the family budget to beer for

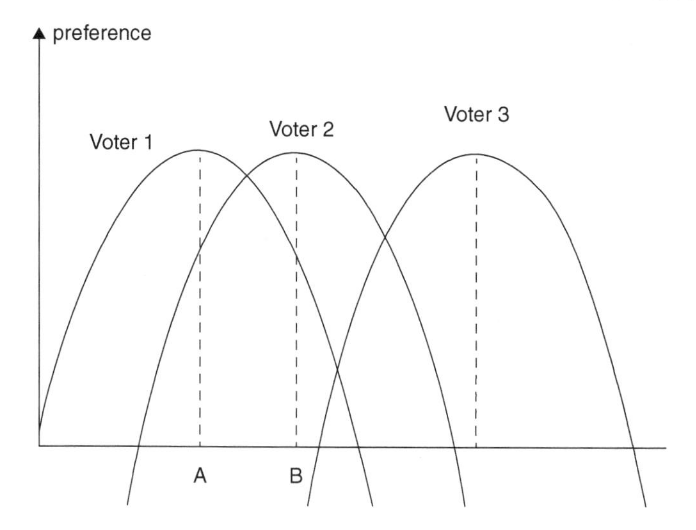

Figure 1.3 Three single peaked preferences for three grocery store "voters"

her husband. Voter 2, with an ideal point at B, on the other hand, might correspond to a husband who knows he'd be in serious trouble at home were he to return from grocery shopping after spending the majority of the family's budget on beer.

We will make a good use of preference curves such as those in Figure 1.3. But before we do so, let us consider some extensions of this representation of preferences. Specifically, suppose there is a third commodity, Z, that can be purchased only outside of the supermarket. If we were now to attempt to represent a person's preferences over X, Y and Z simultaneously in a three-dimensional space by way of extending Figure 1.1, we'd most likely imagine something like a set of nested mixing bowls with their bottoms aimed at the origin of the space, and each smaller or more distant bowl corresponding to a higher level of utility. We refrain from drawing such curves because doing so exceeds our graphic skills. But now imagine a person's budget constraint in this three dimensional space. Rather than a line, that constraint would be a triangle (a *budget simplex*) wherein each vertex of the triangle corresponds to all of the budget being spent on X or Y or Z. Finally, try to imagine what the surface of that triangle might look like as it cuts through various mixing bowls. Some of those bowls will not, of course, touch the triangle because they represent combinations of the three goods that cannot be achieved, given one's budget. And some of them will inscribe circles or some such curve on the triangle as the triangle cuts through them, thereby denoting budget-consuming mixes of X, Y and Z over which the person is indifferent. And unless the decision maker in question has preferences that yield a taste for spending their entire budget on only one of the three goods, we will find that one of the bowls just touches (is tangent to) the triangle. That point of tangency, then, corresponds to the person's ideal allocation of his

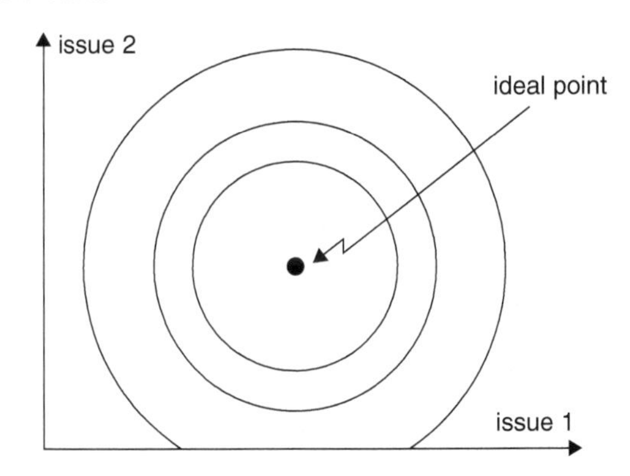

Figure 1.4 Two-dimensional spatial indifference curves

or her budget and, as in Figures 1.2 and 1.3, the further from that point, the less that person will like it. Figure 1.4, then, illustrates these indifference curves on the budget simplex, assuming perfectly round bowls after we lay out that simplex flat on the page.

The reader might ask why we've gone through so much trouble to extend our 2-goods model to 3 goods. Suppose, then, that instead of taking a fixed amount of money from our grocery shoppers when they enter the supermarket, we instead allow them to vote on how much of their budget will be spent on X and Y (and, thereby, how much they can spend subsequently, once released from the store, on Z)—or, more properly, we require that the two candidates take positions on how much will be spent in total on X and Y as well as on the allocation of monies between X and Y. Suppose we also eliminate any reference to beer and baby food and instead substitute such words as "social welfare" and "national defense." And instead of the abstract labeling of the third dimension as good Z, we instead think of it as the negative of a tax rate. Thus, we have arrived at a model—admittedly simple-minded—of an election in which voters must not only choose between different types of public spending, but also on the overall size of the public sector. People with ideal point near the budget simplex's vertex on the Z dimension prefer a small, if not insignificant, state wherein all consumption decisions are left to the private sector; people with ideal points near or at the simplex's vertex on the X dimension prefer massive government spending, provided it is spent on national defense; and people with ideal point at the third vertex prefer that most of society's wealth be devoted to social welfare programs.

Presumably, the majority of us prefer something closer to the middle or at least away from such extremes. For that reason, when making use of such *spatial* representations of preference, we forgo drawing triangles and, as we have done in Figure 1.4, simply denote the axes of the coordinate system along with the indifference curves and ideal points within it. The important thing, though, is to understand how we can move from the economist's usual representation of consumer preferences to those of voters who must make decisions using a more collective (political) institutional arrangement.

Before we sign off on this subject, it is useful to consider some of the forms spatial preferences can take. Figure 1.4 represents those preferences with some nondescript concentric circles, which suggests that the voter in question weighs the two dimensions or issues equally. Circular indifference curves or contours are especially useful for illustrating basic ideas, and are useful when we take advantage of our natural intuitions about geometry and distance to explore a new idea so that our intuition can lead our reasoning. However, consider Figure 1.5a, which represents preferences as concentric ellipses. In this instance, we can say that whoever holds such preferences is more sensitive to changes on the first (horizontal) dimension than the second. And then there's Figure 1.5b, where the elliptical indifference curves are tilted relative to the axes. First, notice that in both Figures 1.4 and 1.5a, a person's preference on one dimension does not depend on what choice is made on the other dimension. So if we arbitrarily fix the value of one dimension, the most preferred value on the second is unchanged (i.e., if, for instance, we draw a horizontal line in either Figure 1.4 or 1.5a, the value of X that corresponds to the tangency of that line to an indifference curve, x^*, is invariant with the height of the line). In this case, a person's preferences are said to be *separable* and their utility can be expressed as $U(x, y) = f(x) + g(y)$. For the case of Figure 1.5b, in contrast, preference on one dimension depends on the value assumed on the other. For this example, the higher we draw a horizontal line across the figure, the higher is the value of X that corresponds to the tangency of that line to an indifference curve (x^{**} versus x^*). Thus, to represent preferences for overall combinations of X and Y we must now write something like $U(x, y) = f(x) + g(y) + h(x, y)$.

Re-Voting at the Philadelphia Convention of 1787: Absent an appreciation of the possibility of non-separable preferences, a naïve reader of James Madison's notes on America's Constitutional Convention in 1787 might occasion considerable confusion, or at least leave one with the impression that the delegates there were indeed a confused lot. Specifically, consider the following recorded votes on the character of the presidency:

June 1: agrees to a seven-year term, by a vote of 5–4–1
June 2: agrees to selection of the chief executive by the national legislature, 8–2 and reaffirms a seven-year term, 7–2–1
June 9: defeats selection of the president by state chief executives (governors), 10–1
June 17: agrees once again to selection of president by the national legislature, 10–0, but postpones decision on seven-year term
July 19: votes 10–0 to reconsider the executive branch; passes by a vote of 6–3–1 selection by electors; defeats 8–2 a 1-term term limit, defeats 5–3–2 a seven-year term, and passes 9–1 a six-year term

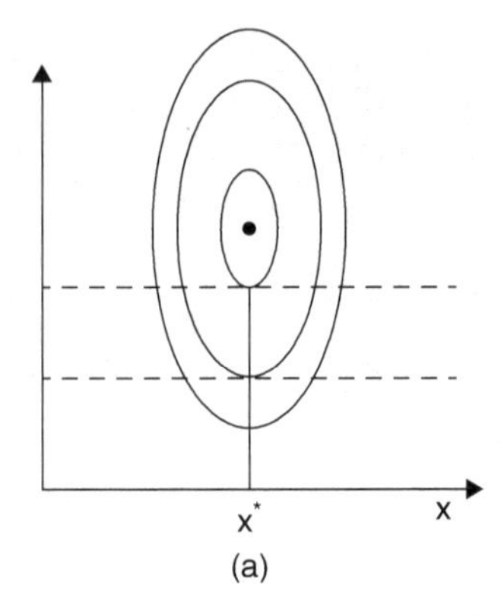

Figure 1.5a Elliptical 2-dimensional indifference curve

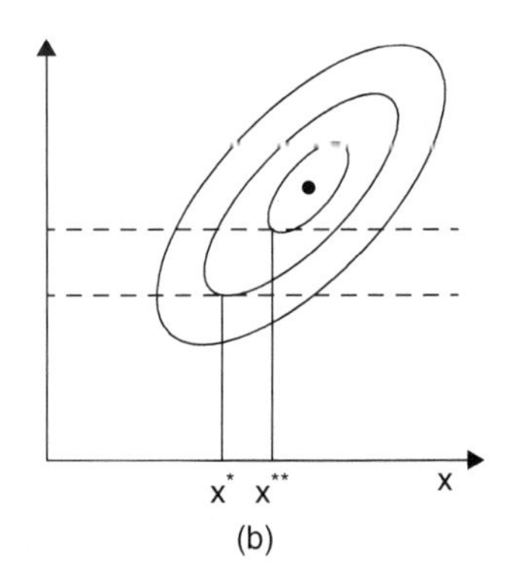

Figure 1.5b Elliptical 2-dimensional indifference curve

July 23: agrees by a vote of 7–3 to again reconsider the executive branch; passes 7–4 selection of president by national legislature and passes 7–3 a seven-year term with a 1-term term limit.
Aug. 24: defeats 9–2 direct election of the president and defeats 6–5 election by electors
Sept. 6: defeats 10–1 a 1-term term limit for the president, and agrees to election of the president by electors via a series of votes refining the Electoral College.

It might be true that the delegates were indeed at times indecisive and uncertain as to the best arrangement when dealing with details. The preceding vote history reveals, though, that the delegates were in fact considering three inter-related issues: The method of selecting a president, the president's term of office, and whether there would be a limit to the number of terms. In addition, decisions were being made on a great many other matters between June 1 and September 6, including the design of the national legislature and the powers of the president. There is little reason to suppose that the delegates, which hardly could be said to not have included some of the greatest political thinkers and engineers at this or any other time, deemed these decisions wholly separable. Thus, a decision on one dimension (issue) might reasonably be expected to impact their preferences on others and, thereby, cause them to reconsider prior decisions.

Institutionally Induced Non-Separability: We might be tempted to think that separable versus non-separable preferences are the product of individual taste, but they can also be institutionally induced. Consider the following clause from Article II, Section 1 of the United States Constitution: "The Electors shall meet in their respective States, and vote by Ballot for two Persons, of whom one at least shall not be an inhabitant of the same State with themselves." The Twelfth Amendment, of course, modified this clause by deleting "for two Persons" and inserted instead "for President and Vice-President" so as to avoid the issues that arose in the election of 1800 when both Aaron Burr and Thomas Jefferson received the same Electoral vote total and it fell to Congress to decide which would be President and which Vice President. For our purposes, though, notice that one's preference for Vice Presidential candidate can be (and generally is) a function of who is nominated for President because the political parties that nominate candidates will quite naturally seek some geographic spread to the two nominees so as to appeal to the electorates' varied sectional interests. The U.S. Constitution sets that preference in stone and dictates that if X is our choice for President, then we cannot choose Y if Y resides in the same state as X. Or, for another example of non-separability induced in part by institutional arrangements, we note that in presidential (as opposed to parliamentary) regimes, while some voters might prefer a unified government in which the same party controls both the executive and legislative branches, there are also those who prefer divided government wherein one party can act as a brake on the actions of the other. One's preference for president, then, might readily depend on who we think will control the legislature and, in the United States at least, whether the same party will control both the Senate and the House.

There are surely other examples of non-separable preferences that are either psychologically or institutionally induced. For example, suppose you must staff a 2-member committee by choosing from a set of 4 candidates, A, B, C and D. It might be that the person you most want to see on the committee is A, because A's preferences most closely match yours. But suppose A as a function of personalities cannot work with either B or C so that any committee that combines A with either of these two people is likely to function poorly or not at all. Thus, if C is chosen first, you most likely would prefer that either B or D be the second serving member. However, if D is chosen to fill the first seat, your preference is that D be joined by A.

Separable versus non-separable preferences do not, though, exhaust the possibilities we might need to consider when describing forms of individual preference. Consider the following example of preferences we call lexicographic:

> **Diamond Rings and the FDA**: We're not sure how many male readers have had the opportunity to shop for an engagement ring, but those of you who have should immediately understand the following preferences among women (hope we're not being too chauvinistic here). There are several dimensions by which to evaluate such a ring—the size (weight) of the stone, the stone's clarity, its cut and the quality of the setting. But if your experience matches ours, you will quickly learn that preferences here can be *lexicographic* wherein the second, third and other dimensions do not come into play in making choice unless the two top alternatives are equivalent on the first dimension. Specifically, cut, clarity and setting are of little note unless the main stone is "big enough." Indeed, if given a choice between a 2-karat stone of average clarity versus a 1-karat stone of superb clarity, not a few women would choose the first stone. After all, how much can clarity count if people don't first at a distance say "wow"?
>
> For another example, it is often argued that America's Food and Drug Administration is too conservative in its approval of new drugs—that effective drugs are available elsewhere in the world long before they are approved for distribution and sale in the United States. Now consider that there are two basic dimensions with which to evaluate any new drug: Its potential effectiveness in treating some disease versus the risks of its side effects. Ideally, these two dimensions should be balanced against each other with a willingness to assume risk a function of a drug's ostensible effectiveness. But consider the incentives of bureaucrats within the FDA. If they disapprove of a drug that later proves to have few side effects, there are unlikely to be any personal consequences—arguments can always be made that further study was necessary before a definitive risk assessment could be conclusively offered. Moreover, if they approve a drug that is effective with no risk, they are unlikely to receive any credit since, after all, they have merely "done what's right." On the other hand, if they certify a drug that proves to have negative or even deadly consequences, there's a good chance that those responsible for the approval will have "hell to pay." Thus, FDA's bureaucrats are likely to be risk averse in the extreme with respect to a drug's side effects to the point that only drugs with no apparent side effects whatsoever are approved. If given the opportunity to certify two competing drugs X and Y from two competing pharmaceutical firms, bureaucrats with lexicographic preferences will consider the matter of relative effectiveness ONLY if both offer equally low risk; otherwise, they will certify neither or the one with no apparent side effects regardless of its relative effectiveness.

We draw attention to lexicographic preferences not because there are advantages to playing analytically with them. Indeed, the opposite is true, but our examples show that not only can such preferences arise "naturally" for psychological reasons, they can also be institutionally induced and, therefore, they are preferences with which we must sometimes deal. Indeed, inducing lexicographic preferences is not the only role institutions can play in determining how to best model preferences in specific circumstances.

A Lesson from Tinseltown: For an example of how the choice of an institution—in this case a voting method—can impact which dimension of preference is most relevant to an individual decision maker's calculus, we note that if an idea is apparent even to those who populate the movie studios of Hollywood—producers, directors, actors and actresses—then the idea must indeed have an element of truth to it. We are reminded then of the ending scene of the movie *1776*, Hollywood's not-altogether historically accurate version of events in Philadelphia at the drafting and signing of the Declaration of Independence. In voting on the Declaration, the delegates abided by the rule of unanimity whereby votes are taken by state and where a single Nay would send the document down to defeat. In the movie version of events, the decision comes down to the Pennsylvania delegation, which, with considerable liberties taken with historical fact, consists of Benjamin Franklin, John Dickinson and Judge James Wilson. Throughout the movie, Wilson is portrayed as a weak personality willing to do the bidding of Dickinson, who is strongly opposed to declaring independence and prefers instead that further efforts be made at seeking reconciliation with England. Thus, with two votes against one for Pennsylvania and a rule of unanimity in effect for the Congress as a whole, the Declaration seems doomed to defeat. Franklin, however, makes the parliamentary maneuver of calling for a roll call vote of his delegation. With Franklin voting Yea and Dickinson Ney, Wilson becomes pivotal "for or against," in Franklin's words, "American independence." With Wilson wavering, Franklin drives home the point of Wilson's pivotal role by noting that "the map makers of the world are awaiting your decision." If preferences over choices are invariant with context, Franklin's parliamentary maneuver should be of no consequence. But by being made pivotal, the basis of Wilson's decision changes. As Wilson himself states the matter, if the delegates were able to vote anonymously within each state, it would be Pennsylvania that would be credited or blamed for having defeated the Declaration; however, under a roll call vote it would be Wilson specifically who did so. As Wilson goes on to explain, if he votes Yea, he will merely be one among many whereas if he votes Ney, he will be remembered as the man who sank American independence. Since his strong preference for anonymity trumps his preference for seeking accommodation with England, Franklin's maneuver changes the basis of Wilson's decision—changes the value (utility) Wilson associates with the alternatives he confronts—and, thus, the final outcome.

This example is not intended to illustrate a situation in which Wilson's core values changed—that somehow Franklin's strategy changed the judge's preferences over some multidimensional issue space, where those dimensions included perhaps one that represented America's alternative relationships with England and another his public visibility. But Franklin's parliamentary maneuver—his switch in voting schemes for Pennsylvania's delegation—did impact the dimensions Wilson deemed relevant to his decision. Only under a voting scheme in which individual ballots are recorded does Wilson's preference for anonymity play a role since only under such a rule are the outcomes "Declaration ratified" and "Declaration not ratified" elaborated to include a specification of how individuals voted. We see here, in fact, yet another door opening to the relevance of game theory—to that of the strategic choice of institutional forms. Hollywood's portrayal of Franklin's genius might not have been historically accurate, but the scene resonates because we know that if that circumstance had in fact arisen, the real Franklin would have understood the strategic possibilities as they were portrayed.

1.4 Collective Versus Individual Choice

To this point, we have focused exclusively on how to represent the preferences of individual decision makers while admitting that our true concerns are collective or political decisions. What, then, of collective or group preferences? After all, everyday discourse about politics is laced with statements or assertions that begin with "Society's interests are __," "The electorate prefers __," "The legislature wants __," "The bureaucracy acted __," "The interests of [country X] lie in __" and so on, as if collective preferences are no less real or tangible than individual ones. We are reminded here of Charles de Gaulle's famous comment that France has no friends, only interests. Here, though, we want to end this chapter on a supremely important cautionary note about attributing preferences to collectivities.

> **The Condorcet Paradox:** Suppose three people hold the following preferences:
>
> Person 1: A preferred to B preferred to C
> Person 2: C preferred to A preferred to B
> Person 3: B preferred to C preferred to A
>
> The question, now, is how to define the *social preference* of these three people as a group. There are, of course, innumerable ways to do so. We could, for instance, simply choose one person at random and define his or her preference as the social preference. Absent any bias in our random selection, such a method seems fair because no person is more likely than any other to represent the group. Alternatively, we could assign 2 points to a first-place ranking, 1 point for a second place ranking and 0 points for a last place ranking and construct the social preference by adding up the scores of A, B and C. In this case, though, such a method is indeterminate, or at least undiscriminating, because each alternative would be awarded a sum of 3 points. Another and seemingly more "democratic" method is to take a majority vote between the alternatives and if X beats Y in a majority

vote, then we would say that X is socially preferred to Y or, equivalently, that the group prefers X to Y. The preceding three preferences, though, point to a general problem with this method. Specifically, note that while C beats A in a majority vote, and B beats C, A beats B. Thus, the social preference is *intransitive*!

The Grandfather, Granddaughter and the Horse: Walking through the village, accompanying his granddaughter and leading the family's horse, the grandfather senses the villagers' disapproval of not affording his granddaughter the pleasure of riding on the horse. So up she goes. But soon thereafter there emerges another sense of disquiet among the villagers: Why is it that such a young girl requires that her elderly grandfather walk while she rides? Not wanting to appear a spoiled, ungrateful child, the girl insists that her grandfather take her place. But nearly immediately the grandfather senses the villagers' disapproval of him riding alone while his sweet granddaughter walks alongside. So up she goes to join him, whereupon the murmurs of disapproval from the village now focus on the horse's burden of having to bear the weight of two people.

The preceding example is but a folky illustration of the more abstract 3-alternative example that precedes it, wherein both illustrate a thing called *the Condorcet Paradox*, named after the 18th-century French mathematician who concerned himself with voting systems and finding a fair method for electing members to the French Academy of Sciences. That our folky example illustrates the same thing as our abstract one can be seen if we suppose that the villagers are of three types:

Type 1: O1 > O2 > O3 > O4
Type 2: O4 > O1 > O2 > O3
Type 3: O3 > O4 > O1 > O2

Where O1 = both ride the horse, O2 = grandfather alone rides the horse, O3 = granddaughter alone rides the horse and O4 = no one rides the horse. In this case, if all three types are represented in the village in approximately equal proportion, the social preference order under majority rule is O1 > O2 > O3 > O4 > O1. The particular paradox here, of course, is that although the individual preferences used to define the social preference in our examples are transitive (and complete), the resulting social preference, at least under simple pair-wise majority rule, is intransitive, in which case we cannot impute a utility function to the group.

Condorcet's Paradox gives rise to any number of important theoretical issues. What, for instance, are the circumstances under which simple majority rule might yield an unambiguous social preference? Are there other ways of applying the idea of majority rule that might avoid the paradox? Do rules other than majority rule also share the property of manufacturing intransitive social preferences out of transitive individual ones? Are there any rules or procedures that guarantee transitive social preferences and, if so, what do they look like?

A Spatial Example of the Paradox: We cannot answer all such questions in this chapter. Presently, then, the Paradox should be taken simply as a cautionary note—a warning against becoming overly anthropomorphic in our approach to politics by inferring or assigning motives, preferences and the like to collectivities regardless of their identity. The reader, though, should not suppose that the Paradox is a mere curiosity and the product of some artfully created individual preference orders. Rather, it is a feature of group preferences with which we must deal in nearly all applications of game theory to social processes. To illustrate this fact, let us return once again to the spatial preferences and the two-dimensional form illustrated in Figure 1.4. This time, though, in Figure 1.6 we draw the indifference contours for three people with ideal points at x_1, x_2 and x_3. Now consider the arbitrarily chosen alternative z_1, through which we draw the indifference curves of all three persons. Notice that the shaded areas bounded by these indifference contours are all points that are closer to a pair of ideal points than is z_1. Alternative z_2, for instance, is closer to the ideals of x_1 and x_2 than is z_1. Thus, under majority rule, z_2 is preferred to z_1. On the other hand, now consider alternative z_3. As placed, z_3 is closer to the ideal points x_2 and x_3 than is z_2. Hence, z_3 defeats z_2 in a majority vote. But finally, notice that z_3 is not in any of the shaded areas corresponding to points that defeat z_1. In fact, z_1 is closer to the ideals x_1 and x_3 than is z_3. Hence, under majority rule, we have the intransitive social order $z_1 > z_3 > z_2 > z_1$.

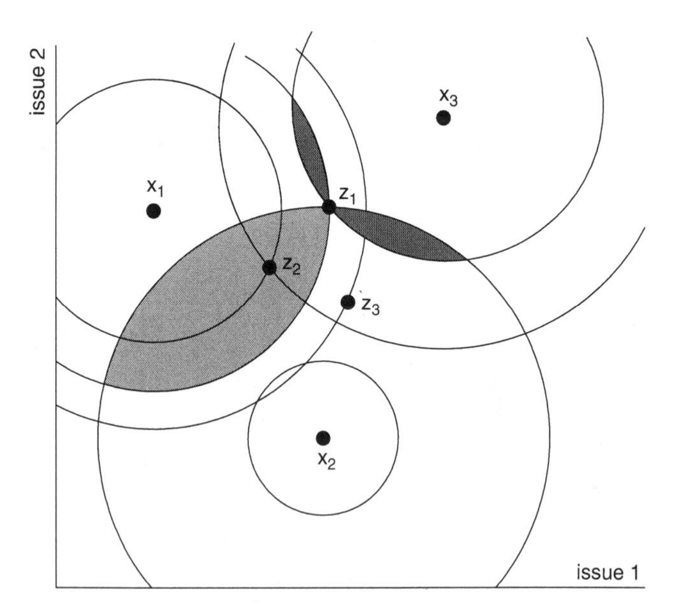

Figure 1.6 Condorcet's Paradox with spatial preferences

This simple example—another manifestation of the Condorcet Paradox—illustrates again the inadvisability of being anthropomorphic about things and, without further analysis, attributing goals, motives or preferences to groups. Barring further developments, we could, of course, assume that the paradox is but an anomalous, albeit unanticipated, characteristic of majority rule. This paradox might merely cause us to question the reverence sometimes associated with outcomes chosen "democratically" by majority rule principles. However, a profoundly important theorem, **Arrow's Impossibility Theorem**, tells us that this paradox is not anomalous or confined to majority rule procedures—it is possible to observe "irrational" social preferences under almost any social process. Arrow's method of demonstrating this fact is to eschew examining each and every rule we might imagine and instead to posit a set of properties or axioms that we think any "reasonable" rule should follow. Without delving into the finer details of things, the axioms Arrow sets forth are, roughly stated, these:

1. The social ordering is complete and transitive;
2. no individual preference order over the feasible outcomes is a priori excluded as a possibility (unrestricted domain);
3. the social preference between any two alternatives never depends on individual preferences regarding other alternatives (pairwise independence of irrelevant alternatives);
4. alternative x is socially preferred to y whenever everyone prefers x to y (the Pareto principle);
5. no individual should be decisive for every pair of alternatives (nondictatorship).

With respect to axiom 2, for instance, there commonly are preferences we prefer to exclude from consideration when making social decisions, such as prohibitions against anti-Semitic or pro-Nazi ideas. Society's current over-indulgence with political correctness is yet another attempt to exclude various preferences from consideration in public discourse. But if we are to fully understand and fully model social processes, then the method whereby certain preferences are excluded ought to be a part of the general rule we are considering. Axiom 3 requires that we can infer the standing of some alternative relative to another by merely looking at individual preferences over those two alternatives, while axiom 4 in effect requires, among other things, that whatever stands highest in the social preference order be Pareto optimal for the collectivity in question. Despite the reasonableness of those axioms, Arrow's theorem establishes that they are inconsistent. Stated differently,

> *For decisions involving three or more alternatives with three or more individuals, at least one of the axioms 1–5 must be violated. Any procedure consistent with axioms 2–5 will allow for intransitive social orderings. Equivalently, the only procedure consistent with axioms 1–4 must violate axiom 5.*

Later we will grapple with the full consequences of this theorem, which is one of the most important in political theory. However, here we want to emphasize the special role played by the transitivity and completeness assumptions in modeling people and the fact that these assumptions cannot play an equivalent role in our discussions of groups. Although it is often convenient to be anthropomorphic and to attribute motives to groups in the same way that we attribute motives to individuals, as our earlier brief discussion of Britain's policies prior to the outbreak of WWI reveals, and as Arrow's theorem precisely formalizes, such linguistic shortcuts are simply that—shortcuts—and not scientifically valid. Thus, although we may choose to use such shortcuts to convey general meaning and although they may be approximately valid when individual preferences are unanimous or at least nearly so, we must keep in mind that any theoretically valid explanation for social processes and outcomes must rest ultimately on an assessment of the preferences and actions of individuals in combination with the institutions (broadly interpreted) within which those individuals and preferences operate.

Arrow's theorem and the associated Condorcet Paradox, though, are not the only problems associated with attributing goals to collectivities. Another way to bring that fact home graphically is to turn to an example wherein a group appears to act utterly irrationally (i.e., in its own worst interest) but where it seems reasonable to assume that the individuals within the group are acting in pursuit of perhaps the most intense of all preferences, that of survival.

> **The Curious Behavior of Herring**: Some time ago the Discovery Channel released a video of a large school of herring swimming casually in beautiful clear water when suddenly it was attacked by a number of blue fin tuna. It seems that herring are deemed quite a delicacy by tuna, at least judging by tunas' enthusiasm for catching and eating as many as possible. The reaction of the herring, though, is surprising. Instead of scattering in every conceivable direction (and indeed there were far more potential directions than there were tuna), the entire school began to swim in a tight swirling ball. The ball, of course, provided a far more inviting target than some widely dispersed cloud of fish, so slowly (or not slowly enough, from the perspective of the herring) the ball began to shrink. The ball itself, moreover, began to appear wholly disoriented and slowly moved toward the surface, which only made it vulnerable to the swooping pelicans above. The video ends when the ball barely exists and the tuna are fully sated.

We are admittedly not in a position to fully account for this odd and seemingly self-destructive behavior on the part of the herring, aside from noting that it appears to contradict Darwin's rules about species survival. In fact, we will later refer to this example as a way to illustrate problems of collective action and social coordination. Here, though, we can use it to illustrate the distinction between individual and collective preference. Were we, for example, to try to explain the behavior of herring with reference to collective preferences and actions, the conclusion that schools of herring prefer suicide and extinction seems inescapable. Surely, since every herring scattering to the wind (or, more properly, ocean current) seems the best choice for all collectively, the school's behavior is consistent with the hypothesis of a group preference for suicide. It seems safe to assume, on the other hand, that while any one herring could give a twit about the concepts of extinction and group survival, each individual fish does value its own skin (or scales) and, if it could, would act accordingly. A careful look at the swirling ball confirms this supposition—specifically, the ball swirls because each fish is doing its damndest to get into its interior. Each herring has but two choices: To swim unilaterally away from the ball and, most likely, into the waiting mouth of one of the surrounding tuna or to try to disappear inside the ball in the hope that the tuna will satisfy themselves by eating only those on the perimeter. Alas, their choice is a Hobbesian one—there is no good choice for each herring, though getting into the interior of the swirling ball would seem the best of two distinctly poor alternatives (since there are a few herring left when the video ends). There is, of course, no issue of collective intransitivity here as in Condorcet's Paradox. After all, there are only two choices for the school—form the swirling ball or scatter and run (swim) like hell. But now we have an example where the presumed unanimous preference of individuals—to survive—is transformed by circumstances into the seemingly irrational preference (if we are to judge by the "school's choice") of maximizing the ease with which it can be eaten.

More will be said of this example later, but note we have just used a word that itself warrants comment—namely, *irrational*. Much has been made of this word and its opposite, *rational*, and much has been said about whether these words have any proper definition in the context of contemporary social science theory. Some will assert that any behavior is rational if it can be conceptualized to follow from some well-defined set of preferences over ultimate outcomes. Rational action or rational behavior is simply any behavior that can be said to follow from our theories and postulates about preference. This definition, though, would seem to render the concept a worthless tautology since we can always ascribe goals to whatever it is we observe. Even the school of herring, for example, can be deemed collectively "rational" if we are willing to postulate for it the goal of suicide. Others then ascribe rationality only to those actions that can be justified by some "reasonable" set of goals or utility functions. But this merely pushes the pebble of contention to a different place in the mud puddle since we are then left with having to define "reasonable." There is no bar of metal sitting alongside the one in Paris that standardizes the measurement of "meter" with which we can measure or distinguish between reasonable and unreasonable. Our preference here, then, is to banish the words *rational* and *irrational* altogether from our lexicon and to instead simply proceed to the task of seeing if we can explain (and predict) social and collective actions with concepts that do not require such words. That is the task to which we now turn.

1.5 Key Ideas and Concepts

decision theoretic
game theoretic
common knowledge
preference
complete preferences
transitive vs. intransitive preferences
risk
utility
indifference curves
expected utility
nondictatorship
unrestricted domain
time discounting
rational
tautological
subjective probability
budget constraint
budget simplex
spatial preference
single peaked preference
separable vs. non-separable preferences
social preference
collective action
Arrow's Impossibility Theorem
Condorcet Paradox
Condorcet winner

Exercises for Chapter 1

1. Portray a utility function for money in which a person is risk averse for small amounts and risk acceptant for large amounts.
2. Portray a utility function for money that "explains" why a person might buy a lottery ticket (and therefore appears risk acceptant) and also purchase a homeowner's insurance policy (appearing risk averse).
3. Consider the following six strict preference orders over three alternatives:

Order 1	Order 2	Order 3	Order 4	Order 5	Order 6
A	A	B	B	C	C
B	C	A	C	A	B
C	B	C	A	B	A

If you are told that a group of people all simultaneously have single peaked preferences over a single issue, can all six preference orders coexist simultaneously; and if not, what are the various subsets of preference orders that can simultaneously describe individual preferences?

4. Assume 6 people have circular indifference contours in two dimensions. Portray their ideal points in a two-dimensional drawing along with the three alternatives A, B and C such that each person has one of the preference orders in problem #3 and no 2 people have the same preference orders.
5. Assume 7 people have circular indifference contours preference orders in two dimensions and assume that for the three alternatives, A, B and C, that B is a Condorcet winner. Locate 7 distinct ideal points consistent with this fact.
6. Suppose 3 people have the following preference orders:

Person 1	Person 2	Person 3
B	C	A
A	B	D
D	A	C
C	D	B

   Assume these people have circular indifference contours in two dimensions. Locate 3 ideal points, x_1, x_2, and x_3, consistent with these preferences.
7. Can a person with a single peaked preference be risk acceptant between two alternatives and at the same time risk averse with respect to two other alternatives?

2 Extensive Forms, Voting Trees and Planning Ahead

2.1 Introduction

Pastor Martin Niemöller, imprisoned by the Nazis for seven years owing to his opposition to Hitler, summarized the way in which a society can slide into an evil authoritarianism with the insightful parable:

> *First they came for the Socialists, and I did not speak out, because I was not a Socialist.*
>
> *Then they came for the Trade Unionists, and I did not speak out, because I was not a Trade Unionist.*
>
> *Then they came for the Jews, and I did not speak out, because I was not a Jew.*
>
> *Then they came for me—and there was no one left to speak for me.*

There are a great many moral lessons to be learned from this quote, but we offer it simply as an example of the need to "look ahead" when making decisions—to attempt to assess the full consequences of immediate actions. It might seem natural, of course, for humans to look ahead, but it is surprising how often, in critical circumstances, they fail to do so or do so only imperfectly. The rise of Nazism aside, at the turn of the 20th century, for example, General Alfred von Schlieffen laid out a plan (called, logically enough, the Schlieffen Plan) for the invasion of France whereby Germany would arrange its armies from North to South with the main thrust proceeding through Belgium and pivoting at the southern tip of that country to catch the French in a grand pincer move southeast of Paris. France's Plan XVII, on the other hand, positioned its armies from South to North, from Alsace to the Belgian border, with a British expeditionary force guarding their flanks to the north. Its plan was a dual thrust through Alsace and Lorraine and a second to the north, possibly through the south of Belgium and Luxemburg if and when Germany violated Belgian neutrality. Various contingencies were allowed for in both plans, with the Germans allowing their armies to sweep to the west of Paris if there was a need to protect their flanks as well as provision for a reallocation of various corps to forestall potential French breakthroughs. The French, somewhat more optimistically, allowed for the possibility that both of their thrusts would be successful. Both plans, however, suffered from two glaring weaknesses. The first was the jointly held assumption that the modern technology of warfare guaranteed that any war would be brief. Thus, neither plan looked very far into the future insofar as how conflict might unfold. And second, each plan assumed that it would work. Thus, what neither plan offered was a specification of what to do in case one or the other failed. And indeed, that is what happened: The Germans suffered an historical defeat to the east of Paris at The Battle of the Marne whereas no French thrust penetrated German resistance to a significant degree. The end result was a four-year war of attrition and millions of battlefield deaths. Perhaps more astonishing still is that Germany under Hitler made an equivalent error in 1941. This time, after somewhat effortlessly defeating France, Hitler turned on Russia, wherein the "plan," Operation Barbarossa, was little more than an all-out attack from the Baltic to the Black Sea under the assumption that the Soviet Union would quickly collapse militarily and politically after experiencing Germany's blitzkrieg tactics. Utterly absent from the plan was any consideration of the possibility that the Soviets would not capitulate but would instead regroup sufficiently to extend the conflict into the harsh Russian winter, whereby, in a replay of Napoleon's disastrous venture, the effect of sub-zero temperatures on unprepared men and material would wreak havoc with any military scheme. Indeed, it seems that Europe's political-military leaders of the 20th century ignored the military maxim "no plan survives contact with the enemy."

The Domino Theory: It is also the case that looking ahead under the wrong assumptions can lead to poor choices in the long run. Much was made, for instance, of the "domino theory" in rationalizing the Vietnam War and America's intense and unsuccessful involvement there . . . the argument being that "if South Vietnam fell, it would be followed by Communist takeovers in the rest of Indochina, then Malaysia, then the Philippines, etc., etc., etc." Events, of course, proved otherwise, since American strategists failed to understand that North Vietnam saw the conflict more as a war of national liberation and a redressing of the consequences of European (read: French) colonialism. But the U.S. was not the first country to formulate and misperceive this domino argument. Consider, for instance, King George III's assertion in 1779 that if the Americans were to succeed in their revolution, then "the West Indies must follow them . . . Ireland would soon follow the same plan and be a separate state, then this island would be reduced to itself, and soon would be a poor island indeed, for reduced in her trade merchants would retire in their wealth to climates more to their advantage, and shoals of manufacturers would leave this country for the New Empire." The King might have been right in the VERY long term, but his comments came before the reign of Victoria and Britain's century-long dominance.

There are, of course, those events that are perhaps impossible to map out in an extensive form so as to facilitate optimal choices, if only because they deal with unique historical events. During World War I again, in 1917, Germany sought a way to free itself from having to defend the Eastern front against Russia. It mattered little that Russia's army was in disarray—resources that could better be used against France and England were nevertheless needed in the East. In response, Germany's strategy seemed a clever one at the time. The Czar had abdicated and a provisional government had taken his place while unrest continued to plague the country. So why not encourage that unrest and a thorough disassembling of the Russian army by sponsoring the country's most radical element, which was itself committed to pulling Russia out of the war—the Bolsheviks. And what better way to aid them than by shipping their leader in exile to Petrograd who, in his Zurich apartment, could do little more than write inflammatory pamphlets and plans—Vladimir Ilyich Lenin. So squirreled away on a sealed train, Germany delivered Lenin to Russia. Little did German planners know, of course, that in 24 years, the cream of German youth would be locked in deadly and, as things turned out, extinction-level combat with Lenin's ideological progeny.

The lesson of these somewhat dramatic examples is that to understand political decision making generally it is essential that we attempt to learn whether or not people look ahead and, if they do, what futures they think alternative choices might yield. This much, of course, is only common sense. But to do this in a theoretically satisfying way—in a way that allows us to generalize what we learn from one situation and apply it to others—requires some structure, and to that end let us consider a few less profound examples of a sort we might address relatively easily:

> **Legislative Pay Raises**: If we know one thing about elected representatives, be they national or regional, American, European or Asian—they love pay raises and having their positions accompanied by all manner of perks. Far be it for members of the U.S. Congress, for example, to relinquish their free parking spaces at Washington's National (Reagan) Airport or to burden themselves (or their staffs) by folding their medical insurance plans into "Obamacare." So suppose a legislature has decided to vote on a raise for themselves (since no one else would do it). The question is, how to do it aside from slipping it in as a part of some obscure bill that seemingly has nothing to do with government salaries? Here we won't worry about the niceties of such things and will confine our example to two simple alternatives: A roll call vote and a secret and simultaneous ballot. To model these alternatives we can suppose there are but two outcomes: The raise passes or it fails. But legislators, being what they are, are likely to see things differently, wherein each of them differentiates among four possible outcomes:
>
> o_1: *the raise passes but I voted against it*
> o_2: *the raise passes and I voted for it*

o_3: *the raise fails and I voted against it*
o_4: *the raise fails but I voted for it*

The reasons for this differentiation should be obvious: Each legislator prefers the raise, but no one wants to vote for it if their vote isn't necessary for passage since doing so puts a target on their back in the next election. Nevertheless, it is not unreasonable to assume that legislators are sufficiently avaricious that the order of the outcomes given above corresponds to their preferences—they prefer the raise, and all the more so if they can secure it while voting against it and the worst possible outcome is voting for a raise that fails to pass. Finally, suppose we want to know who is advantaged and who is disadvantaged by different voting schemes. For example, if your last name is Aardvark, which would you prefer—a roll call vote in which votes are taken alphabetically or a vote in which all choices are, in effect, made simultaneously?

Presidential Vetoes: Suppose, as is commonly the case, that it takes only a majority vote to pass legislation in a legislative assembly and that the political system has a president who can veto any legislation the assembly passes. Commonly, though, national constitutions afford the legislature the authority to over-ride executive vetoes, but different constitutions require different votes for an over-ride to be effective. Some, such as the U.S. Constitution, require that two-thirds of those voting vote to over-ride. Various other constitutions require a three-fifths vote, so that, as in the American case, a special or super majority is required to thwart the will of the president. But there are those national constitutions that allow a simple majority to over-ride a veto and it is widely assumed that since it takes the same vote to approve of the legislation in the first place, allowing a majority over-ride affords a president a toothless power. Our question is whether this inference is necessarily correct.

Horse Racing Against the King: Sun Bin, a military strategist during China's Warring States Period (403–221 BC), was asked by General Tian Ji to advise on how to win a horse race against the King of Qi. At the time the contest consisted of three rounds of three different pairs of horses racing against each other, where the best two out of three was declared the victor. Tian Ji always lost to the king. Sun, however, advised Tian to use his worst horse to race against the king's finest in the first round. In the second round, Tian would use his finest to race against the king's regular horse and in the third round, Tian would use his regular to race against the king's worst. Tian won the horse racing bet that day and both the king and Tian admired Sun Bin's talents. However, as entertaining as this historical account might be, it is incomplete and leaves the door open to a number of possibilities, not all of which results in Tian being victorious over the king.

The Iroquois Confederation: Arguably, the world's oldest extant written constitution (if we allow wampum to be deemed writing) is the one governing the Iroquois Confederation of North America (variously referred to as *The Constitution of the Five Nations* or *The Iroquois Book of the Great Law*). Various dates have been ascribed to it, some as early as 1390. The confederation itself originally consisted of five tribes in present day upstate New York—the Mohawk, Seneca, Onondaga, Oneida and Cayuga—and offered this method whereby the five tribes considered any legislation:

> *First the question shall be passed upon by the Mohawk and Seneca Lords, then it shall be discussed and passed by the Oneida and Cayuga Lords. Their decisions shall then be referred to the Onondaga Lords (Fire Keepers) for final judgment . . . when the Mohawk and Seneca Lords have unanimously agreed upon a question, they shall report their decision to the Cayuga and Oneida Lords who shall deliberate upon the question and report a unanimous decision to the Mohawk Lords. The Mohawk Lords will then report the standing of the case to the Fire Keepers, who shall render a decision as they see fit in case of a disagreement by the two bodies, or confirm the decisions of the two bodies if they are identical . . . if through any misunderstanding or obstinacy on the part of the Fire Keepers, they render a decision at variance with that of the Two Sides, the Two Sides shall reconsider the matter and if their decisions are jointly the same as before, they shall report to the Fire Keepers, who are then compelled to confirm their joint decision.*

Thus, paralleling our previous example, the Onondaga have a veto that can be over-ridden by the same vote as that which passed the proposal in the first place. But in addition, the Onondaga, the smallest of the five tribes, can cast a tiebreaking vote in the event that the Cayuga and Oneida disagree with the vote taken by the Mohawk and Seneca. How might we represent this situation in a way that allows us to analyze voting and the relative voting power of the Onondaga relative to the other four tribes?

Crisis Escalation: One of the inherent dangers of international affairs is the crisis that in a seemingly uncontrolled way escalates to a level of conflict that both sides would have preferred to avoid. The Cuban Missile Crisis of 1962 seemed to be just such a situation, wherein the world was threatened with an all-out nuclear conflict between the United States and the Soviet Union. The conflict was avoided, which should lead us to ask why this case was unlike the escalation leading, for example, to World War I. However, rather than model such situations directly, consider the following artificial scenario: Two persons, labeled 1 and 2, must bid on \$100, with person 1 afforded the opportunity to open the bidding at \$10. If 1 passes, then 2 can buy the \$100 for \$10 and the auction ends. But once the bidding begins, each person, in turn, can bid the price up in increments of \$10. Once one person refuses to bid further, the high bidder wins the \$100. However (and this is the interesting part), not only must the winner pay the amount of his bid for the \$100, but the loser must pay their last highest bid. Thus, if, say, person 1, after a sequence of bids, drops out after person

2 bids, say, $60, then 2 wins the $100 and shows a net profit of $40 = $100 – 60 whereas person 1 wins nothing but must pay his or her last bid, $50. To forestall the possibility of bankruptcy, suppose in addition that we put a cap on the highest possible bid of, say, $200, which, of course, represents a severe overpayment for the $100 prize. The question, then, is whether this simple auction scenario models crisis escalation.

Timing Is Everything: America's presidential election of 1896, held during one of the worst depressions up to that point in the country's economic history, was known even beforehand to be a watershed in American politics. The Democratic incumbent, President Cleveland, was certain not to run, thus leaving the field open for his party's nomination. The Republican's most likely nominee would be William McKinley, a staunch pro-business politician from Ohio and ardent champion of keeping America on the gold standard (i.e., a tight money supply favored by banking and business—J.P. Morgan, in particular). The Democratic party was itself split between proponents of the gold standard versus those who wanted the currency backed as well by silver (i.e., labor, farmers, etc.). But in addition to these two parties, there was a third, the Populists, who also favored moving off the gold standard and whose overall philosophy was far closer to the Democrats than to the Republicans. The Populist party, however, was barely viable, and some of its members believed its best chance of survival lay with establishing an identity separate from the Democrats. On this score, the question was when to hold their national convention: before or after the Democrats? One faction argued that their platform should not focus solely on the issue of free silver but should emphasize its programs with respect to railroad regulation, imposition of an income tax and various pro-labor measures. Their opposite number within the party argued for an explicit focus on the issue of silver versus gold. The first faction wanted the party's convention held early to co-opt the Democrats promulgation of a broad platform, while the "silver faction" preferred a late convention to take advantage of whatever discontent emerged within the other two parties, and perhaps even, to make common cause with "silver-Democrats." As it turned out, the pro-silver faction won the argument, but then confronted the fact that the Democrats nominated William Jennings Bryan, who "stole" much of the pro-silver Populist thunder. Indeed, all that was left to the Populists, then, was to nominate Brian as well and to ultimately disappear as an independent force in American politics.

2.2 The Extensive Form

The common feature of the preceding examples is that each describes a sequence or potential sequence of actions—voting alphabetically, vetoes and veto over-rides, the ordering of horses for a sequence of races, a sequence of tribal votes, and the order with which parties hold their conventions. Each scenario, then, can be described by what we call an *extensive form*. Briefly, extensive form representations of situations attempt to describe in a semi-formal way the decisions people confront when interacting with each other, including who knows what when, whose turn it is to make a decision, and so forth. The components of an extensive form are simple: *decision nodes* that tell us who is making the choice, *branches* that correspond to specific decisions or choices and that lead to subsequent nodes, *chance nodes* that model the uncertainty occasioned by nature's probabilistic moves, *terminal nodes* that specify when a "game" or decision scenario ends, and a specification of the outcome at each terminal node. To this list we will also introduce the notion of an *information set* to describe what a person knows about the prior choices of other decision makers (including the chance moves of nature) when it is their turn to make a decision.

We can begin with our first example, the legislative pay raise, and with the supposition that, initially at least, voting will be an alphabetical roll call vote. Also, to simplify our presentation, assume that there are but three legislators, A, B and C. Figure 2.1, then, portrays the extensive form of this scenario. Thus, each node specifies whose turn it is to vote, each branch specifies the legislator's vote (for or against the raise), and when all voting is done, the terminal nodes are denoted by the final outcome from the perspective of each legislator. Also, to facilitate subsequent analysis, after assuming that a legislator's most preferred outcome is "worth" 2, and least preferred outcome "worth" –1, we also give the utility equivalents of each outcome.

We can now answer the question as to whether you'd prefer that your last name be Aardvark or Zzekial. Look at the problem from A's perspective. If he

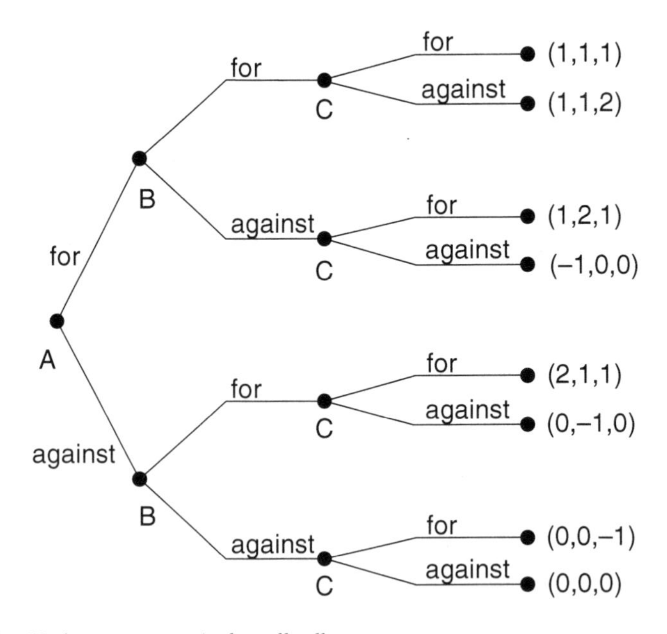

Figure 2.1 Voting on a pay raise by roll call vote

or she votes for the raise, then either B or C can vote against it and still have the raise pass. But if A votes against it, then both B and C must vote for it for the raise to pass. Thus, it seems reasonable for A to speculate that by committing to vote "against," B and C can be forced to vote "for" because their preferences match A's (i.e., to pass the raise even if that means to vote for it). But A can reason even more precisely; namely, A can look to what C might do at each of C's decision nodes. To wit, if C observes both A and B voting "for" or voting "against," then C would be a fool to vote "for"; however, if A and B split their vote, then C is pivotal for the outcome and, given an avaricious preference for more money, is sure to vote "for." Legislator B, not being a dummy (or at least not assumed to be one here), can make the same calculation and conclude that if A votes "for," C can be induced to vote "for" if B votes "against" (thereby securing the most preferred outcome) whereas if A votes "against," then B should vote "for" because C will join to pass the raise. So moving back to A once again, A can reason "if I vote 'for', B will vote 'against' and C will vote 'for'; but if I vote 'against,' both B and C will deliver my most preferred outcome by joining to get their second most preferred outcome by voting for the raise."

This might seem confusing at a first reading, but Figures 2.2a and 2.2b graphically portray the preceding reasoning. Since it is already clear what legislator C will do at each of its decision nodes, we can eliminate the choices that will not be made and thereby produce the reduced extensive form in Figure 2.2a. But now it's clear what B will do at each of its decision nodes—if A votes "for," B will vote "against," whereas if A votes "against," B will vote "for." So if we now take Figure 2.2a and eliminate the choices B will not make, we are left with Figure 2.2b. And from here we see that A's choice is unambiguous—vote "against." So, to answer our question, for this roll call vote at least, it's best to vote first and, thereby, force the other members of the legislature to swallow the bullet in their own self-interest and deliver you your favored outcome.

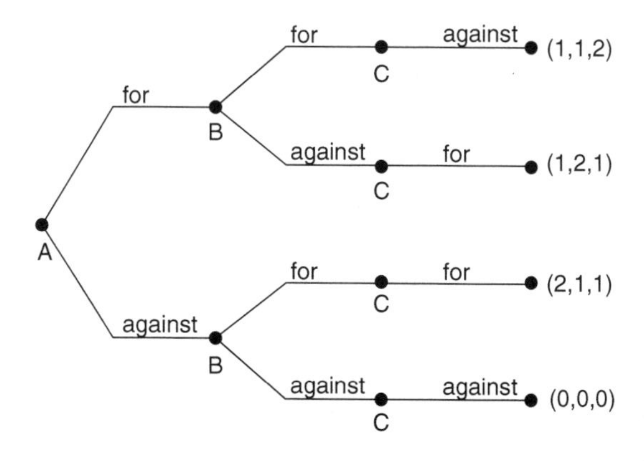

Figure 2.2a First reduction

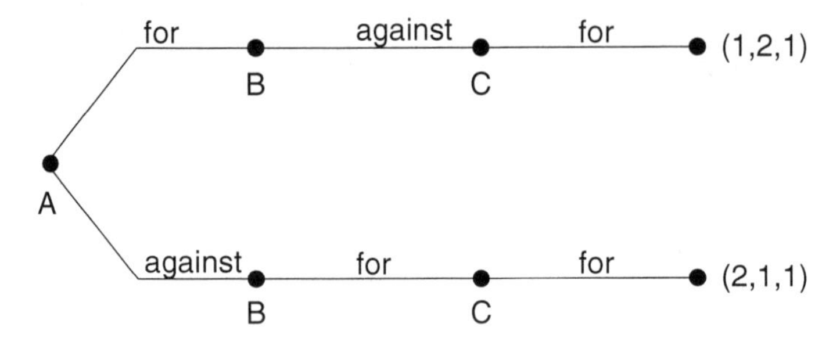

Figure 2.2b Second reduction

For another example of how we approach the analysis of extensive forms, consider this problem: Three committee members must pick one (and only one) candidate from a list of four. The preferences of the committee for the four candidates, A, B, C, and D, are given below, ranked from best to worst. The committee uses a procedure whereby member 1 first vetoes a candidate, then member 2 vetoes a candidate from the remaining three, then member 3 vetoes one of the two remaining candidates. The candidate who is unvetoed is elected.

member 1:	C	B	D	A
member 2:	B	C	A	D
member 3:	A	B	C	D

It might seem reasonable, now, to suppose that a person should veto their least preferred alternative, but to see if this supposition is correct, Figure 2.3a portrays this situation's full extensive form, where the branches label the alternative that the voter in question vetoes. Figure 2.3b shows the first reduction after member 3's optimal responses are determined and Figure 2.3c reduces this form further by eliminating those branches that are not 2's optimal responses, in anticipation of what 3 will do. This last figure reveals that member 1 should first veto alternative B, his second choice, so that C ultimately prevails.

The preceding examples illustrate the approach we ought to take when beginning an analysis of an extensive form. Whenever possible, we should prune that form by seeing if there are any choices we can infer directly. At this point, however, the reader might object with the argument: *What you have said is common sense. However, people typically must act without the opportunity to sit down and draw out extensive forms, in which case there is no guarantee that they will act as you have just described. Indeed, if only a single decision maker fails to act as assumed, your entire analysis is invalid.* This objection is legitimate. But once again, we must keep in mind what the assumption of common knowledge implies—namely, that all persons are aware of the situation's extensive form. If for some reason a person's analytic capabilities are limited—if, for example, they cannot see the future, if they forget the past, or if they are uncertain about various aspects of the situation—then these facts ought to be represented in our extensive form description and be known to everyone involved. The reader

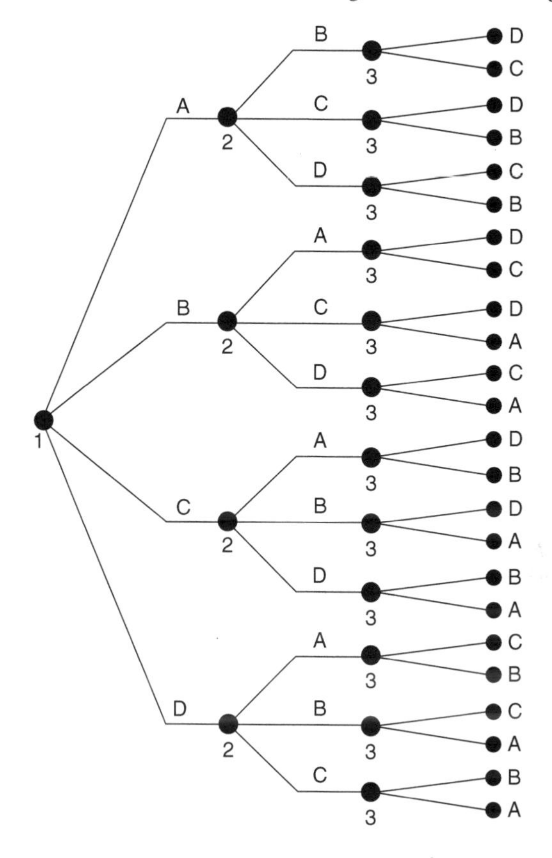

Figure 2.3a Sequential veto

is correct, of course, to believe that such common features of reality can make it difficult to model and analyze a situation. However, before we assume, for instance, that it's a stretch to think that people "work backwards on an extensive form" in the way just described, try playing tic-tac-toe with any average 12-year-old. Odds are, regardless of whether you go first or second, you won't win—especially if that 12-year-old has minimal experience with the game. Why? After all, whoever is assigned X has 9 branches emanating from their first decision node (actually, only 3 if we take account of the symmetry of the cells); the player denoted O then has 8 branches from their first of 9 (or 3) nodes; X then has 7 branches; and so on. Tic-tac-toe surely seems a lot more complicated than our legislative scenario, and yet 12-year-olds with minimal experience seem perfectly able to look ahead to see what you might do in response to whatever they do, and plan accordingly. Now it may indeed be a stretch to suppose that the average member of a national legislator or parliament has the same mental capacity as a 12-year-old, but it's best, as an initial assumption at least, to assume that they do.

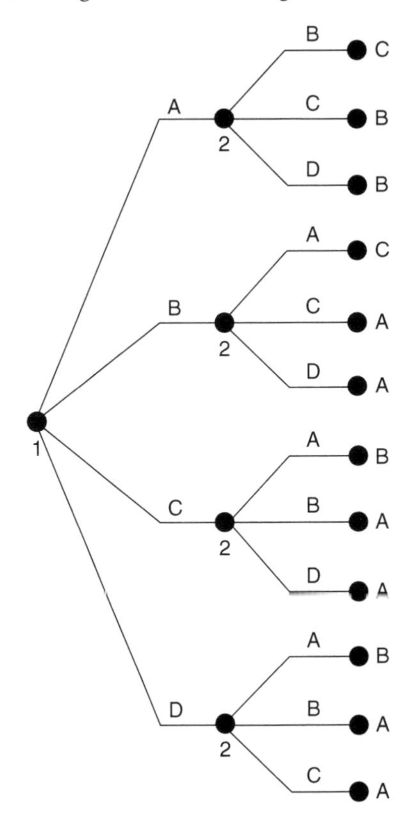

Figure 2.3b First reduction of sequential veto

Continuing with our presentation of extensive forms, we note now that these forms need not be limited to supposing that decision makers can act only once in a situation. People can choose sequentially—first one, then another, then the first again—and, indeed, who chooses and what alternatives they confront can be made dependent on the actions others take. For example, suppose each chamber (H and S) of a bicameral legislature consists of a single member (never let it be said that theorists do not know how to simplify a problem) and that both legislators must pass a bill (choices p and f) before it goes to the president (P), who can approve (a) or veto (v) it. If the president vetoes, the two chambers must vote to over-ride (o) or sustain (s) the veto. If we suppose further that legislative chamber H moves before S, Figure 2.4 portrays the extensive form of this situation. Notice that we allow chamber S to vote even though it knows that chamber H has already killed the legislation. This assumption, of course, allows us to accommodate the possibility that legislators treat the failure of outcomes differently, depending on who can be held responsible for killing the measure.

Of course, if all of politics were as simple as the preceding examples or as readily solvable as tic-tac-toe, there would be little need for game theory—and

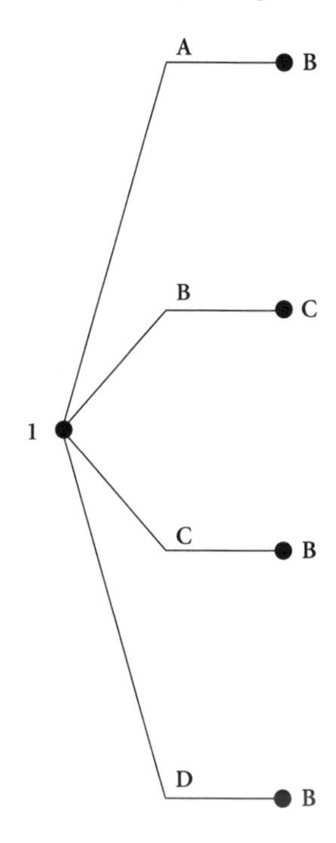

Figure 2.3c Second reduction of sequential veto

perhaps even little need for political scientists. Fortunately, the world can be a tad more complex. For example, suppose in the preceding example that both legislative chambers act simultaneously. That is, suppose the Senate does not know how the House has voted. In that case, we would draw the extensive form as in Figure 2.5.

The dashed envelopes around the Senate's decision nodes—called *information sets*—indicate that when it is that body's time to act, it doesn't know which node within the set it is at: in the first information set it doesn't know whether the House has passed or failed the bill in question and in the second dashed envelope it doesn't know whether the House has sustained or over-ridden the President's veto. Of course, assuming that one legislative chamber must act in ignorance of how the other has voted might reasonably be deemed a far-fetched example, so instead consider our pay raise example again, but this time assume that voting is by secret ballot rather than by a roll call. In this case, it makes little difference who actually casts their ballot first, since the votes and final outcome are not revealed until all votes have been cast. Thus, without loss of generality (be careful when you read those words, since they often disguise more than they reveal) we can again assume in our example that legislator A votes first, B second and C third. However, now we must indicate that B does not know how A voted when it is B's turn to cast a ballot. Thus, in Figure 2.6 we draw an

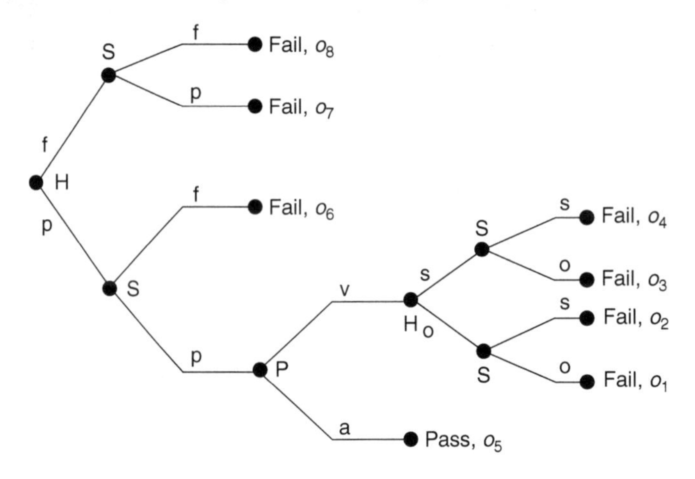

Figure 2.4 Legislation in a 2-chamber legislature with sequential moves

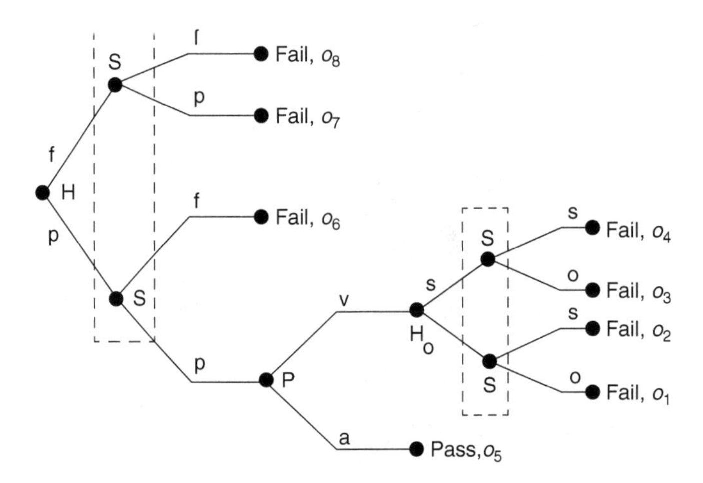

Figure 2.5 Legislation in a 2-chamber legislature with simultaneous votes

envelope around both of B's decision nodes to indicate that although B knows when it is time to vote, B does not know which of the two nodes he or she is at. Similarly, when C votes, C does not know how A or B voted; thus, we draw an envelope around all four of C's decision nodes.

Information sets more often than not make things interesting. Specifically, notice that we can now no longer work backwards up the extensive form in order to advise A how to vote. The envelope around all of C's decision nodes means that C doesn't know, when it is his turn to vote, which node he is at. At some nodes C prefers to vote "for" and at others C prefers to vote "against," but there is no unambiguous choice, short of cheating and looking at what A and B have done (and since the votes are simultaneous, A and B

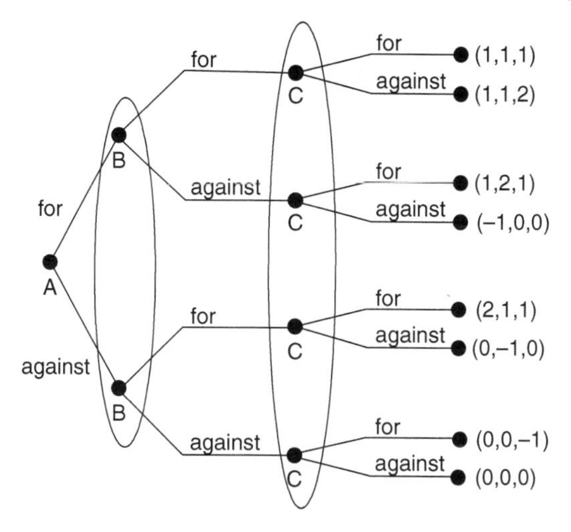

Figure 2.6 Voting on a raise by secret ballot

might not yet have done anything). The same argument applies to B. Not only doesn't B know what A has done, B can't unambiguously infer what C will do since B knows that C will not know how other legislators have voted. We can, moreover, complicate things further by, for instance, letting C spy on either A or B and learn their choice before C votes. In this case, C is in effect choosing between the extensive forms in Figures 2.7a and 2.7b. In Figure 2.7a, C's information sets are divided to indicate knowing whether he is in the upper half of the extensive form (corresponding to A voting "for") versus the lower half (corresponding to A voting "against"). Figure 2.7b, on the other hand, models having C know B's vote beforehand but not A's. It is still the case that we cannot work backwards

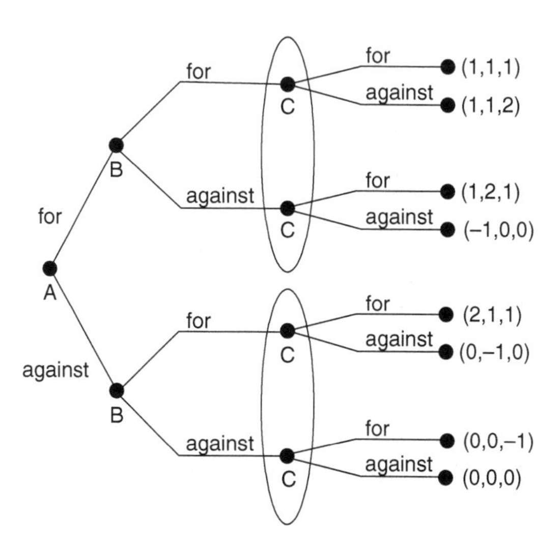

Figure 2.7a C spying on A

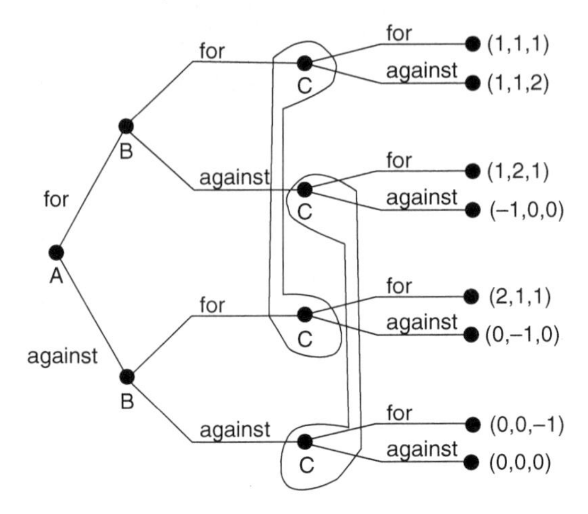

Figure 2.7b C spying on B

up either of these two extensive forms, but once we develop the tools for "solving" for the choices people are likely to make here, we can, for instance, determine whether C should have a preference for whom to spy on.

For a final play on this example, consider Figure 2.8, which assumes that B can decide whether to spy on A or on C.

The reader is free at this point to construct other extensive forms using this example, but as a final example of the use of information sets in extensive forms, consider Figure 2.9. This form illustrates a situation in which persons 1 and 2 alternate in making binary choices, where each observes the choices of the other, but person 1 can only recall the last decision and person 2 can recall only the last two. Because 2 has the better memory, 2's information sets are more detailed and, thus, more numerous than are 1's.

More generally, information sets allow us to distinguish between two classes of interdependent choice situations—games with perfect versus imperfect information.

> *In a game with* ***perfect information,*** *each and every information set encompasses a single decision node. Such games correspond to situations in which all decision makers know the choices that they and others made any time in the past. In a game with* ***imperfect information,*** *at least two decision nodes are in the same information set for at least one decision maker. Such games correspond to situations in which at least one decision maker is unaware of a prior choice made by another decision maker.*

Although information sets increase the applicability of extensive forms, certain rules must be adhered to when we use them. Specifically,

1. All decision nodes contained in an information set correspond to the same decision maker. An information set describes what a decision maker knows

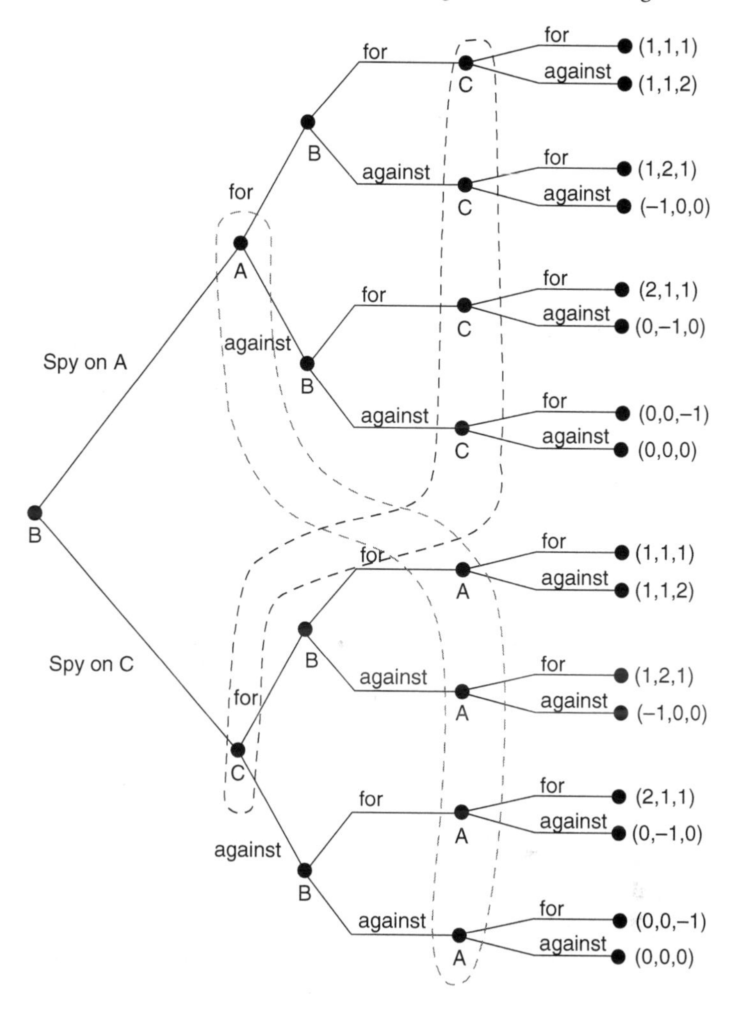

Figure 2.8 B's choice of who to spy on

about earlier decisions when it is his or her turn to act and, thus, each information set describes something about a specific person.

2. Each node in an information set has the same number of branches emanating from it; otherwise, the decision maker can identify which node he or she is at by the number of alternative actions.
3. For the same reason, the labels attached to the branches at each node must be matched by the labeling of branches at any other node in the same information set.

To this list we impose two additional but tentative restrictions on extensive forms. First, we suppose that the branches connecting nodes cannot "double

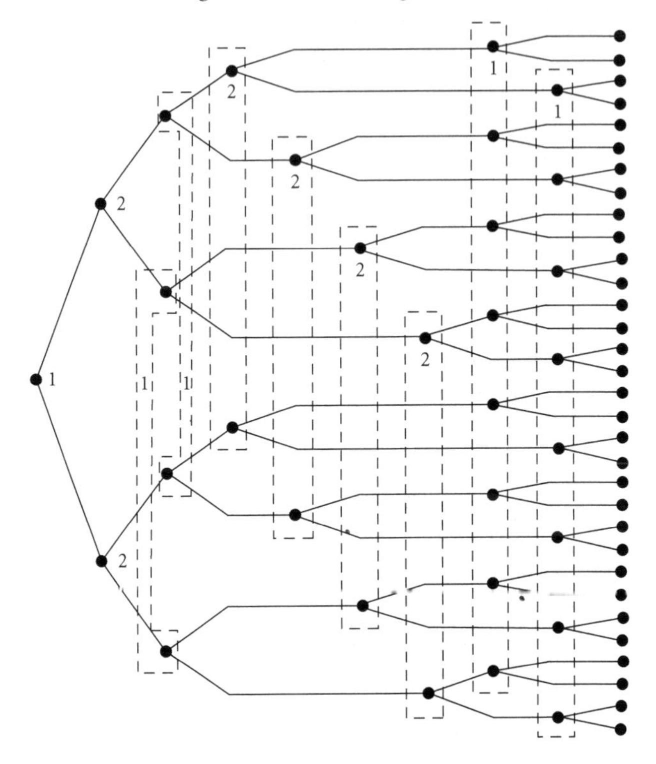

Figure 2.9 Imperfect recall

back" to some previous node. Later we accommodate the possibility that some branches may require that the people replay all or part of the game, but for now it is easier to consider situations in which such possibilities are excluded. A second and related assumption is that all branches lead eventually to some outcome. That is, we suppose that the time allowed to play a game is finite. As with repetition, we later consider situations that, in principle, allow for the possibility of infinite play. Finally, we will suppose that the number of choices at any node is finite. Again, this assumption is not essential, but it does simplify exposition (to say nothing of the graphical portrayal of an extensive form).

With the lessons of our examples in mind, let us now return to the advice offered by Sun Bin to General Tian Ji when formulating a strategy to defeat the King of Qi in horse racing. To offer some notation, suppose the king's horses are labeled Q1, Q2 and Q3, where Q1 is known to be faster than Q2, which is faster than Q3. Let Tian's horses be labeled T1, T2 and T3, where T1 is faster than T2, which is faster than T3, and where overall, the ranking of the horses by speed is Q1 > T1 > Q2 > T2 > Q3 > T3. Sun Bin's advice, now, is predicated on the extensive form shown in Figure 2.10, where the outcome (1,0) means that the king wins a majority of the races and (0,1) means that Tian's horses win a majority of times. That is, the implicit assumption in his advice is that the king first chooses which horse to run in the first race and then which to run in the second (with the third horse, by default, relegated to the last race). If we work down this tree we see that regardless of the king's choices, as long as General Tian can assign his horses to the races knowing the king's assignments, Tian cannot lose. The darkened branches denote optimal choices for Tian, and notice that regardless of how the king chooses, there is a branch for Tian that leads to an (0,1) outcome. There are, of course, other choices, and the reader should be able to draw the extensive form when Tian must move first, when Tian and the king must make their choices simultaneously before each race, and when the order with which each contestant will race their horses must be

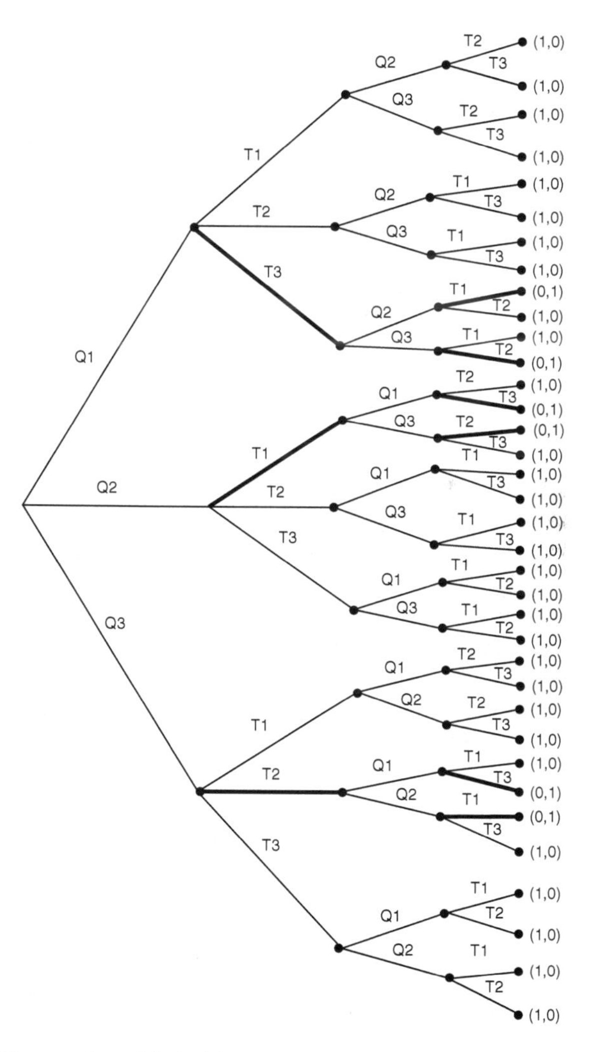

Figure 2.10 Horse racing with the king

decided simultaneously before any race is run. The reader should also be able to show that if Tian must move first, the king always wins. The tools for analyzing the remaining two possibilities, though, will be presented later in this volume, but this much is clear: A king intent on winning the contest, because he is king, will most certainly require that Tian move first. Any monarch who fails to see this beforehand will, most likely, not be king for very long.

If we turn now to our example of presidential vetoes, our intuition that a majority over-ride negates a president's power to veto legislation is essentially correct. But it is correct ONLY if legislative agendas—the order in which motions are considered and voted upon—abide by a certain form. To see what we mean, for purposes of our example let us treat the legislature as a unified entity. Nevertheless, because it is in reality a collection of individuals, it is subject to exhibiting intransitive preferences of the sort discussed in Chapter 1. So suppose there are three possible outcomes, bills A and B plus the status quo Q, and suppose a unicameral (one chamber) legislature holds the intransitive preference $A > B > Q > A$ whereas the president prefers Q to both A and B. Now consider the "extensive form" in Figure 2.11a wherein there is no executive to consider and the legislature votes as follows: First it decides whether to approve of A. If it does, then A is the final outcome, but if it does not, it next considers B. If B is approved, it is the final outcome; otherwise, the status Q prevails. Working backwards in the same way as in our previous example, we see that although B defeats Q, A defeats B and thus A is the outcome. Now, however, consider the agenda in Figure 2.11b, which allows for a presidential veto that the legislature can over-ride if it so chooses. Working backwards up this tree reveals that if A is approved in the first vote and if the president vetoes A, then the legislature will not over-ride the veto since a majority within it prefer Q to A. But if A is rejected in favor of B, then regardless of whether the president vetoes or not, the final outcome is B. So by approving of A in the first vote, the outcome is Q whereas by rejecting A in favor of B, the outcome is B. So now the legislature, which prefers B to Q, should reject A in the first vote to get B.

What we see here, then, is that the insertion of a presidential veto with a majority over-ride changes the outcome from A to B. And since we assume only that the president prefers Q to both A and B, we have an example in which a president might prefer not to have the authority to veto things (if he or she

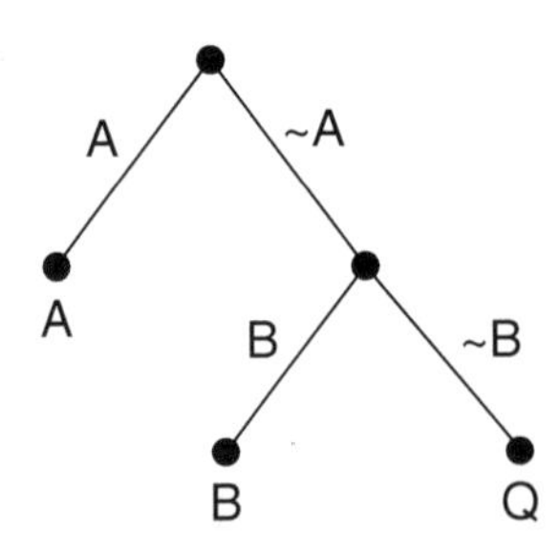

Figure 2.11a Legislative agenda with no presidential veto

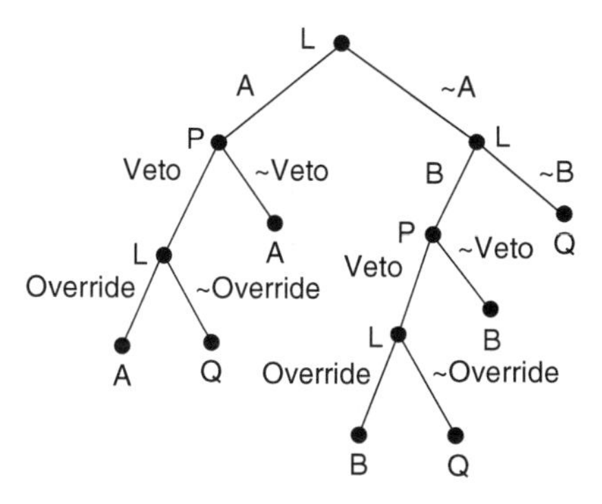

Figure 2.11b Legislative agenda with presidential veto

prefers A to B). There is, though, one other interesting feature of this example. Notice that if we were to rely exclusively on observation and not the preceding abstract analysis, we might conclude that vetoes that can be over-ridden by a majority are toothless. Presumably, presidents prefer not to see their vetoes over-ridden, if only because it gives the appearance of weakness and an inability to control the legislative branch. If so, then because in our example the president knows that a veto yields the same outcome as no veto, unless he or she wishes to publicly state opposition to B, vetoes are never over-ridden because they are never observed. And if never observed, we might incorrectly conclude that a constitutional provision that allows a presidential veto to be over-ridden by a majority affords the president no influence over outcomes.

Turning next to voting within the Iroquois Confederation, even if we again treat the representatives from the individual tribes as unitary actors, the extensive form representation of a simple vote as to whether to adopt A or maintain the status quo Q can be complex. However, because the Mohawk and Seneca lords (representatives) must be unanimous to proceed and because the Cayuga and Oneida lords must also be unanimous, we will treat these pairs of tribes as unitary actors—in which case, the voting tree looks as shown in Figure 2.12 (assuming that if the Mohawk-Seneca side and Cayuga-Oneida side propose nothing, then there is nothing for the Onondaga to rule on).

As complex as Figure 2.12 might seem, the complexity of the situation that results increases considerably if we choose to treat individuals within each tribe as separate decision makers (and in fact, the Confederation Constitution defines subsets of actors within several of the tribes and assigns them specific responsibilities) or if we attempt to model the possibility that the Seneca contemplated discussing matters with the Oneida before conferring with the Mohawk. Minimally, though, Figure 2.12 makes clear that, contrary to what we might come to believe by watching too many Hollywood cowboy movies, the governing body of the Iroquois Confederation did not make decisions by dancing around a campfire and letting the patterns of rocks or bones thrown in the air or the pattern of the smoke arising from their fire dictate decisions. Instead,

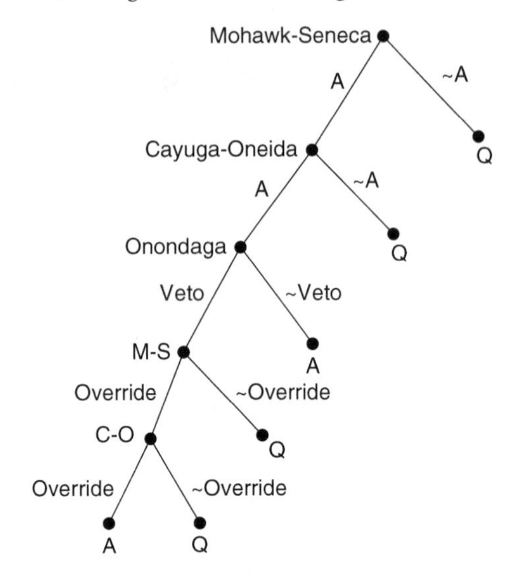

Figure 2.12 Voting in the Iroquois Confederation

they proceeded very much like a contemporary legislature with a well defined, constitutionally prescribed agenda.

Turning now to our discussion of crisis escalation, or rather an auction for \$100, in lieu of actually drawing out an extensive form let us simply examine things logically; which is to say, let us reason back down the implicit extensive form. Suppose the current bid by 1 is \$70, which means that 2's last bid is \$60. Should 2 bid \$80? Person 2 knows that by withdrawing at that point from the auction, he loses the \$60 he last bid and 1 gets the \$100 for \$70. So 2 might reasonably decide to up the ante by bidding \$80 in the hope that 1 will instead withdraw. But is this hope valid? Suppose, then, that we go to the end of that branch of the extensive form where person 2 is the last to decide . . . in this case, whether to pass or bid \$200. If 2 passes, he foregoes the \$100, but since his last bid was \$180, he must still pay this amount. On the other hand, if he bids \$200, he at least gets the \$100, albeit at a cost of \$200. So he bids. Now consider 1's prior move, which is either to pass and incur a cost of \$170 (his last bid) or to bid \$190. But if he bids \$190, he knows 2 will bid \$200 and his cost is \$190. So it seems logical that 1 should pass and let 2 have the \$100 at a cost of \$180. Thus, when 2 must decide whether to bid \$180 or pass, he knows that if he bids the \$180, 1 will pass, so 2 does, in fact, bid. Stepping back once again to where 1 must decide whether to bid \$170 or pass, 1 knows that 2 will bid \$180 if he bids and that he will then pass and let 2 have the \$100. So seeking to avoid an unnecessary expense, 1 passes. Continuing back down the extensive form with this reasoning, consider 1's decision node when he must decide whether to bid \$90 or to pass. Passing allows 2 to walk away with a bargain—\$100 at a cost of \$80 (2's last bid). But if 1 bids \$90, he knows that 2 will bid \$100, and that if he keeps on bidding, he will accomplish nothing other than to force 2 to over-pay for the

$100. So again, if his personal finances are all that matter to 1, he should forego bidding. But this same argument applies to the choice between passing and bidding $70, between passing and bidding $50, and so on. The net result is that person 1 should refuse to bid $10 initially and to let 2 have the $100 for $10. We note incidentally that a simple rule change ensures that no one wins the $100. Specifically, if 1 is allowed to reenter the auction after passing and after 2 opens the bidding at $10, then the situations of players 1 and 2 are now reversed, and we can conclude that both persons will refuse to open the bidding.

Thus, our simple auction DOES NOT model crisis escalation, and so we must conclude that when crises do escalate, they do so for reasons not captured by our example. What might those reasons be? Here we can only speculate, but one possibility is that our model misrepresents people's preferences and that those preferences exhibit a form of reverse altruism. Suppose that while our two bidders might prefer to avoid net losses, *ceteris paribus*, they gain some satisfaction from making their opponent incur losses. For example, suppose 1, looking down the extensive form, knows he can force 2 to pay more than the objective value of the auction's prize by simply refusing to opt out of the bidding and by bidding at least up to $110. If, for whatever reason, 1 discounts the costs to himself and focuses instead on the costs incurred by the opponent, bidding will proceed beyond the value of the prize. Indeed, if monetary value is not the sole consideration and if the primary thing 1 wants to avoid is having 2 incur a net gain from the auction (by having 1 drop out of the bidding early), then surely 1 will bid to $90 or even $110, if not higher. It would seem, then, that a crisis can escalate when the predominant concern of one or both sides to a conflict is that the opponent enjoys a relative gain or a gain at too low a cost. And crises are averted when both sides understand the consequences of such preferences and consciously act to suppress them even if that means one side or the other enjoys an advantage.

An alternative explanation for crisis escalation—one that we are not prepared to consider in this chapter—is that the players have incorrect or imperfect beliefs about each other's preferences or attitudes towards risk. Postponing consideration of such things, we can still exclude one potential explanation for crisis escalation: the mistakes made by players at the early stages of the game. Suppose, for example, that player 1 accidentally bids $10 and that player 2, following the logic of our analysis, bids $20. In effect, the game's "first move" is again player 1's and the existence of 1's initial $10 bid and 2's $20 response simply becomes a description of the situation's status quo environment. Unless player 1 errs again by bidding $30, which will be followed by 2's response of a $40 bid, player 1 should simply admit its error and drop out of the bidding. And, if we sweep aside many of the complexities of the situation, this is perhaps what occurred in the Cuban Missile Crisis when Soviet Premier Nikita Khrushchev ordered his ships to turn back and not run the American blockade (although surely events during that crisis hardly matched such a simple scenario).

With respect to America's 1896 Presidential election and the choice facing the Populists with respect to the timing of their convention, we can ignore the Republicans since they were certain to nominate a pro-gold candidate. Drawing a simplified extensive form is straightforward, then, if we take the bold step of modeling the two parties as unitary actors and add the simplification of the Democrats choosing between a unified versus a factious convention by drafting a narrow versus a broadly based party platform. The bare outline of such a form is portrayed in Figure 2.13. Keep in mind, however, that it is a bold step to model ANY American political party as a unitary actor, and a more careful and satisfactory model of the situation should, at a minimum, divide the parties into their separate factions, which, in turn, makes it difficult to assign preferences to the specific outcomes. Can we, for example, say that the Democrats preferred a unified to a factious convention in the fight over whether to issue a broad platform versus one that focused specifically on the issue of backing the dollar with silver? Or is it better to say that any such preference is the consequence of a variety of sub-games played by the various factions at the convention itself? Similarly, notice that we differentiate outcomes by whether or not the Populists held their convention before or after the Democrats so as to open the possibility that the public's view of things would depend on whether the party is viewed as leading or following the Democrats, which again complicates any assignment of preferences. Thus, Figure 2.13 hardly ends our analysis of this historical event, but any attempt at filling in its details illustrates the analytic decisions that must be made when modeling such real world processes.

For a final example, consider once again our brief discussion in Chapter 1 of the decisions confronting Japan and the United States with respect to America's

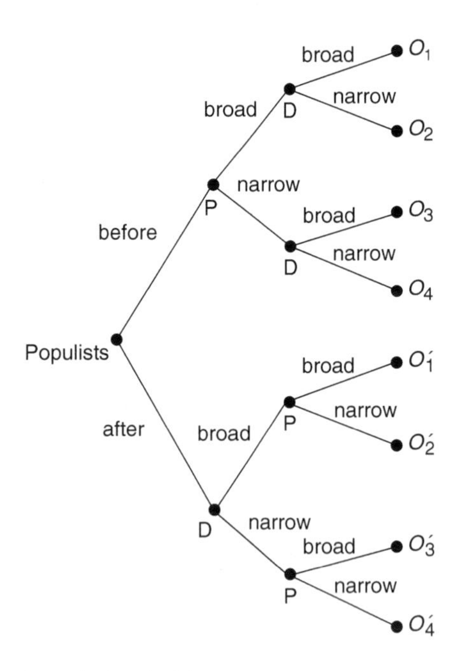

Figure 2.13 Populist Party timing decision

strategy of sending a single plane to drop a second atomic bomb in World War II. Our earlier use of this example was intended to illustrate the difference between decision theoretic and game theoretic reasoning. Here we can consider how to model the situation more completely after adding the wrinkle of supposing that the U.S. has to guess at the likelihood that Japan's strategic command assumed that the U.S. was capable of manufacturing a second bomb. Recall now that in making the decision to fly a lone plane to Hiroshima, the United States was most likely treating matters as a simple decision problem—betting that Japan would treat that bomber as any other lone observation plane. After Hiroshima, however, the United States had to guess as to whether Japan would assume that the U.S. had additional atomic bombs and, if so, whether it would pursue the same tactic as before or whether, in delivering the next bomb, it would shift gears and imbed the plane carrying the next bomb in a bomber group. There are, admittedly, a number of ways to model this situation, so here we take the simplest path. First, we let nature decide whether or not the U.S. has a second bomb. The U.S., of course, knows what "choice" nature has made here, but Japan can only assign the probabilities p and $1 - p$ to the alternatives. And to simplify matters more, suppose that if the U.S. has a second bomb, it will definitely use it. The choices for the U.S., then, are whether to send a single plane (S) or fleet (F) over Japan's cities the next day while Japan must choose between intercepting (I) or not intercepting (not I) whatever appears in its skies. Finally, let us assign the probabilities q to the chances that if the bomb is imbedded in a fleet and if the fleet is intercepted by Japan's air defenses, the bomber carrying the deadly payload is destroyed. In extensive form, then, the situation looks, perhaps, like the one shown in Figure 2.14 (we forgo attaching values to the outcomes, several of which are decidedly negative for Japan).

Clearly, this representation greatly simplifies things. For example, we could include additional intermediate moves on the part of the U.S., such as probing Japan's air defenses with additional (but unarmed) scout planes to test Japan's responses. Japan, in turn, if fully cognizant of the existence of a second bomb and cognizant as well of the fact that the first was delivered by a single plane, might then take into account the possibility that the U.S. would, after Hiroshima, test Japan's air defense tactics by first sending in one or two additional scout planes before deciding on a tactic for delivering the second bomb. Of course, were Japan fully aware of its circumstances, it might attempt to lull the U.S. into complacency by ignoring the first or second lone bomber over its territory after Hiroshima so as to lead the U.S. to believe, falsely, that it was oblivious to circumstances, all with the idea of thereafter intercepting anything over its territory. In other words, we can readily imagine a game where each side attempts to deceive the other.

Manipulation of Beliefs: Lest the reader think we have made too much of the strategic calculations involved with America's use of its atomic arsenal in World War II, consider Zhang Xun, a general of the Chinese Tang Dynasty. Known for successfully defending Yongqiu and Suiyang during the Anshi Rebellion against the rebel armies of Yan, he is credited with

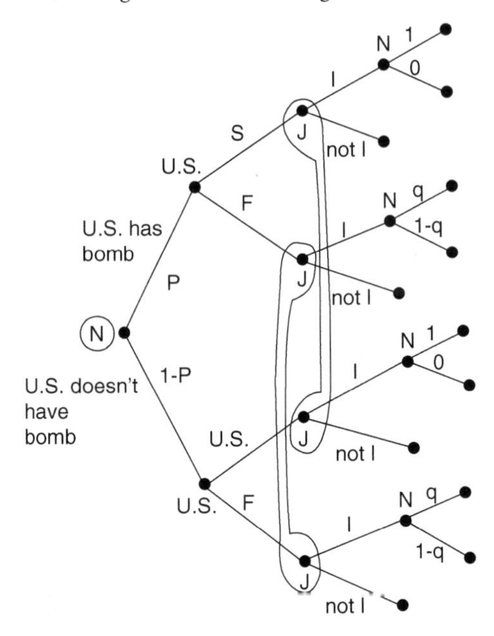

Figure 2.14 Atomic bomb decision

having blocked Yan forces from attacking and capturing the fertile Tang territory south of the Huai River. More precisely, in defending the city of Yongqiu in 756 AD, Zhang's troops had run low on arrows. He ordered his troops to make about 1,000 scarecrows and had his soldiers dress them in their own armor. At night, the soldiers hung the scarecrows from the fortress wall so as to be easily seen, whereupon the rebel commander, seeing the armored black figures at a distance, ordered his archers to shoot and stop a potential ambush. When the scarecrows were pulled back over the wall with about two hundred thousand arrows, the enemy commander realized he'd been duped into resupplying his enemy and ordered his troops not to waste arrows by shooting at black figures coming off the wall. So when Zhang rehung the scarecrows the next night, no arrows were fired at them. On the third night, black figures were again hung from the wall and were again ignored. But these figures were not scarecrows. They were, instead, five hundred of Zhang's best men who proceeded, in a murderous ambush, to inflict heavy casualties on the rebels as they slept.

This classic tale serves as a warning about how to treat uncertainty in situations involving interactive decision making. Zhang's opponent made the fatal error of not fully appreciating Zhang's incentives to strategically manipulate perceived probabilities—in this case, the probability that the "scarecrows" were precisely that. Indeed, a careful reading of Sun Tzu's *The Art of War* would reveal innumerable examples of advice that consists primarily of manipulating the beliefs of one's opponents. But what's interesting from the perspective of this volume is what happens when your opponent knows you are trying to deceive. Consider the rebel commander in Zhang's case. Suppose he knew that there was a chance on any given night that the scarecrows would be real soldiers. What strategy should he follow then, short of quickly exhausting all his resources by firing as many arrows as possible at the hung figures on every occasion? And how might Zhang have acted if he had a higher opinion of the rebel commander's ability to recognize the game they were playing against each other? Could Zhang then exclude the possibility that his opponent was merely "playing possum" the second night by refusing to shoot at the scarecrows and was, in fact, attempting to lure Zhang into a false sense of security when he lowered his best men over the wall the third night? Can one, in other words, deceive someone who knows you are trying to deceive them and can one deceive a deceiver?

Further discussion of situations of this sort and answers to our questions must await the introduction of additional game theoretic concepts and tools as well as a discussion of how we might model beliefs that depend on prior observations when it is known to all decision makers that some or all of them might seek to manipulate beliefs. For the present, we will assume that all decision makers know the game they are playing. Thus, if each decision maker were asked to draw out the extensive form of the situation that confronts them, each would draw the same form. In one interpretation of our example of Zhang Xun, we could say that this assumption is violated because Zhang's opponent failed to include in his mapping of the extensive form any branches in which Zhang chose to hang scarecrows from the fortress walls. Alternatively, we might model the situation in accordance with our assumptions by saying that Zhang's opponent simply put too low a probability on the likelihood of using scarecrows to replenish Zhang's supply of arrows.

2.3 Voting Agendas

As we've already seen in several examples, it is sometimes convenient when dealing with groups of people to collapse their decision nodes into a single node, especially when analyzing formally prescribed parliamentary voting scenarios. Imagine, for example, attempting to represent a vote in the U.S. Congress with 435 members by a fully drawn out extensive form. There may not be an available sheet of paper large enough to do this. At times, of course, exploring such extensive forms in detail may be unavoidable as when individual legislators possess choices unavailable to others or when the distribution of information is not symmetric. But often such detail is superfluous and, as with our example of presidential vetoes (and as long as we keep in mind that collectivities can hold intransitive preferences), it is sufficient to consider only an abbreviated version of the extensive form termed a *voting tree*, such as the ones in Figures 2.4 and 2.5. Voting trees, in turn, can be analyzed in much the same way as a fully drawn extensive form.

To illustrate this along with the impact agendas can have on outcomes, consider a legislature with the intransitive preference $A > B > C > A$ and where all three alternatives are preferred to the status quo Q. In this instance, we can ignore any final votes for passage because the legislature will approve anything that reaches a final vote. Commonly now, Western legislatures will, when confronted with three alternatives, employ an *amendment agenda* whereby two alternatives, say A and B, are first paired against each other and then the winner of that first vote paired against C. Denoting this agenda by {A,B,C}, there are, then, five other possible agendas with these three alternatives: {A,C,B}, {B,C,A}, {B,A,C}, {C,A,B} and {C,B,A}, though of course the agenda {A,B,C} is equivalent to {B,A,C}, {B,C,A} is equivalent to {C,B,A} and {A,C,B} is equivalent to {C,A,B}.

To see, then, how setting the agenda can influence outcomes, consider first {A,B,C}. Assuming again that our legislators look ahead before voting on A versus B, they know that if A passes, it will be paired against C and C will prevail. Thus, the *strategic equivalent* of voting for A is C. On the other hand, if B wins the first vote, it will be paired against C and will again prevail. So the strategic equivalent of voting for B is B itself. Our legislators, then, know that their initial vote is not really between A and B but between C and B, in which case, because a majority prefer B to C, B prevails as the final outcome. More interestingly, however, notice that if we reproduce this analysis with the agenda {A,C,B}, the final outcome is A whereas if we use the agenda {B,C,A}, the final outcome is C.

This simple example not only illustrates how voting trees are analyzed, it also illustrates the power an agenda setter can have on outcomes. Indeed, in this example at least, the agenda setter is a virtual dictator of the outcome. For another example that illustrates the power of an agenda setter as well as the difference between *voting strategically* (i.e., looking ahead to see what happens at different branches of a tree) versus *voting naïvely*, consider the example we have just considered in which we assumed—implicitly, by presuming that our voters look ahead—that everyone is strategic, and suppose instead that they are naïve. In this case, the agenda {A,B,C} yields the outcome C rather than B—the initial vote yields A, in which case the final vote, which pits A against C, yields C. Similarly, the agenda {A,C,B} yields B rather than A and the agenda {B,C,A} yields A rather than C.

Lest the reader remain unconvinced that setting agendas is perceived by those who do so as an important element of political strategy, we note that perhaps the classic instance of agenda setting in American history, albeit in a less formal context than a rigorously defined legislative agenda, occurred in the framing of the U.S. Constitution in 1787. Preparing for the Constitutional Convention weeks, even months, prior to its opening, James Madison arrived at the convention several days early and, in consultation with other members of his state's delegation, drafted his Virginia Plan, which outlined his ideal constitutional form. In the end, Madison did not get all that he wanted, but because the Virginia plan was the sole well-defined proposal presented to the Convention when it opened, it largely determined the course of the Convention's deliberations. "By preparing a plan for discussion that was neither too elaborate nor too vague, and that was an open challenge to the Articles of Confederation, he seized the initiative in behalf of the reform minded nationalists that was never relinquished from the first day to the last" (Clinton Rossiter, *1787: The Grand Convention*, NY: Macmillan, 1966: 171). However, the other members of the convention, which included the likes of Benjamin Franklin, James Wilson, Gouvernor Morris, Roger Sherman and Alexander Hamilton, were neither fools nor naïve. So while they largely allowed the Virginia plan to guide their discussions, they not only made provision in their procedural rules for a reconsideration of prior votes, as we noted in the previous chapter with respect to the particulars of presidential selection and terms of office, they did revote on things that appeared to have been resolved earlier. The potential utility of being flexible in this way is perhaps best illustrated by these three preference orders over four alternatives, A, B, C and D.

voter 1	voter 2	voter 3
B	C	A
A	B	D
D	A	C
C	D	B

The social preference order under majority rule for these four alternatives is the intransitive relation A > C > B > A > D > C and B > D. Now consider the amendment agenda {A,B,C,D} with naïve voting. In this case B defeats A, C defeats B and D defeats C. However, notice that A, which is where we began in the voting, is unanimously preferred to D, which is where we ended. It would surely be foolish, then, if a committee such as the Philadelphia delegates, realizing that such possibilities existed, did not make provision for a reconsideration of prior votes.

The preceding two examples also illustrate why it is sometimes advisable to set one's procedures in stone, including the order in which motions are voted upon (e.g., in the order in which they are introduced), so that voters can plan their strategies beforehand and not cycle endlessly over the choice of rules and procedures or the choice of rules for choosing rules, and so on. Regardless of whether voting is strategic or naïve, in our 3-alternative example, the choice of agenda inherits the intransitivity of the alternatives themselves. Specifically, with strategic voting, because A > B > C > A, then {A,C,B} > {A,B,C} > {B,C,A} > {A,C,B} whereas with naïve voting, {B,C,A} > {A,C,B} > {A,B,C} > {B,C,A}. Thus, our example reveals why it is important to have constitutions wherein rules and institutional forms are set somehow in stone before it is known to what alternatives they will be applied. If, in our example, the committee must first vote on an agenda or an agenda setter, then the committee might never reach a decision.

Our examples, though, should not be taken to mean that all voting decisions are susceptible to agenda manipulation. Specifically, suppose the legislature's preferences are not wholly intransitive and that there exists a motion to be voted on at some point in the process that is a Condorcet winner—an alternative that defeats all others in a pair-wise majority vote. It is straightforward to see now that *if all legislators are strategic and if they employ any form of a binary agenda (an agenda that only pits pairs of alternatives against each other at each decision node), then the Condorcet winner will necessarily prevail regardless of how the alternatives are ordered or organized in the agenda.* To see why this is true, suppose C is a Condorcet winner. Then there must be somewhere on the agenda a final decision node that pits C against some other alternative that C defeats. Thus C is the strategic equivalent at that node. But then C will be the strategic equivalent of the node that precedes that one, and so on and so forth. In other words, C will "bubble up" to become the strategic equivalent of at least one of the branches of the first node and, thereby, the branch chosen by a majority of the legislature. It might seem, then, that any legislative choice that entails a Condorcet winner will be uninteresting and not worth analyzing. This is not necessarily true, though, if the legislature employs an agenda other than an amendment agenda and if some legislators for various reasons are unable to vote strategically.

Pliny the Younger and Puerto Rican Independence: For an example of this possibility, we can turn to a classic example of *agenda manipulation* offered by antiquity in the records of the Pliny the Younger (AD 61–112). As recounted by him, several servants had been accused of murdering their master and the question was how to vote on their guilt and punishment. One option was to put the three alternatives up for a simultaneous *plurality vote* wherein the alternative receiving the most votes dictates the outcome:

A: innocent and set free
B: guilty and banished
C: guilty and put to death

A majority of those voting felt that the servants were guilty, but of those a minority believed there were extenuating circumstances and that at worst the servants should be banished. However, since the number believing they were guilty and that extenuating circumstances were of no consequence outnumbered all others that believed in their guilt, Pliny presumed that C would prevail in a plurality vote. Pliny, however, also believed the servants innocent and wished, at a minimum, to avert a death penalty. He understood, moreover, that those who believed the servants guilty and fully appreciated the extenuating circumstances had as their second choice setting the servants free. Thus, he foresaw that there were but three preference orders approximately evenly represented among those voting:

$A > B > C$
$B > A > C$
$C > B > A$

With each group approximately evenly matched in numbers, the majority preference is transitive, with B the Condorcet winner. So Pliny proposed a binary agenda in which they first voted on the guilt or innocence of the servants and then, if they were convicted, on their punishment. The outcome thus was changed to B—a not inconsiderable change from the perspective of the servants. We should note here that Pliny's analysis is incomplete and, in the process, raises an interesting question. Specifically, his analysis ostensibly assumed that voters were naïve under the plurality rule and, in particular, that those who most preferred B did not see the potential value in voting for A instead of their first choice. The question then becomes whether plurality rule is guaranteed to yield a Condorcet winner if one exists when voters are strategic. Answering that question, though, is one we must postpone until we develop some additional analytic tools in subsequent chapters.

However, lest the reader think we must revert to examples from antiquity to illustrate the potential role of agendas, consider the status of Puerto Rico. Without delving into complexities, Puerto Rico is classified as a commonwealth of the U.S. whereby it is neither a state nor an independent country. There are those who favor statehood as well as those who favor independence, and in a recent poll (*The Daily Koss* poll of November / December 2011), 41% of respondents preferred statehood, 35% preferred the status quo and 4% preferred independence, with 20% undecided. On numerous occasions, the U.S. Congress has authorized the residents of the island to vote on their status, but here the question is how should a popular referendum be organized. One alternative is to have all three alternatives on the ballot simultaneously and the outcome decided by plurality rule. The alternative, and the one suggested by the Congress and approved by Puerto Rico itself, is to have a two-stage referendum: In the first stage, the vote will be between maintaining the status quo versus making a change in Puerto Rico's status. If making a change wins, voters will choose between independence and statehood. It seems evident that if a vote were taken simply between the status quo and statehood, the outcome would be close, although most observers agree that the preference for statehood has been gaining strength. Those in favor of statehood, then, should most likely prefer a simple three-way plurality format. In contrast, opponents of statehood, and notably those who prefer the status quo, should prefer the two-stage format. Because those favoring independence are strongly opposed to statehood and because statehood is virtually certain to win if paired against independence, pro-independence voters can forestall statehood and live to fight another day by voting against change in the initial balloting. In fact, in the November 2012 vote, 52.4% voted as being dissatisfied with Puerto Rico's current status, while 44.7% were satisfied with the status quo. However, on a separate question, only 44.9% registered a preference for statehood, with 26% not registering any preference. If one assumes, then, that those who strongly prefer statehood indicated that preference, then clearly the two-stage procedure ensures maintenance of the status quo because the pro-independence voters would not vote for making a change.

Agenda Manipulation with a Few Unsophisticated Voters: Far be it for us to suggest that our elected representatives are unsophisticated when voting on legislation. However, there may be times when they are unable to vote strategically for fear their constituents would not understand their vote, especially if their vote can be misrepresented by their opponents in the next election. For example, consider the following three preference orders and suppose each group holding these preferences is sufficiently numerous so that the social preference is the intransitive order A > Q > B > A,

$A > Q > B$
$B > A > Q$
$Q > B > A$

Suppose B is not initially on the agenda so that A wins in a simple majority vote over Q, but that you are intent on defeating A and having the status quo prevail. One tactic, then, is to introduce B so that A and B are first paired, with the winner set against Q. Given the above preferences, though, this tactic will not work if everyone is strategic: If A prevails against B, it will defeat Q, ultimately, whereas if B wins in the first vote, it will lose to Q. So if everyone is strategic, they will see that the first vote is really between A and Q, and A will again prevail. However, suppose for some reason that those who hold the preference B > A > Q are unable to vote strategically. In this case, they will join with those who hold the preference Q > B > A to defeat A in the first vote, after which B loses to the status quo Q. Thus, those who prefer the status quo can get their way if they can find an alternative to A—or, more precisely, an amendment to A—that forces some legislators to vote sincerely.

The 17th Amendment to the U.S. Constitution: Can such a thing happen in a legislature of otherwise sophisticated members? In fact it has, and in quite dramatic fashion. For years, opponents of direct election of U.S. Senators thwarted Senate approval of the 17th Amendment to the U.S. Constitution by precisely this tactic. Keep in mind that up to the passage and ratification of this amendment, the Constitution allowed individual state legislatures to decide how Senators were chosen, and a good share of states, though not a majority, still appointed their Senatorial representation by the legislature without reference to a popular vote. In the attempt, then, to expand the domain of direct election and make the Senate "democratic," an amendment, A, was introduced that proposed simply that all members of the Senate be subject to direct election. It would seem that A was sure to pass because a significant majority of states already directly elected their representatives. After all, if you have to take your chances before an electorate that can fire you at the next election, then why shouldn't everyone have to endure the same threat? Of course, there were those Senators appointed by their state legislatures who liked things as they were and so, to defeat A, they proposed an amendment to the amendment designed to put an end to things. Specifically, they proposed the seemingly reasonable requirement that the federal government oversee all elections to federal office. However, as progressive as this requirement might seem as a means of guarding against electoral fraud, it was anathema to the South, which had used state and local control of elections to exclude blacks from voting. Thus, while Southern Senators—the product of states controlled by one party—had no objection to direct election, their worst nightmare was federal supervision of their Jim Crow electoral systems. We might suppose that all others favoring direct election could see the tactic of the amendment for what it was and vote strategically to defeat it. But if there is one thing an incumbent politician fears, it is the possibility that voters will fail to see the virtue of their voting record. Thus, a good enough share of pro-17th Amendment Senators felt boxed in by a concern that their constituents would not understand a vote against a provision that seemed pro-reform and strengthened the teeth of the proposed constitutional Amendment.

The tactic of thus amending the 17th Amendment postponed its ultimate approval for upwards of ten years, but such a tactic is not always guaranteed to work. In the 1970s the U.S. Congress debated a piece of legislation—The Common Sites Picketing Bill—that organized labor strongly supported. The tactic employed in this instance by opponents of the bill was to propose an amendment that actually strengthened it. Their calculation was that the pro-labor members of Congress would feel compelled to vote for the amendment, but that it would so radicalize the bill that the more conservative supporters of the original bill would vote against the amended version when considering it for final passage. The tactic, though, failed. Aware of the consequences of passing the strengthened bill, organized labor set about to inform their membership of the threat while simultaneously coordinating with their supporters in Congress, thus freeing those supporters to vote strategically for the weaker version of the bill.

Amendment agendas and plurality voting agendas are not, however, the only possibilities. Indeed, unless there are but two or three proposals on the floor of the U.S. Congress, amendment agendas are NOT used, despite the academic literature's nearly exclusive focus on them. For example, suppose the Congress must consider a bill, B, as originally reported out of committee, an amended version of that bill, B_A, a substitute bill, S, an amended version of that substitute, S_A, and the status quo Q. Then the rules of the U.S. Senate and House of Representatives dictate that first the substitute bill should be refined—a vote must first be taken on S versus S_A. Then, rather than take the winner of that vote and pair it against the next alternative, the bill reported out of committee must be refined with B pitted against B_A. The winners of the first two votes are then pitted against each other, and the winner of that third vote then pitted against Q in a vote for final passage. Alternatively, consider voting on U.S. Supreme Court appointments. The first vote, then, would be to approve or disapprove of the President's nominee. If that person is approved, the voting ends whereas, if disapproved, the Congress takes up consideration of the President's next nominee, and so on until someone is approved. Agendas of this type are referred to as *sequential elimination agendas.*

The potential complexity of U.S. Congressional agendas suggests that they leave room for considerable strategic manipulation, both in terms of the labeling of motions and in the actual voting. They also suggest that there is a considerable premium put on parliamentary expertise and skill. It would be foolhardy, though, to suppose that such skills are limited in their application to this specific legislative body and that they do not apply broadly to legislative bodies elsewhere. Indeed, it would be foolhardy to suppose the expertise and skill do not apply to virtually any voting body wherein decisions of some consequence are made.

2.4 Games and Subgames

Agendas represent only one class of games relevant to the study of politics, but the conclusions we arrive at with them using backward reduction suggests that we should try to generalize this process so that we can apply our method of analysis more broadly. To this end, we introduce the concept of a *subgame.*

> *Beginning at a decision node, a* ***subgame*** *is a part of an extensive form that is itself an extensive form and that, when detached from the original form, does not divide any information set.*

A subgame, then, begins at a particular node and includes all branches and nodes that follow it in the original extensive form and that by some path are connected to that node. For example, the legislative voting form shown in Figure 2.5 has three subgames: The "game" beginning at the node denoted H_O, the "game" beginning at the node P, and the full extensive form itself. The version of this scenario shown in Figure 2.4, in contrast, has seven subgames—the three just noted in Figure 2.5 as well as the four "games" beginning at the four S nodes. These additional subgames are not subgames in Figure 2.5 because separating them breaks information sets.

Next, we introduce the notion of a *strategy* for extensive forms:

> *A* ***strategy*** *is a rule that tells a decision maker what choice to make at each of his or her information sets.*

The concept of a strategy is critically important in the development of game-theoretic reasoning and analysis, and we will have reason to elaborate on its definition with a number of examples in subsequent chapters. For present purposes, we can think of a player's strategy informally as a "plan for how to play the game"—a specification, laid out beforehand, for what choices will be made under every possible contingency (information set). A typical strategy, then, might read as follows: "If I arrive at information set I_1 because my opponent(s) [or nature] chose y_1, then choose x_1, but if instead I find myself at information set I_2 because my opponent(s) [or nature] chose y_2, then choose x_2." Moreover, if the game continues beyond these choices, then the strategy under consideration, if it is to be complete, would also specify choices at any subsequent information sets that might be encountered. For example, then, when playing poker a player might decide to "play conservatively," which might entail such choices as not bluffing and folding whenever a hand appears weak. Alternatively, a player might choose an aggressive strategy of bluffing on occasion and never folding unless confronted by an opponent who appears to be acting overly aggressive. And a more complex strategy might allow for switching between conservative and aggressive play, depending on one's success (e.g., how much money they've won or lost to that point)

With the notion of strategy, or at least its rough outlines in mind, we can then move next to the idea of an *equilibrium*. In later chapters we will expand on this definition, but for now the following is sufficient:

> *An* ***equilibrium*** *is a vector of strategies, one for each decision maker involved in the extensive form, such that no person has a positive incentive to switch unilaterally to some other strategy.*

Continuing with our poker example, suppose each player around the table has, for whatever reason, chosen a strategy for how to play cards that night, where one of them decides to be aggressive in a wholly naïve (stupid) way and commits to bluffing at every opportunity. Unless his opponents are comatose, such a strategy is unlikely to be successful for very long because it will soon become evident to everyone that he is raising the pot regardless of whether his hand is weak or strong. Bluffs will begin to be called every time, and most likely the value of the chips before our player will soon approach zero. Clearly, then, such a strategy cannot be part of an equilibrium since there is a clear incentive to abide by a different and more sophisticated scheme.

Once again we will revisit these definitions, but we are now in a position to define a *subgame perfect equilibrium*, which formalizes the essential ideas behind working backward up an extensive form.

> *A* ***subgame perfect equilibrium*** *is a vector of strategies, one for each person, such that, when we look at any subgame, that vector implies an equilibrium in that subgame.*

Continuing with our poker scenario, we can, if we want, view every hand of poker and all that follows as a subgame—once the winner of the previous hand is determined and the chips reallocated accordingly, the next hand might be said to begin the evening anew (aside from the fact that some people may have more chips and others less than they did earlier). The all-out naïve strategy of being aggressive, then, dictates that our player in question bluffs in the next hand and all that follow. Suppose, though, that our player has come to understand that always bluffing is just plain dumb. Thus, being aggressive cannot be part of any equilibrium in what follows—in the subgame that consists of all subsequent hands.

As noted above, we will have reason to return to the notion of strategy and the representation of strategic situations in greater detail, but for the present it is useful to see the distinction here between an equilibrium and a subgame perfect one. Consider, then, the following simple scenario: If person 1 (who moves first) chooses L, the outcome (2,2) prevails and the game ends. But if he chooses R, then person 2 gets to choose between L and R. If 2 chooses L, the outcome (3,1) prevails whereas if he chooses R, the outcome (0,0) prevails. Working backwards up the implied extensive form yields (L, L) and the outcome (3,1) as the unique subgame perfect equilibrium. However, the strategy pair (L, R) and the corresponding outcome (2,2) also satisfy our definition of an equilibrium. Thus, the notion of a subgame perfect equilibrium can be viewed as a refinement of the notion of an equilibrium.

We appreciate, of course, that terms such as *subgame* and a *subgame perfect equilibrium* might seem as far removed from politics as any mathematical jargon. However, the preceding two sections are concerned with precisely these topics—*pruning a voting tree from its end to its beginning is equivalent to finding a subgame perfect equilibrium.* To see more precisely what we mean by this, notice that we have quite briefly introduced several new terms here—the notions of a strategy and of an equilibrium. We will, in fact, have reason to elaborate on the meaning of these words throughout this volume, but for the time being, let us simply define an equilibrium in the context of extensive form games that points the way to generalizing the ideas implicit in the method of working backward up such a form. Specifically, if, in addition to identifying the strategic equivalents of nodes, we also keep a careful record of the implied choices of each decision maker, we could then describe strategies for action at each node that are in equilibrium in the following sense:

> *No person, after learning what any other person intends to choose, has any incentive to make a different choice at any decision node.*

That the sequence of choices arrived at by working backward up an extensive form are in equilibrium follows from the fact that we have constructed the path from the first node to the predicted outcome by looking at each person individually when it is that person's turn to act and by retaining only those choices that are not inferior. Thus, the choices of all persons, taken together, are in equilibrium in the sense that no single decision maker has a positive incentive to make a different choice when we hold constant everyone else's choices (strategies).

For yet another way to illustrate these ideas, let us consider another card game—blackjack—played against a Las Vegas casino. Of course, this isn't strictly a "game" in the game theoretic sense since a casino dealer's choices are predetermined and publicly stated: If the dealer's cards sum to anything less than 17, draw another card and continue to draw until a sum of 17 or greater is realized. The only free choices, then, are yours, and here we can outline some of the components of an equilibrium strategy. Ignoring the possibility of card counting and finite decks, if the dealer's up (face) card is 5 or less, we know that it plus the down card cannot exceed 16, so the dealer must draw one or more cards before he or she reaches 17. There is, then, a good chance the dealer will go over 21 and bust. So even if your two initial cards sum to, say, 14, its best to hold pat and not ask for another card—only drawing a 3, 4 5, 6 or 7 will improve your hand, which, *ceteris paribus*, has a less than even chance of happening (5 card types out of 13 possibilities). On the other hand, suppose the dealer's up card is a king, queen, jack or ten. There's now a better than even chance the dealer's down card is a seven or greater, in which case staying pat with two cards totaling 14 or 15 will, on average, lose so you might now want to consider drawing another card and hoping it takes you above 16 but not over 21. A strategy, then, will, among other things, tell you what do to when holding cards of certain values while confronting a specific up card on the part of the dealer. Indeed, if you ask, Las Vegas casinos will either give you or direct you to where you can buy a little card that tells you what the odds favor doing for all possible contingencies. That card, then, describes a strategy. And unless the casino is using a single deck (they now generally use six) so that each hand is essentially independent of whatever happened before—i.e., there is little chance of you gaining an edge by card counting—every new hand begins a new subgame consisting of that hand plus all that follow. Since you are not likely to be able to improve on the probabilistic calculations summarized by that little card, committing to abiding by its advice for the rest of your Las Vegas experience constitutes a subgame perfect equilibrium.

Turning from cards to the specifics of voting agendas, notice that a node corresponding to any final vote in a binary agenda tree that is not connected to any other node by an information set is a subgame since we can "split it off" without splitting any information sets. Although nodes in a voting tree may merely be shorthand in large committees for the simultaneous choice of a great many persons, as long as we don't break any information sets, we have not violated the definition of a subgame. And second, since everyone possesses a well-defined best choice at each final decision node of a voting tree, the only choice that can be part of a subgame perfect equilibrium is voting for one's preferred outcome or alternative from the alternative outcomes offered at that node (ignoring the modest complication that voters might be indifferent between alternatives). Since a subgame perfect equilibrium must have voters voting optimally in every subgame, we can replace all final nodes with their strategic equivalents. But this same argument now applies to any node that precedes any final nodes. Hence, a subgame perfect equilibrium requires that voters choose optimally between strategic equivalents at such nodes, in which case pruning a voting tree back and substituting strategic equivalents for voting nodes determines that equilibrium. Stated differently, *a subgame perfect equilibrium requires that all voters vote for the strategic equivalent ranking highest in their preference order, and this is precisely what we assume when we suppose voters are strategic and we prune a voting tree from bottom to top, substituting strategic equivalents for eventual outcomes.*

Figures 2.1, 2.2a and 2.2b illustrate this process of finding subgame perfect equilibria for our roll call vote on a legislative pay raise, while Figures 2.3a, 2.3b and 2.3c illustrate the same thing for a committee that uses a sequential veto. In the case of the roll call vote, Figure 2.1 is the entire extensive form (which is itself a subgame, albeit an all inclusive one). The four individual nodes corresponding to C's choice points with their two possible decisions to vote for or against constitute four subgames which, after eliminating C's dominated choices to find the strategic equivalents of those nodes, yields the subgame in Figure 2.2a. And that subgame, after we eliminate B's dominated choices and substitute instead the strategic equivalents for B's choices, yields a subgame in Figure 2.2b in which only legislator A has choices. Similarly, Figure 2.3a gives the extensive form of our sequential veto example, and here again each of legislator 3's twelve decision nodes plus their associated branches correspond to twelve subgames. After eliminating 3's dominated choices and substituting the choices he or she would make at each node for the nodes themselves, we arrive at Figure 2.3b, wherein the extensive forms beginning at each of 2's four decision nodes are now subgames. Doing the same thing with legislator 2 as we did with 3 by substituting strategic equivalents for those four subgames, we arrive at the subgame in Figure 2.3c that corresponds to legislator 1's decision node. Identifying 1's optimal choice solves for the subgame perfect equilibrium.

The reader should be able to see now with these examples how the notion of subgame perfect equilibria applied to voting trees applies equally well to any extensive form that models a game of perfect information—a game in which no two decision nodes share the same information set. It is at this point, though, that we begin to run into problems. Instead of a roll call vote, suppose legislators must decide whether or not to implement a pay raise for themselves by a secret ballot. We have already seen, however, what happens then to the situation's extensive form; namely both of B's two decision nodes are within the same information set and all four of C's decision nodes are contained in a single information set. There is, then, but one subgame here and no opportunity to work backwards from final outcomes to individual choices unless, in violation of the definition of a subgame, we break information sets. In other words, we cannot, as we already know, work backwards on such an extensive form to deduce optimal choices. Thus, we require something in addition to the notion of subgame perfection to prune non-binary voting trees or to prune any extensive form for which we cannot identify uniformly best choices at each person's information sets. The development of such an idea is the focus of our next chapter.

This is not to say, however, that as soon as we see information sets in an extensive form we must abandon any application of subgame perfection. Indeed, precisely the opposite is true because it may still be possible to eliminate some branches and nodes so as to simplify the analysis. To see what we mean, consider the following example:

> **Example**: Suppose three voters (1, 2, and 3) use a single secret ballot vote and plurality rule to choose a final outcome from the set {A, B, C}. Suppose, however, that voter 1 can break ties if each of the three alternatives receives one vote. If, then, voter 1's preferences are A > B > C, then Figure 2.15 describes the extensive form after we have eliminated 1's choices in the event of a tie and substituted instead the strategic equivalent of A for

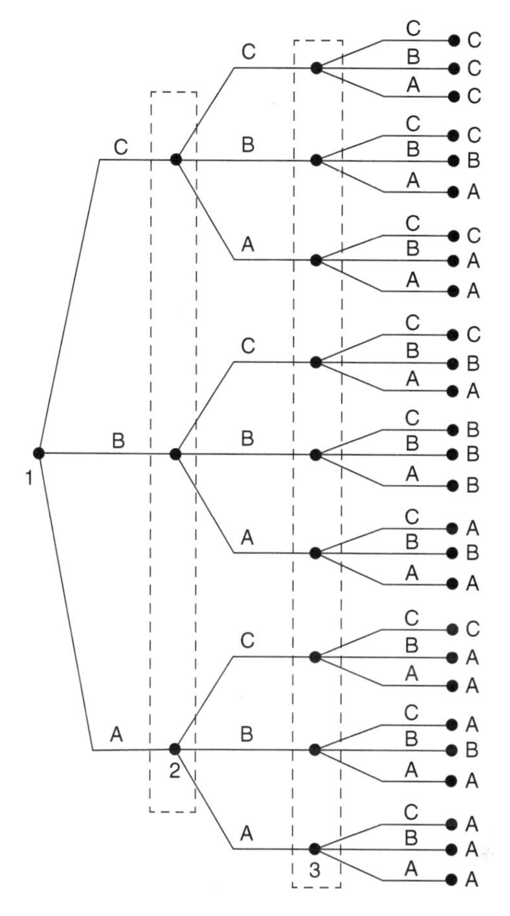

Figure 2.15 Secret ballot with tiebreaking by voter 1

the outcome (since 1 would always break the 3-way tie in favor of his most preferred alternative).

In addition, we can see from this extensive form that voter 1 should, in the initial secret ballot, always choose A, since doing so ensures the selection of A if voter 2 votes for A or if 2 and 3 disagree. In other words, voter 1 has a strategy—voting for A—that is best regardless of what the others do. Thus, while doing so breaks voter 2 and 3's information sets, we can nevertheless focus our analysis on the bottom third of the extensive form in Figure 2.15 because only the branch corresponding to voter 1 voting for A is relevant. Beyond this we cannot go. Suppose preferences now are the usual ones that give rise to the Condorcet Paradox; that is, voter 2 prefers C to A to B and 3 prefers B to C to A. Because 1 is decisive if all three voters choose differently, and because voter 3 wants to avoid allowing voter 1 to dictate the selection of alternative A, voter 3 does not have a strategy that is best under all circumstances. In particular, if 2 votes for B, then 3 should vote for B, but if 2 votes for C, then 3 should vote for C. For the same reason, voter 2 does not have a uniformly best choice. Nevertheless, the preceding reasoning has simplified our analytic task in the event we wish to pursue an analysis of this voting situation (which we do in Chapter 6).

2.5 The Centipede Game: A Word of Caution

To this point everything might seem routine and unexceptional. However, consider the following scenario introduced by Robert Rosenthal in 1981 ("Games of Perfect Information, Predatory Pricing, and the Chain Store," *Journal of Economic Theory* 25,1: 92–100): In a classroom experiment, two piles of money are set on the table in front of two students now labeled players 1 and 2 (but don't hold your breath to be offered the chance to play this game anytime soon). One pile contains $40, the other $10. Player 1 has the first move and can take the larger of the two piles, in which case player 2 takes the smaller pile and the game ends. But if 1 "passes," the piles are switched and doubled! Player 2 can then take the larger pile ($80), whereupon 1 gets the smaller pile ($20) and the game ends. But, once again, if 2 passes, the piles are switched and doubled yet again and it is now player 1's choice of taking the larger pile ($160), giving 2 the smaller one ($40) and ending the game—or passing. Suppose this situation is allowed to proceed until player 2 must choose between taking and getting $320 with 1 getting $80 or "passing" and getting $160 and player 1 getting $640. This circumstance is portrayed in extensive form in Figure 2.16.

If we now work backwards up this extensive form, beginning with player 2's last choice in order to compute the subgame perfect equilibrium, it is evident that 2 should take the $320 rather than pass and receive $160. Thus, player 1 knows if he reaches the third decision node (his second) and if he passes, 2 will take and 1 will get $80. On the other hand, if 1 takes, then he gets $160. So player 1 should take. Working backwards and repeating this argument for each of the nodes leads to the seemingly incontrovertible conclusion that each player should take at each node and, therefore, that 1 should take at the first node, thus saving the experimenter a considerable amount of money.

But now consider the following problem: We've concluded that 2 should take at his first decision node because player 1, preferring, as he does, more money to

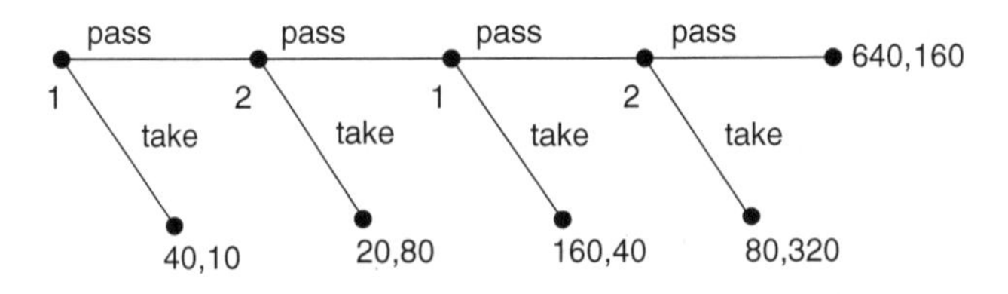

Figure 2.16 The centipede game

less, will take at his second node, and so on. But shouldn't player 2 ask himself, "If 1 is reasoning as I am, then how did I ever get to choose anything—shouldn't 1 have taken his money off the table at the very first opportunity?" We can, of course, only speculate as to what 2 might think at this point. Player 2 could, for instance, hypothesize that 1 simply made a mistake that is unlikely to be repeated and that it's best to grab the money now and run. Alternatively, he could speculate that 1 has a broken arm and can't pick up the money or even that 1 is supremely irrational and has no appreciation for money whatsoever. The most interesting possibility, though, is that 1 is trying to signal a willingness to let things slide for a bit so that both players can maximize their take from the experimenter. It wouldn't be the first time the subjects of a classroom experiment turned things into a game against their professor as opposed to against each other. But notice now that each of these possibilities leads to a different decision as to what to do if player 2 ever gets to choose anything. If 1 merely made a mistake, 2's best choice is probably to assume that it won't happen again and that he best take the $20 now. But if 1 has a broken hand and can't somehow pick up the money, then 2 should pass at all times until the final node. Finally, if 1 is trying to signal something, then 2 has a harder decision to make; namely, at what point does he think 1 will decide the time for cooperation is over and will take the money? To make things more complicated still, suppose we run this experiment out to n nodes, where player 2 gets to choose between taking and passing on the even numbered nodes. Since the maximum payoff is increasing to $40(2^{n-1})$ and the minimum to $10(2^{n-1})$, allowing n to equal 12 gives 2 the opportunity to end the game with a considerable amount of money for each player ($163,840 for 2 and $40,960 for 1). Surely, if both subjects are of a mind to try to signal cooperation early in the experiment, their willingness to do so will depend on n, the size of the carrot not only at the end of the stick but also on the carrots along the stick.

The critical thing to notice now is that none of these alternative thought processes is represented in our extensive form. Nor does our extensive form at this time tell us or allow us to deduce how n might impact behavior if the experimental subjects contemplated signaling to each other. In other words, Figure 2.16 leaves something out of the description of things and is not truly representing the complexity of the decisions confronting our two players. Needless to say, we would not have so explicitly offered this example so early in this book if we did not have some idea as to how to treat such circumstances. We aren't fools who would reveal this quickly the likelihood that there are fundamental social processes about which we have no idea how to model. Selling a book, or at least inducing those who read it to continue reading, is as much of a game as any other.

Lest we end this chapter on a seemingly inconclusive note, let us return to where this chapter began and to the observation that it seems counterintuitive at times to suppose that even in historically critical times, people act in accordance with any well thought out extensive form in mind. And here let us consider one other historical event in which the chief protagonists almost certainly failed to lay out the extensive form of the situation that confronted them, with dire consequences for those persons. We refer here to the August 1991 coup against Mikhail Gorbachev, wherein a group of Communist Party hardliners sought to thwart the seemingly inevitable disintegration of the USSR or at least to usurp Gorbachev's reconstruction of the Soviet system. As part of their strategy, several leaders of the coup—notably Oleg Baklanov (head of the Soviet military-industrial complex), Valery Boldin (Gorbachev's chief of staff), Oleg Shenin (Central Committee Secretary of the Communist Party), General Valentin Varenniko (commander of Soviet land forces)—flew from Moscow to Gorbachev's Black Sea vacation villa, intending to isolate and then intimidate Gorbachev into signing a decree declaring a state of emergency and authorizing the arrest of Russian President Boris Yeltsin. What appears to have been the case, however, is that none of the coup plotters, including its leaders back in Moscow (notably KGB chief Vladimir Kryuchkov, Interior Minister Boris Pugo, Soviet Vice-President Gennady Yanayev, and Defense Minister Dimitry Yazov), had developed contingency plans in the event that Gorbachev would refuse to cave in and that Yeltsin would act to oppose the coup. Nor, of course, had the coup plotters foreseen the refusal of a number of critical Soviet military units to participate in their plans. The coup, thereafter, disintegrated and, as we now know, three months later the Soviet Union itself passed into history. It is evident, then, that Kryuchkov et al. had at best an incomplete, if not an utterly naïve and stupid, extensive form in mind when implementing their scheme. This, however, should not be taken to mean that drawing out an extensive form in the attempt to understand these events is a fruitless or meaningless activity. Indeed, exactly the opposite is true. In hindsight and with further reflection it is almost always easier to imagine branches to a situation's extensive form description that were either ignored or unrecognized by decision makers at the time. Monday morning quarterbacking isn't reserved for football fans; it is perhaps no more extensively engaged in than by historians. But what drawing out an extensive form with hindsight does in this case is provide the historian with a research agenda—an agenda that includes attempting to answer questions such as: Did decision makers perceive the choice of ____, and if not, why not? And if they did perceive those choices, what consequences did they foresee from them that they wished to avoid? What actions did they imagine as being available to __ and what was the extensive form they believed others perceived? In the case of Moscow's coup plotters, what did they think would happen and what choices would be left to them if, as happened, Gorbachev refused to capitulate? And if Gorbachev had done what they demanded, how might Yeltsin and the Soviet military have responded? That is, understanding the whys and wherefores of that coup requires that, in the context of laying out a more encompassing extensive form than might have been in anyone's head at the time, we explore choices not made and perceptions not revealed by the actual unfolding of events. It is only by knowing what choices were not made, and why, that we can begin to explain the choices that were made.

2.6 Key Ideas and Concepts

extensive form
decision nodes
terminal nodes
information sets
imperfect recall
roll call voting
backwards reduction
perfect vs. imperfect information
voting agendas
voting tree
agenda setter
amendment agendas
sequential veto
agenda manipulation
sophisticated vs. unsophisticated voting
strategic voting
centipede game
subgame
strategy
equilibrium
subgame perfect equilibria

Exercises for Chapter 2

1. Suppose persons 1 and 2 make a sequence of binary decisions, first 1, then 2, then 1, then 2, and so forth, and suppose that all decisions are observed by both persons. Portray an extensive form to represent a situation in which 1 has perfect memory but 2 can only recall the last move of 1.
2. Assume there are three states (1, 2, and 3) of equal size, that the polls close first in state 1, then in 2, and then in 3, and that the election winner must capture a majority (two) of the states. Draw an extensive form representing the decision of a single citizen in state 3 who must decide whether to vote and for whom to vote in a two-candidate election. Assume that this person treats the decisions of all other voters and the outcomes of the balloting in other states as probabilistic choices by nature. (Assume that ties within states never occur.)
3. Draw the extensive form of the agenda "A versus B, the winner against C" for a three-person legislature in which legislator 3 observes 1's choice, but in which no other legislator observes any other choices.
4. A defense department bureaucrat oversees two related programs: X and Y. Recent cost overruns force him to "bury" these costs in the accounting of one of these programs. An agent from the government accounting office

that you oversee will, at the end of the year, review the bureaucrat's records. You can assign the agent to one of the following activities: (A) audit the personnel records of both programs; (B) audit the inventory of both programs; (C) audit the expenses of parts suppliers for both programs. If your agent detects the overrun, you earn +10; otherwise, you earn −10 whereas the respective payoffs for the bureaucrat are −10 and +10. The probability that the auditor detects the overrun given the focus of the audit and given the program in which the overrun is buried is as follows:

	X	Y
A	.5	.2
B	.3	.6
C	.1	.8

Portray this situation in extensive form.

5. Referring to problem 4, suppose you must make your decision before the bureaucrat chooses the program in which to bury the cost overrun and that the bureaucrat can pay you two units to learn beforehand how you intend to assign your auditors. Describe the extensive form for this situation.
6. Assume three people (1, 2, and 3) must choose one candidate from the list of A, B, C, and D. Assuming that all persons observe all prior choices, draw the extensive form of a procedure in which first 1 deletes a candidate, then 2, and then 3, and in which the candidate that remains is elected.
7. If free competition reigns in an industry with two firms, each firm sells 20 million units of that industry's products at a net profit of $1/item. But if they collude to set a higher price, each sells 15 million units at a net profit of $2 each. If one firm defects to the lower competitive price, its sales soar to 35 million units while the other firm sells nothing. Before each firm sets its price (which they do simultaneously) Senator Billie Bob proposes a licensing agreement whereby each pays a tax of $.20/item to produce the product at the fixed cartel price—ostensibly to insure that "destructive competition" does not "leave hard-working Americans unemployed." Construct this situation's extensive form, where each firm must first approve or disapprove of the licensing arrangement, which goes into effect only if both firms agree to it.
8. You are a contestant on a game show. There are three doors, labeled A, B, and C. Behind two of the doors are prizes worth nothing, and behind the third door is a prize worth $10,000. The game show host knows which door contains the valued prize, but you don't. You get first move, at which time you must choose one of the three doors. The game show host gets the second move, at which time he must open one of the doors that you didn't select, revealing the contents behind that door and leaving the other two doors unopened. You get the third move, at which time you can choose to stay with your initial choice or switch your choice to the remaining unopened door. After you make your final choice, the door you selected in move 3 is opened, and you win the prize behind that door. Your preferences are to maximize your expected dollar payoff whereas the preference of the

game show host is to keep the suspense going as long as possible. Under what circumstances, if ever, should you switch doors after the game show host opens a door?

9. Assume the following preferences by a five-member committee, all of whom are sophisticated voters.

1	2	3	4	5
A	B	C	C	B
B	C	A	A	C
C	A	B	B	A

You are voter 1. Each voter receives a payoff of $100 from his or her first choice, $50 from a second choice, and $0 from a last choice. You and voter 2 are competing to set the agenda, but who will do the setting depends on who can "buy" voter 4. You and voter 2 must announce how much of your winnings you will pay, after which voter 4 will choose between you and 2 to be the setter. Because of your sterling character, an indifferent voter 4 will choose you. To simplify matters, suppose that you and 2 must each bid $75, $25, or $0.

   a. Assuming that money and utility are equivalent, what is your bid, and what is the final outcome?
   b. How does your answer to part (a) change if 4 chooses 2 when indifferent?

10. Suppose the value of some object is $x_1 + x_2$, where both x_1 and x_2 are random variables that can each take on a value of 0 or 1. Suppose person 1 observes x_1 but not x_2 whereas person 2 observes x_2, but not x_1. Portray in extensive form an auction in which 1 and 2, after observing x_1 and x_2, respectively, bid .5 or 1.5 for the object by submitting their bids in sealed envelopes. Assume that the person submitting the highest bid wins the object, but must pay whatever he or she bid, and assume also that, in the event of a tie, a coin toss determines the winner.
11. Portray a congressional voting tree in which, with the alternatives B (bill), A (amended bill) and Q (status quo) already on the agenda, a predesignated legislator must decide before the actual balloting whether to introduce a substitute bill S. Assume that, regardless of whether or not the substitute is introduced, Congress must first decide whether to amend the original bill B.
12. You, player 1, and your opponent, player 2, each have three horses. Let your horses' speeds be denoted by a_1, a_2, and a_3, and let your opponent's horses' speeds be b_1, b_2, and b_3. Suppose their speeds are as follows: $b_1 > a_1 > b_2 > a_2 > b_3 > a_3$. Suppose three races are to be run in sequence and that each of you must decide which horse to enter in each race (no horse can race twice). You are allowed to decide, though, which horse to enter after the previous race is run. You and player 1 must choose a horse simultaneously for the first race, but your opponent chooses first (and you observe this choice) for the second race. The person who wins two or more races

wins the stake of $1,000. However, since your horses are known beforehand to be slower on average, you are paid $250 to participate. Portray the situation's extensive form and deduce equilibrium strategies. Is there a determinate final outcome?

13. Suppose two countries, 1 and 2, each have second-strike nuclear capabilities in that they can retaliate after suffering a first strike or they can launch a second strike after sustaining a retaliation. Nature picks a country at random, whereupon it, say 1, can launch a preemptive attack (p) against the other's military capabilities. If 1 attacks, 2 can retaliate (r) or capitulate (c). But if 2 retaliates, 1 can choose between launching (l) its second-strike (which 2 no longer has) that moves 2 back to the stone age or it can do nothing (~l). On the other hand, if 1 does not launch a preemptive attack (~p), then the first move is 2's, with 1 and 2's roles reversed. The game ends if neither country attacks, after one country capitulates, or after one country launches its second strike following a retaliation. Assign some "reasonable" payoffs to the outcomes and determine the eventual outcome. What are the preferences that make a preemptive attack inevitable?

3 The Strategic Form and Nash Equilibria

3.1 Introduction

In their entertaining and informative book *The Art of Strategy*, Avinash Dixit and Barry Nalebuff (New York: W.W. Norton, 2009) recount a contest between two competing teams, euphemistically labeled Red and Blue here, on the TV series *Survivor* in which 21 flags are planted in the sand. Each team, beginning with Red, must in sequence, take 1, 2 or 3 flags (a team cannot choose to take no flags, nor can it take 4 or more), and the team taking the last flag, whether it be alone or in combination with one or two others, is declared the winner. We leave it for the reader to consult Dixit and Nalebuff for the details on how this contest actually unfolded in the TV series; here we note simply that if one considers this contest an extensive form game, team Red should, if possible, maneuver things so that Blue is confronted with 4 flags on its last move. In this case, regardless of whether Blue chooses 1, 2 or 3 flags, Red will win by taking the remaining 3, 2 or 1 flags. But that means Red should maneuver to leave Blue with 8 flags on its move before that, since with 8 flags, regardless of Blue's subsequent choice, Red can ensure that Blue confronts 4 flags on its last move. Moving further back on the extensive form, this implies that Red should maneuver to have Blue confront 12 flags before that, 16 before that and 20 before that. In other words, Red should choose 1 flag on the first move and counter any selection by Blue thereafter so that the number of flags chosen by Blue + Red on each successive pair of moves sums to 4.

To this point, then, the example once again illustrates the value of backward reasoning—of looking ahead and setting one's plans accordingly. However, suppose we change the scenario slightly so that each team, after discussing the situation, must select one member to implement whatever plan or scheme is decided upon. The question then is: Is there a reason for the other members of each team to be physically present while their scheme is implemented by the selected member? That we've asked this question suggests the answer is No. Indeed, suppose the Red team agrees upon the following instructions for its representative:

> Initially choose 1 flag. Thereafter, if Blue chooses x flags ($x = 1, 2$ or 3), then choose $y = 4 - x$ flags.

For Blue, the problem seems hopeless. As long as Red abides by the above rule, there is little Blue can do to win the contest. Nevertheless, it would be prudent to prepare for the possibility that Red makes a mistake (which it did in the actual play of the game); in which case, its representative should abide by the following rule:

> If confronted by 20, 16, 12, 8 or 4 flags, it matters little what you choose. But if confronted by any other number of flags, choose enough flags to set the resulting number for Red at 16, 12, 8 or 4.

With this rule, if Red errs, Blue's representative can turn the tables and ensure a victory for itself. The critical lesson to be drawn from this discussion, then, is that once the game is analyzed and a correct rule identified for each team, it matters little whether the entire team or merely each team's representative is there to play. Each team, in fact, could replace its representative with a robot, programmed to implement the agreed upon plan. We call that plan a *strategy*, and this chapter examines how situations described as games of strategy might be formulated and analyzed.

3.2 Strategies and Simultaneous Choice

The reader might recall that we introduced the notion of a strategy in the previous chapter when defining subgame perfect equilibria, and our notion of a strategy here is identical to the one we discussed earlier. Presently, though, we want to focus on the fact that in addition to underscoring the importance of thinking ahead and affording us the opportunity to introduce the notion of a strategy for playing a game, the preceding example from *Survivor* also suggests an alternative representation of strategic situations that often is an especially convenient way to think about, model and analyze strategic choice. Consider again the situation that confronts our three legislators voting on a pay raise from the previous chapter, but where their votes must be cast simultaneously. It is true that an extensive form with appropriately drawn information sets can represent this circumstance, but since legislators cannot then condition votes on how others vote, let us instead represent this situation as in Figure 3.1.

		Legislator C			
		For		*Against*	
		Legislator B		Legislator B	
		For	*Against*	*For*	*Against*
Legislator A	*For*	3,3,3	3,4,3	3,3,4	1,2,2
	Against	4,3,3	2,2,1	2,1,2	2,2,2

Figure 3.1 Voting on a pay raise with simultaneous balloting

Notice that this figure conveys all the information contained in our earlier extensive form, provided we keep in mind that each legislator must choose whether to vote *for* or *against* in ignorance of what other legislators choose (or, equivalently, that they all choose simultaneously). In such instances, a tabular form seems unexceptional. But now, let us return to the original scenario in which legislators cast their ballots sequentially (alphabetically) in a roll call vote and see if we can't do the same there—namely, represent the roll call scenario in tabular form. To see that we can, suppose we take the following somewhat odd perspective: Each legislator must hire an agent who will vote for them, conditional on what those agents observe before acting. Moreover, suppose those agents act precisely as instructed. In this case, legislator A, who votes first, cannot condition any decisions on what B and C do, and thus can tell an agent only one of two things:

$s^A{}_1$: Vote *for* (the raise)
$s^A{}_2$: Vote *against*

Legislator B's agent, in turn, will observe A's vote and thus can be given one of four instructions:

$s^B{}_1$: Vote *for* regardless of how A votes
$s^B{}_2$: If A votes *for*, vote *against*, but if A votes *against*, vote *for* (vote opposite of A)
$s^B{}_3$: If A votes *for*, vote *for*, but if A votes *against*, vote *against* (vote the same as A)
$s^B{}_4$: Vote *against* regardless of how A votes

In other words, legislator B's agent will confront one of two circumstances: either legislator A will have voted *for* or *against* the raise. For each of these circumstances B has two choices—to vote *for* or *against*. In effect, then, the instructions B can give an agent take the form:

$s^B{}_i$: If A votes *for*, then vote ____; but if A votes *against*, then vote ____.

Since there are two ways to fill in the first blank, and two ways to fill in the second, there are 4 distinct alternative instructions for B to choose from. Finally, consider legislator C. A generic instruction that C can give an agent takes the following form:

$s^C{}_i$: If A votes *for* and B votes *for*, then vote ____; but
if A votes *for* and B votes *against*, then vote ____; but
if A votes *against* and B votes *for*, then vote ____; but
if A votes *against* and B votes *against*, then vote ____.

Legislator C's instructions to an agent, then, must anticipate four possible contingencies corresponding to its four decision nodes and their associated information sets, and since there are two ways to fill in each blank at each contingency, there are $2 \times 2 \times 2 \times 2 = 16$ possible instructions. Finally, notice that if we take any triple of instructions—one for A's agent, one for B's and one for C's—we can specify which outcome results and what payoffs the individual legislators realize. That is, we can calculate the vector

$$U(s^A_i, s^B_j, s^C_k) = (u_A(s^A_i, s^B_j, s^C_k), u_B(s^A_i, s^B_j, s^C_k), u_C(s^A_i, s^B_j, s^C_k))$$

where $u_A(s^A_i, s^B_j, s^C_k)$ denotes the payoff to legislator A and is stated as a function of the instruction given to each legislator's agent, and so on. Indeed, we can now write down a $2 \times 4 \times 16 = 128$ cell table that tells us what payoffs prevail for each combination of instructions. This table is what we show in Figure 3.2 (recall that we assign a payoff of 2 if the raise passes and the legislator in question votes against, a payoff of 1 if the raise passes and he votes for, a payoff of 0 if he votes against and the raise fails, and a payoff of –1 if the raise fails but he votes for it).

Several questions should concern the reader at this point. The first is: What did we gain in this instance from going through the effort of constructing such a table when we can decide what each legislator should or would do using the original extensive form? Admittedly, the answer in this case is "nothing." In other words, sometimes it's silly to go through the exercise of rewriting an extensive form in tabular form. The second and perhaps more important question, though, is: Aren't the legislators giving something up in terms of their

	Legislator A votes *for*				Legislator A votes *against*			
	Legislator B chooses:				Legislator B chooses:			
C:	*f,f*	*f,a*	*a,f*	*a,a*	*f,f*	*f,a*	*a,f*	*a,a*
f, f, f, f	1, 1, 1	1, 1, 1	1, 2, 1	1, 2, 1	1, 1, 2	–1, 0, 0	1, 1, 2	–1, 0, 0
f, f, f, a	1, 1, 1	1, 1, 1	1, 2, 1	1, 2, 1	1, 1, 2	0, 0, 0	1, 1, 2	0, 0, 0
f, f, a, f	1, 1, 1	1, 1, 1	1, 2, 1	1, 2, 1	0, –1, 0	–1, 0, 0	0, –1, 0	–1, 0, 0
f, f, a, a	1, 1, 1	1, 1, 1	1, 2, 1	1, 2, 1	0, –1, 0	0, 0, 0	0, –1, 0	0, 0, 0
f, a, f, f	1, 1, 1	1, 1, 1	0, 0, –1	0, 0, –1	1, 1, 2	–1, 0, 0	1, 1, 2	–1, 0, 0
f, a, f, a	1, 1, 1	1, 1, 1	0, 0, –1	0, 0, –1	1, 1, 2	0, 0, 0	1, 1, 2	0, 0, 0
f, a, a, f	1, 1, 1	1, 1, 1	0, 0, –1	0, 0, –1	0, –1, 0	–1, 0, 0	0, –1, 0	–1, 0, 0
f, a, a, a	1, 1, 1	1, 1, 1	0, 0, –1	0, 0, –1	0, –1, 0	0, 0, 0	0, –1, 0	0, 0, 0
a, f, f, f	2, 1, 1	2, 1, 1	1, 2, 1	1, 2, 1	1, 1, 2	–1, 0, 0	1, 1, 2	–1, 0, 0
a, f, f, a	2, 1, 1	2, 1, 1	1, 2, 1	1, 2, 1	1, 1, 2	0, 0, 0	1, 1, 2	0, 0, 0
a, f, a, f	2, 1, 1	2, 1, 1	1, 2, 1	1, 2, 1	0, –1, 0	–1, 0, 0	0, –1, 0	–1, 0, 0
a, f, a, a	2, 1, 1	2, 1, 1	1, 2, 1	1, 2, 1	0, –1, 0	0, 0, 0	0, –1, 0	0, 0, 0
a, a, f, f	2, 1, 1	2, 1, 1	0, 0, –1	0, 0, –1	1, 1, 2	–1, 0, 0	1, 1, 2	–1, 0, 0
a, a, f, a	2, 1, 1	2, 1, 1	0, 0, –1	0, 0, –1	1, 1, 2	0, 0, 0	1, 1, 2	0, 0, 0
a, a, a, f	2, 1, 1	2, 1, 1	0, 0, –1	0, 0, –1	0, –1, 0	–1, 0, 0	0, –1, 0	–1, 0, 0
a, a, a, a	2, 1, 1	2, 1, 1	0, 0, –1	0, 0, –1	0, –1, 0	0, 0, 0	0, –1, 0	0, 0, 0

Figure 3.2 Legislative pay raise in strategic form

strategic options by handing things over to an agent? The answer to this question is No—at least not if we've laid out the instructions to the agents properly. Specifically, if we ask what is it that a legislator might learn by foregoing the use of an agent and being physically present to cast his or her ballot, the answer, of course, is: The choices of those legislators who voted before them. But the instructions given to agents already take this information into account by conditioning an agent's choice on what the agent observes. Thus, there is nothing a legislator can observe to condition a choice on that an agent cannot also observe. It may be that a legislator might want to postpone making any decisions in order to think about the problem longer. But what is there to think about? Presumably we have drawn our extensive form to represent all possible contingencies, and unless a legislator is a slow thinker (which is hardly out of the question), nothing is lost by passing off to an agent the mechanical task of casting one's vote.

Representations of strategic environments as in Figures 3.1 and 3.2 are termed *strategic forms* and the instructions that might be given to an agent are called *strategies*, where, once again, a strategy is a specification of how to play the game—of what choice to make at each and every information set a player might encounter in the situation's extensive form. To this conceptualization, however, we can add one additional wrinkle. Specifically, under the assumption that the explicit extensive form underlying a situation's strategic form is complete—that all moves by opponents and by nature are fully represented by the form's decision nodes, branches and information sets—then when analyzing a strategic form we can assume that each player chooses their strategy in ignorance of what strategies all other players have or will choose—*or equivalently, that all players choose their strategies simultaneously.*

This formulation might seem a counterintuitive way to analyze or predict what people do, especially when accompanied by our assertion that a game in strategic form should be analyzed under the assumption that the players choose their strategies simultaneously. Consider, though, the game of tic-tac-toe. Once again, if one is playing against a 12-year-old with some minimal experience, more than likely you will be unable to do better than guarantee yourself a draw. In other words, its seems safe to assume that both you and your pre-teen opponent are playing the game "optimally." Yet, just as it is unreasonable to suppose that either of you have explicitly mapped out this game's extensive form with its hundreds of decision nodes and branches, it is also unreasonable to suppose that either of you have written down its strategic form, since tic-tac-toe has literally billions of strategies (most of them quite horrible). Yet even 12-year-olds are easily bored with this game. The reason, of course, is that the human brain is marvelously complex. We are only now beginning to understand its capacity, but one thing is clear: The human brain has little difficulty looking ahead two or three moves in a relatively simple game and deducing what paths lead to bad outcomes and which lead to better ones. It is also quite capable of recalling patterns of play so as to avoid bad patterns in the future and to pursue good ones instead. Nevertheless, suppose you are told that you have no choice:

Absent the opportunity to play tic-tac-toe endlessly, the 12-year-old in your company will reveal himself to be a spoiled crying brat. You know beforehand how boring that task will be and doubtlessly you'd love to find a way out of it. So suppose you can hire your own 12-year-old to take your place and play against your 12-year-old opponent—someone you haven't been able to beat at this game—and suppose the cost of hiring this "agent" is a trip to the ice cream store. Surely this would be more enjoyable than experiencing an endless series of draws against your youthful opponent. Moreover, because you have not been able to defeat her, you know your "agent" will employ the same strategy you would choose, thereby keeping your 12-year-old opponent no less entertained than he would be if playing against you. So why not let her take your place while you go off and do something more productive? The only problem you might confront here is establishing the right incentives so that your "agent" plays in your self-interest (i.e., she doesn't become so bored herself that she loses, owing to inattention)—but that is a problem we can address later in this volume.

In the case of tic-tac-toe, then, the idea of hiring an agent is utterly reasonable provided you have some confidence she will employ whatever strategy you'd have chosen had you been forced to be there. Moreover, from your perspective, the hiring of this youthful stand-in is equivalent to choosing your strategy simultaneously with your opponent or at least in ignorance of precisely the strategy your opponent will pursue. But now consider the game of chess or Go. Chess and Go, of course, are far more complex and are not ones that any existing computer has yet mapped out fully in extensive form. We know, though, that even here we see the ability of the brain to learn and store some good, albeit incomplete, strategies. It may be that masters of either game can look further ahead than the rest of us, but even masters are aided by complex classifications and categorizations of moves drawn from the experiences of others. With the advantage of a "library" of moves stored away in the neurons and synapses of their brains, they can, in effect, look ahead dozens of moves. That the full extensive form for either game has not (yet) been constructed and stored in some computer capable of deducing optimal strategies depending on whether one moves first or second explains, in fact, why they remain a test of skill. But we can be certain that as soon as that extensive form is mapped out for either game and it is no longer theoretically possible to defeat a computer, then the sales of chess sets or Go boards will plummet. The question here, though, is whether you'd be willing to hire an agent to play such a game for you. The answer depends, we suppose, on the pleasure you derive from simply testing your skill against an opponent (who presumably doesn't have a computer immediately at his or her disposal). But suppose all you care about is winning and that you somehow can hire someone who plays at least as well as you do, and perhaps better. Again, then, it seems that hiring such a person as one's agent is not a bad idea. It is equivalent, in fact, to choosing a strategy before your opponent even makes a move and, indeed, perhaps even before you know which side of the board you'll be playing.

Of course, we hire agents of various types all the time to make choices for us. When confronted with a lawsuit, for instance, we hire a lawyer under the assumption that whoever we hire has a better understanding of the law than we do and will be able not only to counter the unfolding strategy of any opposing lawyer better than we can but also may be able to devise a strategy that is more sophisticated than anyone else's. If we hire a plumber to fix a leak or a doctor to cure an illness we are in effect hiring an agent to "play a game against nature" under the assumption that our agent will play with better skill than we can. The notion of hiring an agent to play a game for us here, though, is of a slightly different sort. Saying that once we have chosen a strategy we could employ anyone or anything to implement the actual choices is not quite the same as telling our spouse, "Honey, why don't you do the grocery shopping today, but just make sure you get that snack I like." Your instructions to your spouse do not correspond to a strategy because they are incomplete. They do not say what to do if the store is out of your favorite snack. Instead, a strategy in the game theoretic sense is akin to saying something like "Honey, when you do the grocery shopping today, get me a box of ____, but if they are out of that, get me ____ instead, and if they are out of that then get whatever you might like." In other words, to repeat what we've said before about the strategic form, a strategy is a complete plan, and once we have selected one, there is no need to afford our "agent" any flexibility in making decisions and there is nothing to be gained by doing the shopping ourselves.

Speaking of grocery stores, let's return to our earlier example of buying groceries in the peculiar fashion of voting on the proposals offered by two competing candidates. Here one's strategies as a voter are simple: Vote for candidate 1 or for candidate 2, where presumably you prefer to vote for the candidate who proposes a basket of goods closest to your ideal—closest to what you'd purchase if allowed to shop in the usual way. Actually, a fully specified strategy here is: "If candidate 1 proposes a basket of groceries that is closer to your ideal than is the basket proposed by candidate 2, then vote for 1; otherwise, vote for 2." The outcome is then the proposal of the winning candidate. But here you also have to plan for the possibility of being one of the candidates. Your strategies now are more numerous and consist of all possible allocations of the budget among the goods on the store's shelves. Indeed, even if we simplify things and assume there are but two distinct goods for sale, the positions along the budget constraint are most likely easier modeled as infinite in number—a segment of the real line. There is, then, no practical way to portray all choices you might make as a candidate in a finite table, so instead the strategic form must take a more abstract algebraic form. Thus, if we normalize the budget constraint to be on the line between 0 and 1, a strategy would look as follows:

> If chosen to be a voter, then vote for the candidate whose 'platform' is closest to your ideal; and if chosen to be a candidate, then adopt the policy position __ $\in [0,1]$ as your platform.

In this strategy, the unspecified position, denoted __, is yet to be determined. The strategic form is then defined by a function that, in effect, tells us how each voter votes for every pair of platforms, that adds up those votes to determine which policy position wins (or ties) within each pair, and that identifies the winning candidate based on their chosen positions. But as with tic-tac-toe, once given all this information, there is no reason why you cannot hire an agent to implement the strategy you set out for yourself.

If the reader is still confused by or uncomfortable with the idea of analyzing strategic form games with the presumption that players choose strategies simultaneously or in ignorance of each other's choices, then perhaps it is useful to differentiate between "choices" and "moves." When playing chess, Go or tic-tac-toe, one *moves* sequentially. The moves in such games, then, are contingent on what one observes and individual players may not be able to predict with certainty an opponent's moves beforehand. But none of this precludes a player from *choosing* a strategy on how to play the game beforehand—a specification of what moves to make as the game unfolds. Thus, in our legislative pay raise example, legislators *move* (vote) sequentially, but for the analyst nothing is lost by assuming that legislators *choose* a strategy beforehand.

3.3 Nash Equilibria

Suppose, now, that we have a game in strategic form and wish to predict what choices the players will make. Before doing so, however, we should first review some of the assumptions we've explicitly or implicitly made about the situation when laying out a strategic form.

> Our *first* assumption is that our strategic form is complete. There is no possibility of unforeseen circumstances, including one or more players actually changing the game before we act. All possible circumstances must already be represented either as choices and choice nodes available to the players or to nature. And if there is the chance that one or more players can change the game, that possibility as well must be represented.
>
> *Second*, we assume that all players are completely informed about the game—about the strategic form. Any uncertainty must also be represented in what we assume about the players, their preferences, choices and information sets.
>
> *Third*, in the actual play of the game, we assume that the players choose their strategies simultaneously or in ignorance of what others have chosen (which is not to say that a player cannot make good and even precise guesses about one's opponents).
>
> *Finally* (and here's the biggie), we assume that everything stated above is *common knowledge*, where here common knowledge is not some imprecise use of language but a specific assumption about what people know. Specifically, people know not only the game others perceive they are playing, but they know that everyone else knows what they know, that you know that they know what you know, and so on and so forth.

Clearly, these are heroic assumptions, but they are perhaps not quite as heroic as they might seem. So lest the reader tune out, thinking, "The theory I am about to be taught cannot be about anything real," we ask that you postpone definitive judgments to see if we can sustain these assumptions while modeling things of interest.

We begin, then, by noting that if a person's information is not perfect in the sense that his or her information sets encompass two or more decision nodes, then backward reduction solves the extensive form only if that (and every other) player has an unambiguous best choice at each such set. But if this requirement is not satisfied—as in the example of voting on a pay raise by secret ballot—the "he-thinks-that-I-think" regress reappears and confounds any attempt to reduce an extensive form by backward reduction. This does not mean, however, that the notion of rational or self-interested choice is undefined. Consider again our earlier discussion of plurality rule from the previous chapter with three alternatives and three voters and the dilemma the voters might encounter with deciding whether to vote for their most preferred alternative or to vote strategically for a second ranked alternative (see Figure 2.15). Recall that in that example, voter 1's preference is A > B > C, 2's is C > B > A and 3's preference is B > C > A. Also, recall that because voter 1 is empowered to break ties, he or she should always vote for A: If 2 and 3 fail to vote for the same candidate, A wins whereas, if 2 and 3 make the same choice, it is irrelevant how 1 votes. And because both 2 and 3 rank A last, they do in fact have an incentive to avoid voting differently. Of course, absent some mechanism for coordinating their actions, there is no guarantee that 2 and 3 will vote identically, but suppose that voter 2 believes, for whatever reason, that 3 will vote for his or her most preferred alternative, B, whereas 3 believes that 2 is risk averse and, preferring to avoid a three-way tie between A, B, and C, will vote for B also. Thus, B prevails as the final outcome. Notice, moreover, that given these beliefs, neither voter 2 nor 3 has an incentive to unilaterally change his or her vote, since a *unilateral* switch to C yields A via a 3-way tie that voter 1 breaks in favor of A. Put differently, voting for B on the part of voters 2 and 3 terminates the "he-thinks" regress. For example, if 2 reasons that he should vote for B because 3 will vote for B, and if he believes that 3 can anticipate his reasoning, then he should conclude that 3 will, in fact, vote for B as originally conjectured. Of course, we could have just as easily set beliefs so as to have voters 2 and 3 choose C, in which case C terminates the regress. But more later on the possibility of multiple termination points.

The preceding discussion illustrates one of the most important concepts in game theory—that of a **Nash equilibrium**—conjunctions of strategies that terminate "he-thinks ..." regresses. To illustrate this idea with another example, consider the following fanciful two-person scenario: Two people must each, in secret, choose an integer number between 1 and 10. If the sum of the numbers chosen equals 10, each player wins an amount equal to the number they chose; otherwise, they win nothing. With the assumption that this game is played only once, each player then has ten strategies, one of the integers in the interval [1,10]. There are, then, ten possible combinations of numbers that yield both players a positive payoff, such as (1,9), (2,8) and so on. So suppose one person, for whatever reason, chooses 7 while the other chooses 3. Then clearly, once these numbers are revealed, neither has a *unilateral* incentive to switch to a different number. Thus, the strategy pair (7,3) as well as (3,7) are Nash equilibria. On the other hand, if one chooses 7 and the other chooses, say, 4, then clearly their strategies are not in equilibrium—at least one player (and in this case, both) has an incentive to choose a different number. And insofar as any he-thinks-that-I-think regress is concerned, if I intend to choose 7 because I think you will choose 3 and if you plan to choose 3 because you think I plan to choose 7, etc., then we in fact choose 7 and 3 respectively, and the he-thinks regress ends.

Before formalizing this idea, we want to give special emphasis to the word "unilateral" in our discussion. Suppose we change the preceding game so that if the numbers chosen sum to 10, each player wins the sum of squared numbers chosen. Thus if (1,9) or (9,1) prevails, each player wins $1^2 + 9^2 = 82$ whereas if, say, (3,7) or (7,3) prevails, their mutual payoffs are $3^2 + 7^2 = 58$. Clearly, both players now prefer that a combination of 1 and 9 prevail to any other combination that sums to 10. Nevertheless, strategic choices such as (3,7) remain equilibria because neither player has an incentive to switch *unilaterally* to any other number since doing so lowers that person's payoff (in this case to zero).

To define the idea of a Nash equilibrium more formally now so we can apply it generally, recall that a strategic form consists of a set of decision makers, N, a set of strategies for each decision maker, $S_1, S_2, \ldots, S_n$, and a set of functions, $u_1, u_2, \ldots, u_n$, which takes any combination of strategies by all n persons and maps each combination to utility payoffs. Letting s_i be a generic strategy for person i, and letting $u_i(s_1, \ldots, s_i, \ldots, s_n)$ denote i's payoff from the outcome that results from the joint strategic choice of $s_1, \ldots, s_i, \ldots, s_n$, then,

> ***Nash equilibrium:*** *A particular n-tuple of strategies, say* $(s_1^*, \ldots, s_i^*, \ldots, s_n^*)$, *is a Nash equilibrium if the following is true for every person i in N and for every* s_i *in* S_i:
>
> $$u_i(s_1^*, \ldots, s_i^*, \ldots, s_n^*) \geq u_i(s_1^*, \ldots, s_i, \ldots, s_n^*) \qquad (2.1)$$

Thus, the n-tuple of strategies $\mathbf{s}^* = (s_1^*, \ldots, s_i^*, \ldots, s_n^*)$ is a Nash equilibrium if, once at $\mathbf{s}^*$, no person has an incentive to shift *unilaterally* to some other strategy.

Although our verbal definition conveys the general meaning of this idea, the formal definition of a Nash equilibrium shows that each individual strategy that is part of such an equilibrium is a **best response** to the equilibrium strategies of the remaining relevant decision makers—it is a strategy that leads to the most preferred outcome from among those outcomes made feasible by the strategies of the remaining decision makers. That is, if a person can infer the equilibrium strategies of these other persons, then the person's choice problem reduces to an elementary decision problem of choosing the best response. *A Nash equilibrium, then, is a set of strategies—one for each player—such that each strategy in the set is a best response to all the others. Resurrecting the notion of utility maximization, a Nash equilibrium strategy maximizes each person's utility or expected utility, given that every other person is choosing a Nash equilibrium strategy.*

The particular virtue of a Nash equilibrium, of course, is that it terminates "he-thinks-that-I-think" regresses. If (a^*, b^*) is such an equilibrium, and if I (person 1) think that my opponent will choose b^*, then I should choose a^* because it is my best response; and if my opponent believes that I will choose a^*, then he should in fact choose b^* because it is his best response. However, despite the fact that it is designed to resolve "he-thinks" regresses, we should be careful before we assert that a Nash equilibrium "solves" all of our problems, since this idea does in fact sweep a great many problems under the rug. Specifically, we should ask:

1. What of the possibility of multiple Nash equilibria? And if there is more than one equilibrium in a game, will we have difficulty deciding which equilibrium offers the most reasonable prediction about choices?
2. What guarantees the existence of a Nash equilibrium? Is it possible that we will need to broaden its definition in order to ensure universal existence?
3. Is the argument for focusing on such equilibria suggested by the preceding discussion—namely that they terminate "he-thinks-that-I-think" regresses—compelling?
4. Throughout this discussion we have assumed that all persons share the same information about the situation. What is the relevance of the notion of a Nash equilibrium when this condition is not satisfied?

Recall, moreover, that we already introduced an equilibrium notion in the last chapter, that of a subgame perfect equilibrium, and thus we should also ask about the relationship of this notion to that of a Nash equilibrium. We can, in fact, address this issue immediately to show that although these two ideas are necessarily closely related in that any subgame perfect equilibrium is also a Nash equilibrium, it is not the case that every Nash equilibrium is subgame perfect. Suppose, for a truly simple example, that two players must sequentially choose A or B and where the person choosing second observes the other player's choice. If they both choose A, they both receive a payoff of 1; otherwise they both get 0. The sole subgame perfect equilibrium, then, is, unsurprisingly, for both to choose A. In strategic form, though, the game looks as in Figure 3.3 (note that because column chooser acts after row chooser, column chooser has four strategies, contingent on what he observes. The strategy AB, for example, denotes, "If row chooser picks A, then choose A, but if row chooser picks B, choose B").

Thus, while (A, AA) is clearly a Nash equilibrium, it is also the case that (B, BB), for example, satisfies this concept's definition. In this case, of course, we

	AA	AB	BA	BB
A	1, 1	1,1	0,0	0,0
B	0, 0	0, 0	0,0	0,0

Figure 3.3 Nash equilibria that are not subgame perfect

can readily preclude (B, BB) as a reasonable prediction about choices by merely eliminating (weakly) dominated strategies—strategies that are never better than some other strategy and that are sometimes worse. But the fact remains that (B, BB) satisfies the definition of a Nash equilibrium, and so we need to be concerned that there are other games, not as simple as this one, that yield Nash equilibria that are not subgame perfect but that are also not quite so easy to dismiss. However, before we address this issue as well as the questions posed above, let us first consider a few examples that illustrate some possibilities:

A 2-candidate, 3-voter, 1-issue Election: Earlier in Chapter 1 we introduced the concept of *single peaked preference* when discussing, in the context of shopping in a grocery store, the distinction between modeling markets and modeling political decisions. So suppose we consider a situation in which there are only three shoppers acting as voters, two candidates (who do not vote or who simply vote for themselves and cancel each other out) and one "issue" that can be framed as "X units of good 1 and Y units of good 2" where each voter holds a different preference on their ideal mix between the two goods. This situation, now, is portrayed in Figure 3.4, which merely reproduces Figure 1.3. Suppose finally that each candidate must announce a policy—a value for X and Y—that, perhaps somewhat unrealistically, requires a level of expenditure that precisely matches the revenues collected from the three voters. Notice now that if any one of the candidates chooses the *median ideal point* (voter #2's ideal in this case), that candidate cannot be defeated in a majority vote. If that candidate's opponent chooses a policy to the left [or right] of 2's ideal, then 2 along with the voter to the right [or left] will prefer the candidate advocating 2's ideal. The best an opponent can do is also advocate 2's ideal, in which case the election will be a tie and determined perhaps by a coin toss. Moreover, once

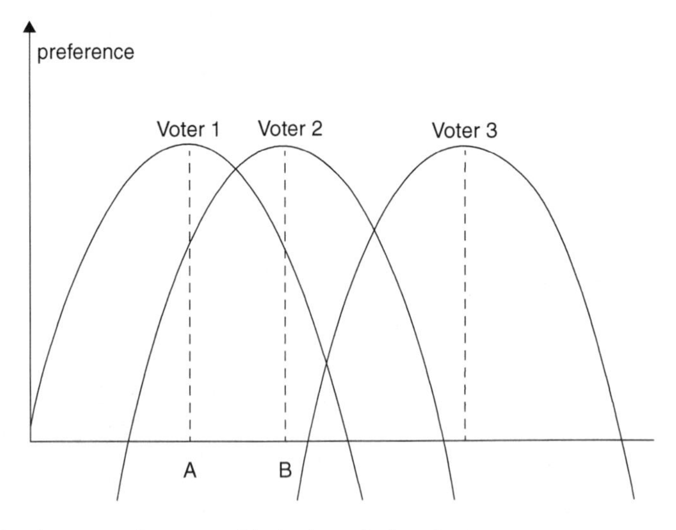

Figure 3.4 A 3-voter electorate with single peaked preferences

both candidates are 'at' this median preference, neither has an incentive to defect unilaterally since to do so is to throw the election to the opponent.

It should be evident at this point how readily the preceding example can be formulated to correspond to a simple 2-person game. The strategies of the players—the two candidates—can be identified as a point on, say, a [0,1] interval where the different points on that interval correspond to alternative choices of X and Y, where the payoffs to the candidates are +1, 0 and –1, depending on whether the candidate in question wins, ties or loses the election. And finally and most critically, the joint choice of the median voter's preference is the game's Nash equilibrium. This example gains special importance, though, if we note that there is nothing special about three versus n (odd) voters, provided that all preferences are single-peaked. From the definition of a median, at least half of the electorate most prefers it or positions to the right, so if one candidate is at the median and the other is to the left of the median, the candidate at the median wins. Since this argument is symmetric with respect to positions to the right of the median, and since the election is a tie if both candidates adopt the median, the median is the unique equilibrium platform in a two-candidate election. We have, then, the following result about one form of majority rule:

> ***The Median Voter Theorem:*** *In two-candidate elections that concern a single issue, if both candidates know the distribution of citizen preferences on the issue, if each candidate's strategy consists of a position on the issue, if citizens know the candidates' strategies, if all citizens have single-peaked preferences on the issue, and if no constraints are placed on the candidates' strategies with respect to the issue, then the electorate's median preference is both a Condorcet winner and the Nash equilibrium to the corresponding two-candidate election game.*

The Median Voter Theorem is important because it reveals the strong centralizing tendency of simple two-candidate plurality-rule elections. It suggests, for example, that people who complain about the fact that political parties in the United States often fail to offer the electorate meaningful choices on important issues misconstrue the purpose of those elections. To the extent that the median voter result models real two-candidate or two-party elections, that purpose is not necessarily to provide meaningful choices; rather, it is to select public policy in accordance with majority rule principles and to assure the rejection of radical candidates. Thus, those who prefer elections in which the main competitors offer distinct platforms other than the median preference must give good reasons for rejecting the median preference as a reasonable policy outcome.

Admittedly, it requires a considerable stretch of the imagination to argue that the Median Voter Theorem constitutes a model of any real election. We will in this volume, then, return to this theorem and assess its generality when we begin to impose less severe assumptions about such things as voter information and the nature of the issues relevant to the choices voters make. But first let us consider some additional examples that illustrate the character of a Nash equilibrium.

A Game with an Undesirable Equilibrium: In Chapter 1 we briefly described the notion of an honor code as it is employed in China as well as in America's military academies. Our purpose then was merely to illustrate a situation in which people's fates are interdependent, but now we can examine such codes in greater detail in order to understand more precisely the individual incentives they establish and why regimes might employ them. So consider a scenario in which one of two friends has committed a violation of the code. Suppose Friend 1 is the initial culprit and thus has the options of Confessing outright to the violation or Not Confessing. His counterpart, Friend 2, who (we assume) knows of the violation, can either Report or Not Report the violation to the appropriate authorities. Figure 3.5 portrays this situation as a simple 2 × 2 game.

	Report	Not Report
Confess	–Z, A	–X, –B
Not Confess	–Y, C	0, 0

Figure 3.5 Honor code strategic form

Somewhat arbitrarily here we denote the payoffs to both friends as 0 if no one confesses and no one reports on the violation. Thus, all remaining payoffs are relative to this outcome. Turning then to the unassigned payoffs, we can reasonably assume that all lettered amounts for Friend 1 are positive so that the minus signs denote a negative payoff. Thus, all payoffs to the primary culprit are negative and most likely satisfy $Y > Z > X$. Thus, the worst outcome for this person is to not confess but to have his "friend" report him. And if he confesses, he prefers that his friend not report him, in which case the onus of his violation falls at least as much on his friend as on him (hence the negative sign in front of B). His friend, in turn, least prefers not reporting an offense that is ultimately confessed to. And presumably, the most preferred outcome is to report an offense that is confessed to. The final character of this scenario, then, depends on the payoff C—in effect, whether one feels guilt when ratting on a friend. If one feels guilt and C is correspondingly negative, there are two equilibria: (Confess, Report) and (Not Confess, Not Report), in which case without further considerations, we cannot say which equilibrium will prevail. But suppose this system was implemented to forestall violations of the code or to uncover those violations when they in fact occur. One solution is to somehow attempt to remove the psychological onus of ratting on a friend, perhaps with appeals to one's loyalty to the institution of which you are a part in lieu of any loyalty to individuals. And indeed, by rendering C positive, one makes Report a *dominant* choice—one that's better for the person involved regardless of whether the other confesses or not. Thus, (Confess, Report) becomes the unique equilibrium.

	Confess	Not Confess
Confess	−5, −5	A, −10
Not Confess	−10, A	0, 0

Figure 3.6 Honor code with two culprits

Now suppose both persons are guilty of a violation of the code. Perhaps Friend 1 brought to the apartment some forbidden reading material (e.g., pornography) and both of them read it. Both are then equally guilty, and it might seem in this case that an honor code system might not work. To see if this is true, consider the situation portrayed in Figure 3.6.

Owing to their symmetric situation, the payoffs here are themselves wholly symmetric (assuming it is no less of an offense to read pornography than it is to purchase or own it). But the question here is what value should be placed on the payoff A. If there is a sense of guilt at ratting on one's roommate as well as oneself (if $A < 0$), there are the same two equilibria as before (even though only one of them, (0,0) is Pareto optimal now). However, suppose the powers-that-be confer a positive reward on anyone who confesses to a jointly committed violation of the code when the "partner in crime" remains silent. In this instance, confessing becomes the dominant choice for both persons and the outcome (Confess, Confess) prevails.

The interesting feature of this scenario is that both persons have an unambiguous incentive to confess so that their interests and (simultaneous) joint actions lead to an outcome both prefer to avoid. In fact, this game has a name—the Prisoners' Dilemma—and its applications to the study of politics and economics are extensive. It can be used to understand market failures, the logic of government with its coercive powers, as well as a variety of social processes from group formation to revolution to why voters in a democracy might prefer to tie their own hands and require term limits of the politicians whom they elect. Those applications, though, must come later in this volume. Here we are interested solely in using the Dilemma to illustrate the fact that although an outcome can correspond to a unique Nash equilibrium—in this case one supported even by dominant strategies—there is no reason to assume that the players will prefer that equilibrium to some other outcome. It also suggests that it may be foolhardy to suppose that once the players of a game understand the situation that confronts them, they will not attempt to change the game in some fashion so as to ensure against the selection of undesirable outcomes—the game itself may be unstable. Indeed, in Chapter 5 this is precisely what we suggest as the motivation for establishing governments and writing national constitutions.

If games with undesirable equilibria might cause us anxiety, games with multiple equilibria can be even more perplexing in terms of what to predict about people's choices. Consider the decidedly draconian example in Figure 3.7.

\$100,000, \$100	Death to both players
Death to both players	\$100, \$100,000

Figure 3.7 A serious coordination problem

Presumably the utility of death is some significant negative number (we demur from suggesting a specific value here), but it is evident that this game has two equilibria. Unfortunately for the participants, they can guarantee realizing one of these equilibria only if they violate the rules of playing a strategic form game—only if they somehow find a way to coordinate without the opportunity to communicate in any way. Indeed, here the existence of Nash equilibria only seems to make the he-thinks regress especially aggravating. Imagine you are row chooser, in which case you might reason "Even though it doesn't preclude the possibility of my demise, I might as well choose my first strategy since it promises me the greatest payoff." But, thinking further, you can reason "… but if my opponent thinks the same, we both die. So perhaps I best settle for \$100 and let him have the \$100,000." Unfortunately, thinking further, you must consider the possibility that your opponent is thinking in precisely the same terms, in which case if he chooses to forego the big payoff, you once again find yourself at an outcome you clearly prefer to avoid. Indeed, absent the opportunity to communicate and coordinate in some way, this is clearly a game you prefer not to play and if forced to play, you are hardly likely to take solace in being told that the game has two quite reasonable Nash equilibria.

Chile in 1970, Korea in 1987 and Taiwan in 2000: To see that the problem of coordination can impact both political elites as well as ordinary citizens, we note that in the 1964 Chilean presidential election, Liberals and Conservatives, thinking they were unlikely to find a viable candidate to compete against Salvador Allende of a united Left (Socialists and Communists), endorsed the centrist candidate, Eduardo Frei. Frei won the election by a comfortable margin of 56.1% of the vote against 38.9% for Allende. In the 1970 Chilean presidential election, however, Salvador Allende won the plurality rule election with fewer votes than he received in 1964—36.2%—whereas the centrist Radomiro Tomic received 27.8% and right of center Jorge Alessandri received 34.9% of the vote. It is not unreasonable to suppose, though, that had either of Allende's competitors dropped out of the race, Allende would not have secured a majority. A similar outcome occurred in South Korea in 1987. Although there is no reliable polling information to precisely place the candidates on any left/right ideological space, it is commonly believed that the presidential winner, T.W. Roh, was actually the Condorcet loser. If either Y.S. Kim or D.J. Kim had dropped out of the race, Roh could not have won. Finally, in 2000 in Taiwan, according to a survey conducted by the Election Study Center of the National Chengchi University, the plurality winner, Chen Shui-Bian, with 39.3% of the vote, was the Condorcet loser by a slim margin. Based on the revealed preferences of respondents, the plurality loser, Lian Chan, with 23.1% of the vote, was actually the Condorcet winner. And although many attempts were made between Lian and the runner up to Chen Shui-Bian, James Soong, to avoid a three-way race, they failed to reach a compromise.

Doubtlessly many similar examples can be found of plurality losing candidates who could have won an election if one or more of the other unsuccessful competitors had dropped out of the race or if the supporters of such candidates had coordinated their vote. To illustrate matters without supposing that our example here matches the above cited circumstances in detail, suppose an electorate consists of voters with the three types of preference orders among candidates A, B and C that we discussed earlier when introducing the notion of a Nash equilibrium; namely:

type 1	type 2	type 3
A	B	C
B	C	B
C	A	A

Suppose further that 40% of the electorate holds preferences of the first type, and 30% holds preferences of the second and third types. Thus, if everyone votes sincerely, A is the plurality winner. Notice, though, that B is the Condorcet winner—B defeats A in a head to head contest 60% to 40%, and defeats C head to head 70% to 30%. The problem confronting supporters of B and C, then, can be portrayed as a simple 2 × 2 game as in Figure 3.8 (the reader should confirm once again that regardless of how voters of second and third types vote, voters who most prefer A should always vote for A).

	Vote for B	Vote for C
Vote for B	B wins	A wins
Vote for C	A wins	C wins

Figure 3.8 Coordinating in a plurality rule election

From the perspective of supporters of B and C, then, there are two equilibrium outcomes to this game: B wins or C wins. However, to achieve either of these outcomes, those supporters must coordinate their actions at the voting booth. This coordination is secured by default if either B or C declines to run, as occurred in Chile in 1964. But absent that action—absent a coalition between B and C and their supporters—voters are themselves confronted with the difficult task, without the benefit of any roundtable negotiations, of deciding who will unilaterally pull out of the contest.

One of the more interesting aspects of politics is learning how people coordinate (or fail to do so) in choosing one equilibrium over another. Indeed, as our example suggests, the failure to coordinate can yield an outcome that does not correspond to an equilibrium so that once that outcome is revealed, some subset of decision makers would, if given the opportunity to do so, prefer to unilaterally change their strategies. And again, although we will pay considerable attention in subsequent chapters to learning how coordination might be achieved, our objective here is to simply note that even if a game has Nash equilibria, we cannot assume, *a priori*, that any one of them will ultimately prevail.

3.4 Mixed Strategies

One need not conjure complex games to illustrate circumstances in which Nash equilibria are not unique or where players, in violation of a game's rules, might prefer to coordinate their choices. However, before we look further into the matter of how players might coordinate in a game when equilibria are not unique, we should first contend with the opposite problem—that of the potential non-existence of equilibria.

> **Penalty Shots in Soccer and the Normandy Invasion:** In one of his less memorable movies, *Escape to Victory*, the actor Sylvester Stallone is cast in the role of the World War II reserve goalie of an Allied prisoner of war soccer team formed by the Nazis to compete against a German national team in a game set up to serve as entertainment and propaganda for the Nazis. The game ends, though, with the Germans being awarded a dubious penalty kick that Stallone must defend against. The agonizing decision he must make as his German opponent lines up to kick the ball, all highlighted by Hollywood's flair for dramatizing things, is whether to dive right, dive left or remain in the center of the net. Of course, as is always the case with the heroes of Tinseltown, Stallone guesses correctly, blocks the shot and, in the melee of the celebration of their on-field victory, the prisoners/Allied soccer players escape among the crowd.
>
> Of course, soccer fans need not subject themselves to some grade-B movie to know the uncertainties associated with penalty kicks. For either player—goalie or shooter—to guess "wrong" results in a bad outcome. Clearly, even if we form an overly simplified model of such situations and allow a goalie to only guard "left," "right" or "center" and the shooter to aim only "left," "right" or "center," the corresponding game cannot have a Nash equilibrium as we have thus far conceptualized the idea. If the goalie chooses "left," the shooter should choose "left" or "center" (since his or her "left" is the goalie's "right") whereas if the goalie comes to believe that the shooter will thereby choose "left," he should alter his choice of the side of the goal to defend to "right." Things grow even more interesting, moreover, if we try to imagine what might be in the mind of, say, a goalie in such

a situation. A goalie knows that the penalty shooter has an incentive, via various feints and facial moves, to try to deceive him. Suppose the shooter quite explicitly indicates a shot to the left. This might seem to call for an automatic response of moving to guard the net's right (again, since his right is the shooter's left). But now the goalie might reason further, "Is he merely trying to deceive me? Is the shooter indicating a shot to the left so that I will guard the right, in which case he will actually shoot to my left?" Of course, the shooter has an equivalent problem since the goalie is free to also attempt to deceive. The result is that an infinite he-thinks-that-I-think regress emerges for both players and the ultimate outcome may be more a function of pure chance than any brilliance on the part of either player (unless, of course, one is a script writer for Hollywood).

One needn't be a soccer fan to appreciate the situation confronting a goalie or penalty shooter in such situations. Anyone familiar with the Allied invasion at Normandy in WWII knows of a similar strategic dilemma confronting both the Allies and Germans. For the Allies, the task was to convince the Germans, and Hitler in particular, that the actual invasion would occur at Calais rather than Normandy and that any troop landing at Normandy was merely a feint. For the Nazis, the choices were whether to defend at Normandy, at Calais or, spreading their resources thin, at both places simultaneously. The history of the invasion entailed an extensive and complex series of actions designed to deceive the Nazis, including putting the general the Germans thought most likely to lead the invasion, Gen. George Patton, in charge of several divisions spread out across the English countryside that consisted primarily of rubber tanks and cardboard trucks. Apparently several of Hitler's generals were well aware of the Allies' incentives to deceive, but Hitler, while vacillating in his beliefs, seemed less able to handle such strategic complexity. It is also interesting to note, however, that even one of Germany's most renowned generals, Erwin Rommel, in charge of coastline defenses, visited his family on June 6th in the belief that the weather would cause the Allies to postpone the invasion—which, of course, meant that a less-than-perfect weather forecast made it a perfect day to invade. Had Rommel immersed himself in the self-evident "he thinks that I think" regress, he might at least have exhibited more anxiety about blowing out his wife's birthday cake candles. So while the relevant actions on both sides occurred over a period of time as distinct from the few seconds in which shooter and goalie confront each other in a soccer penalty shot, the strategic issues here are nearly identical and one can readily imagine he-thinks-that-I-think regresses preoccupying the minds of strategic planners at least on the English side of the English Channel in the days leading up to "The Longest Day" of June 6th.

To grapple with the issues occasioned by the ostensible absence of a Nash equilibrium, consider the simple 2 × 2 game in Figure 3.9 and notice that regardless of which cell we consider, one player or the other has an incentive to change their strategy. In response to this fact, suppose person 2 (column chooser)

	b_1	b_2
a_1	4, 0	2, 5
a_2	1, 4	3, 2

Figure 3.9 A game with no pure strategy Nash equilibrium

reasons that since the "he-thinks" regress cannot be resolved, he might as well let a coin toss choose his strategy. But before doing so, suppose he reasons further and realizes that person 1 (row chooser) can anticipate this maneuver, in which case 1 would conclude that he or she can maximize their expected utility by choosing a_1 ($E(a_1) = 4(1/2) + 2(1/2) = 3$ whereas $E(a_2) = 1(1/2) + 3(1/2) = 2$). But if 1 chooses a_1, then 2 should rethink the idea of tossing a coin and choose b_2 instead. But if 1 concludes that 2 will choose . . . and so on. In other words, tentatively deciding to toss a coin fails to resolve 2's dilemma because it fails to terminate the he-thinks regress.

Despite the fact that tossing a coin fails to resolve matters, notice the conceptual innovation here. To this point we've focused on choices that specify the selection of one strategy or another with certainty. However, such choices are merely special cases of a more general type of strategy called a *mixed strategy*. First,

> *if s is a strategy for a player that specifies a specific action at each of that player's information sets, then s is a* ***pure strategy***.

Thus, aside from the random moves of nature, a determinate outcome prevails if all players choose pure strategies. A mixed strategy, on the other hand, is defined thus:

> *If the set of strategies available to a player is $S = \{s_1, s_2, \ldots, s_m\}$, then a* ***mixed strategy*** *for that player is a lottery over S,* $\mathbf{p} = (p_1, p_2, \ldots, p_m)$. *The player is said to choose the strategy* $\mathbf{p}$ *if he uses this lottery to determine which pure strategy he will implement in the actual play of the game.*

A mixed strategy, then, chooses a particular pure strategy by some random device. And although the toss of a fair coin does not terminate the "he-thinks" regress in our example, we want to ask whether it is possible that some other lottery might do so. That is, is it possible that there exists a *mixed strategy equilibrium* in which each person chooses a pure strategy in accordance with some lottery in such a way that if these lotteries are common knowledge, no one has any incentive to use a different lottery?

That the answer to this question is "Yes" is an important result in game theory. To see this in the context of the previous 2×2 example, consider this reasoning: Persons 1 and 2 begin with the supposition that 1 (row chooser) will choose a_1 with probability p and a_2 with probability $1 - p$, and that 2 (column chooser) will choose b_1 with probability q and b_2 with probability $1 - q$. For these probabilities to terminate the "he-thinks" regress, notice that, from 1's perspective, the choice between a_1 and a_2 is effectively a choice between two lotteries defined by the probabilities with which 2 chooses b_1 and b_2 (the chance q of 4 and the chance $1 - q$ of 2 versus the chance q of 1 and the chance $1 - q$ of 3). However, if 1 uses the lottery $(p, 1 - p)$ to choose between a_1 and a_2 and if there is some chance of choosing a_1 as well as a_2 (if $0 < p < 1$), then the lotteries offered by a_1 and a_2 should yield the same expected utility. Otherwise, if $u_1(a_1) \neq u_1(a_2)$, where we now let the u's denote expected utility, then person 1 can improve his expected payoff by switching from $(p, 1 - p)$ to choosing with certainty the strategy that yields the greater expected return—$(p, 1 - p)$ can not be an equilibrium strategy. That is, the expected payoff from using $(p, 1 - p)$ is $pu_1(a_1) + (1 - p)u_1(a_2)$, and unless $u_1(a_1) = u_1(a_2)$, this expected return is increased by increasing or decreasing p. Thus, in our example we must have

$$u_1(a_1) = 4q + 2(1 - q) = 1q + 3(1 - q) = u_1(a_2)$$

which solves to yield $q = 1/4$. Similarly, if 2 has no incentive to shift from $(q, 1 - q)$, where $0 < q < 1$, it must be that

$$u_2(b_1) = 0p + 4(1 - p) = 5p + 2(1 - p) = u_2(b_2)$$

which requires that $p = 2/7$. Hence, we have established that the pair of lotteries $((2/7, 5/7), (1/4, 3/4))$ constitute a Nash equilibrium.

American Football: It is not difficult to contrive examples in which the chair of a committee prefers to have ties broken in an agenda by lottery rather than be empowered to break ties because fellow committee members will, if he is empowered to break a tie, act to avoid such a possibility. This fact seems paradoxical, but only if we ignore the interdependent nature of choice and the fact that others will respond in their own self-interest to any alteration in someone's voting power. A similar paradox often materializes in the American game of football. Those who follow such games closely at the professional level will be familiar with the following seemingly paradoxical event: A team's star but somewhat aged quarterback, whose record over the years of passing the ball marks him as a future member of the Hall of Fame, is suddenly injured and replaced by a recently recruited fresh-out-of-college backup whose professional experience is essentially non-existent. Suddenly, rather than see the team rely on running plays, the frequency with which the backup quarterback passes increases. Average yards gained from a completed pass may drop, but to everyone's surprise, the new wet-behind-the-ears quarterback's completed pass percentage increases above that of the aged veteran. Thereafter sports commentators begin speculating on whether the team confronts a "quarterback controversy," whether the star ought to be permanently retired or whether the team's coach is failing to do his job by not forcing that retirement.

Before we jump to conclusions as sports commentators often do, consider this example: Suppose two candidates, 1 and 2, must decide whether to allocate the final few days of the campaign in either state A or B, since the polls tell them that these two states are pivotal. Suppose the candidates' respective probabilities of winning the election are as shown in Figure 3.10.

	A	B
A	0.5, 0.5	0.9, 0.1
B	0.9, 0.1	0.6, 0.4

Figure 3.10 Probabilities of winning the election

If we make calculations of the same sort as before, we find that the mixed strategy equilibrium is for both candidates to campaign in A with probability 3/7 and in B with probability 4/7. Now suppose that prior to implementing these strategies, a new poll reveals that if candidate 1 (row chooser) alone goes to state B, his probability of winning is actually 0.7 and not 0.9, as originally thought. It might seem that this reassessment should cause 1 to increase the likelihood of visiting A. However, the equilibrium strategy for 1 now is to go to A with probability 1/5 and to increase the likelihood of visiting B from 4/7 to 4/5. So the inference that a decrease in the value of visiting state B should lead to a lower likelihood of visiting B is erroneous, because it neglects the fact that 2 (column chooser) will also adjust to this change (2 should now increase his likelihood of campaigning in state A from 3/7 to 3/5).

Returning to American football, what sports commentators are unlikely to consider is that the teams opposing the backup have adjusted their strategies by relaxing somewhat on their passing defense. Football is a game where it is reasonable to suppose that coaches and quarterbacks employ mixed strategies—when confronted with equivalent circumstances, they will "mix their plays" so that it is impossible for the opposition to guess with certainty which play will be run at any specific instance. Sometimes the quarterback will attempt a pass; in other, seemingly identical, circumstances, he will call for a run. At the same time, defenses will also be mixing since otherwise the offense can take full advantage of any regularized pattern in the defense. And once mixed strategies enter the domain of strategic choices, we can witness things that seem paradoxical.

For purposes of a numerical example, consider Figure 3.11, in which payoffs are stated in terms of yards gained (a positive gain for the offense is a negative "gain" for the defense). Presumably, if the defense incorrectly guesses what type of play the offense will choose, the play will be successful for the offense; otherwise, the offense's gain is at best small. The mixed strategy equilibrium for the payoffs in Figure 3.11, now, is ((4/5, 1/5), (2/5, 3/5)).

Now, however, suppose the payoffs for the rookie replacement are changed as in Figure 3.12 so that the yards gained from a pass against a run

	Run defense	Pass defense
Run	5, –5	10, –10
Pass	20, –20	0, 0

Figure 3.11 Yards gained with the star quarterback

	Run defense	Pass defense
Run	5, –5	10, –10
Pass	10, –10	0, 0

Figure 3.12 Yards gained with the rookie backup quarterback

defense drops from 20 to 10. Now the mixed strategy equilibrium is ((2/3, 1/3), (2/3, 1/3)). Thus, the rookie now calls a pass play more frequently than does the veteran (1/3 versus 1/5) despite his lower efficiency. Such a result is paradoxical only if we ignore the fact that the defense is also responding to the lower efficiency of the rookie. Specifically, notice that the frequency with which the defense now guards against a pass drops from 3/5 to 1/3, thereby accounting, in part for the increased use of the pass on offense (and accounting as well for any "quarterback controversy" that sports writers might invent since our rookie is successful with his passes two thirds of the time as against two fifths for the veteran—two thirds of the time the rookie passes, he does so against a run defense whereas the veteran does so two fifths of the time).

Aside from offering seemingly paradoxical results, the introduction of mixed strategies raises a new set of questions, chief among them being whether they resolve the issue that occasioned their introduction in the first place. That is, do mixed strategies necessarily resolve the issue of existence? The answer is provided by von Neumann and Morgenstern in 1947 and subsequently generalized by John Nash in 1950:

Every strategic form game in which all persons must choose from finite sets of pure strategies has at least one Nash equilibrium in either pure or mixed strategies.

Requiring a finite number of pure strategies is important (although not for all games). For a game without an equilibrium, consider this example: Choose an integer, and whoever chooses the largest integer wins. Since there is no integer greater than all others, there is no equilibrium. Or suppose two people must each choose a number in the interval [0, 1]. If they choose the same number, they tie. Whoever chooses the larger number wins, with the exception that if one person chooses 1 and the other chooses a lower number, the second person wins. This game also has no equilibrium, pure or mixed, since for any mixed strategy (i.e., probability density function) over the interval (0,1) the opponent can find another mixed strategy that pushes everything closer to 1 without putting any weight on 1. The problem here is akin to walking half the distance to a wall, then walking half the distance of the remaining distance, then walking half the distance of the remaining distance (one quarter of the original distance), and so on. No matter how many such increasingly small steps one takes here, you never reach the wall, and someone else, in principle, can walk half of whatever distance remains and get still closer to the wall than you are.

Ignoring these anomalous cases, the curious reader might ask how it is that mixed strategies solve, at least mathematically, the issue of existence of equilibria. We will not offer here a formal mathematical proof, but we can convey the general idea that underlies that proof. To begin, consider a function $f(x)$ that maps values of x in the interval (0,1) to values in that same interval. In fact, imagine a two dimensional coordinate system limited to the interval (0,1) on both the horizontal (x) axis and vertical ($f(x)$) axes. Now put your pencil somewhere on the vertical axis at $x = 0$ (i.e., pick a value for $f(0)$ in the interval [0,1]) and, without lifting your pencil from the paper, draw a line, curved or otherwise, across to $x = 1$ and where that line never goes above $f(x) = 1$ or below $f(x) = 0$. What has to be the case now is that your line will cross the diagonal line running from (0,0) to (1,1) at least once—the only way to avoid that line is to do what we just told you not to do, namely lift your pencil from the paper or let the function you've drawn take on values outside the interval (0,1). Thus, for at least one value of x, say x^*, it must be the case that $f(x^*) = x^*$. The value x^* is called a *fixed point* of the function $f(x)$—a value of x that f maps onto itself. There are, as one might suspect, a number of theorems in mathematics called *fixed point theorems* which establish conditions on f and on a function's domain and range such that fixed points necessarily exist. Our little exercise with pencil and paper merely illustrates the simplest of those theorems and one set of conditions to which those theorems apply (in this case that f be continuous and that its domain and range be both identical and closed convex sets). The proof that a Nash equilibrium necessarily exists then proceeds as follows (after simplifying notation by limiting discussion to a two-person game): Let the function $g_1(q)$ define person 1's best response to 2's choice of the strategy q, where q can be any mixed strategy, including one that puts all weight on a single pure strategy. Similarly, let $g_2(p)$ denote person 2's best response to 1's choice of p. Thus, we can define the function $h(p,q) = (g_1(q), g_2(p))$, which maps the strategy pair (p, q) to a new pair in accordance with the preferences of the individual players and the game's structure. Notice now, though, that the domain and range of h are identical, and that by allowing for mixed strategies, we have rendered h's domain and range closed convex sets (i.e., if a player has, say, m pure strategies, then the set of all possible pure and mixed strategies is an m dimensional simplex or, in the case of $m = 2$, the interval [0,1]). All that remains, then, is to show that h is a continuous function, in which case the existence of a fixed point, as in our paper and pencil exercise, is assured. We skip this last step, but since it is true, there must exist a (p^*, q^*) such that $h(p^*, q^*) = (p^*, q^*)$. Such a strategy pair is, by definition, an equilibrium. Hopefully, then, the reader can see not only the underlying logic of the proof of the existence of Nash equilibria when we allow mixed strategies, but also the role mixed strategies play in allowing us to apply an otherwise seemingly unrelated theorem in mathematics to establish something that might be relevant to the study of politics.

However, as elegant in its simplicity this proof might be, we are nevertheless left with a second question: Why should a person use a mixed strategy when, as our analysis of the example of Figure 3.12 shows, any lottery over the pure strategies yields the same expected payoff provided the opponent chooses in accordance with his equilibrium strategy? The answer is provided in the last clause of the question. If for the game in Figure 3.12 person 2 (column chooser) switches to a pass or run defense with certainty, or to any lottery other than (2/3, 1/3), and if, as we have assumed, row chooser (player 1) can "get into 2's head" because he knows as much about the game as does 2, then 1 can take advantage of 2's decision, which in this case at least, reduces 2's expected payoff. Thus, not only does 2 have no positive incentive to switch from (2/3, 1/3), but switching can even be dangerous.

This defense of mixed strategies as a solution to strategic form games should not be taken to mean that we deem them a universal solution to the treatment of games without pure strategy equilibria. There are times when they make sense (e.g., football and perhaps even in the selection of military tactics) but there are other times when relying on them to solve a game is unbelievable. Consider again the three-voter, two-dimensional spatial example we've used to illustrate a majority rule spatial game without a Condorcet winner (see Figure 1.6). Because that winner does not exist—indeed, because there is no alternative that is not beaten in a majority vote by something else—there cannot be, as we note earlier, a pure strategy Nash equilibrium to the two-candidate election game. Our difficulty now lies in the fact that if we take such spatial examples as preliminary models of electoral competition in plurality rule systems, we have a rather difficult time bending our brains around the hypothesis that candidates abide by mixed strategies. Even if we could compute such a strategy for a specific election (and indeed, such a calculation is at best truly complex), we suspect that if we had the audacity to show up at a candidate's campaign headquarters, computer in hand, and attempted to show them the mixed strategy they should use in preparing a candidate's campaign policy proposals, we'd quickly and justifiably be shown the door. Campaigns are dynamic events wherein our representation of candidate strategies as points in some multidimensional issue space can be too much of an abstraction. The Median Voter Theorem bypasses this problem because it identifies a circumstance in which there is a pure strategy equilibrium and thus some reason for supposing that campaign dynamics would not upset that equilibrium. This isn't to say that other features of real elections will not intervene to forestall convergence to the median, only that the median as a Nash equilibrium will remain a point of attraction just as gravity exerts its influence on an airplane in the air. But if no such winner exists, then these dynamics, including the maneuvering that occurs within parties between campaigns, appear more critical and the simple static view of the Median Voter Theorem is most likely out of place.

In discerning what role mixed strategies can play in our analysis of things, we should keep in mind that as with almost any modeling enterprise, whether it be in the natural or social sciences, the efforts of the researcher are as much an art as a science. There are no hard and fast rules for when mixed strategies make sense and when they don't, and part of the "art" is knowing when to state that a problem has an indeterminate solution, or a solution that can be identified only after we have specified other aspects of the situation under investigation. For an example of what we mean here, consider another game without a pure strategy equilibrium, *The Colonel Blotto Game*. In this game we have two equally sized and equipped armies that must be allocated over three battlefields. Whoever allocates the most armies to a battlefield wins that battlefield, and whoever wins two battlefields wins the war. The strategy space of both players is, then, equivalent to a three-dimensional budget simplex, with each dimension measuring the share of one's army allocated to a particular battlefield. As it turns out, this game has a mixed strategy solution—indeed, it has several—and one such solution can be described as follows: Take the simplex and lay it flat on the table; inscribe a circle inside the simplex that touches all three sides and erect a cylinder up from that circle; drop a cone into the cylinder so that its tip touches the simplex and its sides fit snugly around the rim of the cylinder. The final step is to normalize the volume of space bordered by the cylinder and cone to 1.0 and interpret that volume as a probability density function. That function is a mixed strategy equilibrium strategy. Pretty horrific, huh? But notice this: An American Presidential election can, if we so choose, be conceptualized as a 50-battlefield (state) Colonel Blotto Game in which the different battlefields have different weights (Electoral College votes). The simplest model now is to suppose that whoever allocates the most resource (e.g., time, money, etc.) to a state wins that state. We shudder, of course, at trying to imagine what the mixed strategy equilibrium looks like for such a game. But we shudder doubly so if, after computing such a strategy, we imagine how we would be greeted at a candidate's headquarters if we were to try to convince them that they need to "spin the big spinner" and choose an overall allocation in accordance with our computations. And if an appeal to America's Electoral College seems too country-specific, suppose the candidates are debating policies that, one way or another, redistribute wealth—a tax reform, a government subsidy to some class of industries, a new or revised social welfare program. In doing so, perhaps they've conceptualized the problem by dividing society into clusters that seem to make substantive sense (e.g., those living in poverty, those who are 'merely' poor, the lower middle class, the upper middle class, the rich, and the super rich). If this allocation is part of an election campaign, what they have here is but another version of a Colonel Blotto Game—in this case, perhaps even with 3, 4 or more "armies" in contention. For the present, then, we will simply regard mixed strategies as a potential solution to the task of predicting people's choices in strategic environments, and leave any final determination as to whether such a solution makes substantive sense to the problem at hand or whether it is a mere mathematical trick and that some other more substantively meaningful approach needs to be considered.

3.5 Mixed Strategies and Domination

The analytic value of the concept of a mixed strategy depends, then, on the care we exercise in modeling a situation and whether or not we think it reasonable to suppose that players might actually contemplate allowing chance to dictate their actions. If the strategic form accurately reflects the strategic environment, then mixed strategies are merely a mechanism for choosing a pure strategy in such a way that opponents cannot take advantage of their knowledge. On the other hand, if our strategic form is but a crude approximation to reality, then the notion of mixed strategy may make little or no sense. However, even if we believe that dynamics are relevant, it may still be the case that a more limited static analysis can assist us in understanding strategic imperatives. To see what we mean by this, it is first useful to introduce the notion of domination. Briefly,

> *The strategy s' **weakly dominates** another, s", if s' does at least as well as s" in all contingencies (against all possible strategies by other decision makers), and if it sometimes does better. And if all these conditions hold strictly (i.e., if s' is strictly better than s" in all circumstances, then s' [strictly] **dominates** s".*

In a simple one-vote agenda, voting for one's preferred alternative dominates a contrary vote. If one's vote is not decisive, then it does not matter what choice is made so we cannot say voting sincerely here strictly dominates voting in any other way, but in the event that one's vote is decisive, then it is always best to vote for the preferred alternative (assuming that one's vote does not count negatively). For the simple 2×4 game in Figure 3.3, A dominates B, but not strictly, whereas if the payoffs in the lower right cells corresponding to (B,BB) and (B,BA) were less than 0, then A would strictly dominate B.

With respect now to election models, it is fortunately the case that the sort of election scenarios that Figure 1.6 models, multi-dimensional spatial elections, have strategies that dominate others. For example, if candidates are concerned solely with whether or not they win the election, then for two-candidate elections, platforms located far from all ideal points are dominated by ones that are closer to those points. And since it seems unreasonable to suppose that anyone would choose a dominated strategy, it seems unreasonable to suppose that a candidate, regardless of the election campaign's dynamics, would act differently. The difficulty here, however, is that multi-issue election models of this type fail to have a strategy that dominates all others (since those examples have no Condorcet winner), and thus, the elimination of strategies that are dominated will not present the candidates with unique choices. Nevertheless, our argument, if formalized, establishes something of substantive significance. With respect to our spatial conceptualization, candidates will not choose platforms far from the "main body" of voter preferences—indeed, as we argue more fully in the next chapter, unless voter preferences are skewed in strange asymmetric ways, they will not choose platforms far from the median preference on each issue (assuming, of course, that the relevant assumptions of the Median Voter Theorem about perceptions and candidate mobility are satisfied, as well as the assumption that voter preferences correspond to distance from some ideal in the issue space). Thus, despite the fact that our representation of an election is too abstract to allow us to assert the relevance of mixed strategies, the simple idea of dominance offers us the beginnings of an insight into the general form of the strategies candidates are likely to limit their attention to.

In the next chapter, we give greater structure to the idea of "not far" and apply the notion of domination to say something more specific about spatial elections by way of putting strict bounds on the strategies candidates are likely to consider. At this point, we simply want to emphasize that the notion of domination is a general one and its application is hardly confined to election models. Pruning voting trees also illustrates the elimination of dominated strategies because, in the case of voting in agendas, for instance, a strategy that specifies voting for something other than the alternative one prefers at the final nodes of any binary voting tree is dominated by a strategy that specifies voting the other way. The elimination of dominated strategies, moreover, can simplify an analysis considerably. Earlier we noted that the strategic form of our legislative pay raise example yields a 2 × 4 × 16 strategic form. However, since legislator C's choice is evident when it comes time for him to vote, any strategy that has him voting for the raise when A and B have already voted to pass it, or voting against the raise when either A or B have voted against it, is dominated by strategies that have C voting the opposite. Pruning the extensive form back at C's decision nodes eliminates precisely these strategies. However, eliminating dominated strategies is a process we can repeat as many times as possible, so if, after eliminating C's dominated choices in the strategic form in Figure 3.2, we then turn to legislator B and eliminate dominated strategies for him as well in the reduced form of Figure 3.2, we will learn that legislator A's dominant strategy is to vote against.

Thus, whenever we confront a game in strategic form, the first thing to be done is to check whether it has dominated strategies. If any player has such a strategy, then, after its elimination, we should check to see if other players now have dominated strategies.

Example: Consider the two-person strategic form in Figure 3.13a, and notice that person 2 (column chooser) does not have a dominated strategy. However, 1 has such a strategy, a_2 (dominated by a_1). After eliminating a_2 to get the form in Figure 3.13b, b_2 is now dominated by b_1. Eliminating b_2 to get the form in Figure 3.13c, a_3 and b_3 are dominated by a_4 and b_4,

	b_1	b_2	b_3	b_4
a_1	3, 3	2, 2	4, 3	3, 4
a_2	2, 0	1, 3	0, 2	2, 0
a_3	3, 4	4, 2	2, 2	0, 3
a_4	4, 3	2, 1	3, 1	4, 2

Figure 3.13a A 4 × 4 game with dominated strategies

	b_1	b_2	b_3	b_4
a_1	3, 3	2, 2	4, 3	3, 4
a_3	3, 4	4, 2	2, 2	0, 3
a_4	4, 3	2, 1	3, 1	4, 2

Figure 3.13b First reduction of Figure 3.13a

	b_1	b_3	b_4
a_1	3, 3	4, 3	3, 4
a_3	3, 4	2, 2	0, 3
a_4	4, 3	3, 1	4, 2

Figure 3.13c Second reduction of Figure 3.13a

	b_1	b_4
a_1	3, 3	3, 4
a_4	4, 3	4, 2

Figure 3.13d Third reduction of Figure 3.13a

respectively. But now, as Figure 3.13d shows after the elimination of a_3 and b_3, the strategy a_4 dominates a_1, so b_1 subsequently dominates b_4, leaving us with a unique Nash equilibrium and a unique prediction about choice.

The argument that legitimizes this sequential elimination of dominated strategies appeals once again to the assumption of common knowledge. If, in our example, 2 knows that a_2 is dominated, then both 1 and 2 know this strategy can be ignored, and both persons can focus on the strategic form in Figure 3.13b. This same argument, repeated, eventually eliminates all but (a_4, b_1).

The Paradox of Power: For another example of the application of Nash equilibria to politics as well as the method of the sequential elimination of dominated strategies, consider this proposition: Any increase in the number of seats that a party controls in a legislature or parliament is advantageous to the party, and thus the number of seats it controls is a measure of its "power." We should, of course, refine this proposition to exclude possibilities in which one party gains control of seats while some other party secures a majority, because it is then obvious that the party's gain is offset by what happened elsewhere. Let us then consider an example in which one party gains seats at the expense of other parties, but no other party gains seats. Specifically, suppose the distribution of seats in a 100-member, three-party parliament is (32, 33, 35) and that this distribution changes to (35, 32, 33). Once again, intuition suggests that party 1 would be delighted with this turn of events. However, intuition here once again commits the error of confusing decision theoretic with game theoretic reasoning.

To see what we mean we must first specify a voting rule for the parliament, so suppose all decisions are made by a plurality vote. This means that, in a three-party parliament, any two parties can dictate the outcome, but if the parties vote differently, the largest party predominates. Now suppose the parliament is considering three alternatives, A, B, and C, and that preferences are those occasioning the usual illustration of the Condorcet Paradox; namely:

Party 1:	A	B	C
Party 2:	C	A	B
Party 3:	B	C	A

Consider a secret ballot in which each party has three strategies: vote for A, for B, or for C. If seats are distributed (35, 32, 33), then the game's strategic form is as shown in Figure 3.14a (for example, if party 1 chooses A, if 2 chooses C, and if 3 chooses B, then A prevails because party 1 has more votes than either 3 or 2; however, if 2 and 3 both vote for C, C prevails regardless of what party 1 chooses). Notice that for party 1 (row chooser), voting for A (weakly) dominates voting for B or C. However, as things stand in Figure 3.14a from party 2's perspective, while voting for C (weakly) dominates voting for B, we cannot eliminate voting for A via domination. Similarly for party 3, while voting for A is dominated, we cannot yet eliminate voting for B or C. However, once we eliminate party 1's choice of B and C as well as 2's choice of B and 3's choice of A, we get the reduced game in Figure 3.14b, and here we see that voting for C is a dominant choice for parties 2 and 3. Hence, alternative C prevails as the final outcome. Thus, the "strongest" party realizes its least preferred alternative, and the "weakest" party successfully secures its ideal preference.

Clearly, then, party 1 in our example should prefer a reversal of roles, so that some other party (notably 3) controls a plurality of seats. Quite directly, then, our example demonstrates that "power" is not a simple variable measured by the accumulation of some commodity—in this instance,

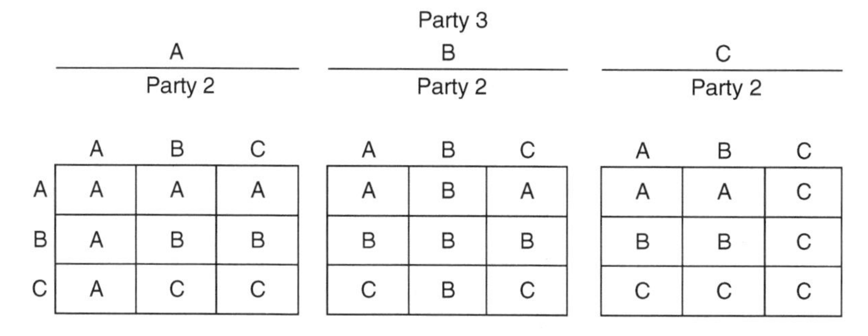

	Party 3: A			Party 3: B			Party 3: C		
	Party 2: A	B	C	Party 2: A	B	C	Party 2: A	B	C
A	A	A	A	A	B	A	A	A	C
B	A	B	B	B	B	B	B	B	C
C	A	C	C	C	B	C	C	C	C

Figure 3.14a A 3-Party legislator with the seat distribution (35, 32, 33)

	B: A	B: C	C: A	C: C
A	A	A	A	C

Figure 3.14b Reduced form of Figure 3.14a

> parliamentary seats. Instead, our example shows that the empirical determinants and measures of power—conceptualized as the ability to secure favorable outcomes—can be defined only in the context of the strategic environment of those who wish to exercise it.

As simple as this example is, it illustrates something of profound importance in politics; namely, power is not always what it seems to be. While having the authority to break ties, to issue vetoes over legislation, to dismiss various government officials, and perhaps even to set the agenda might all seem dimensions of power, if by "power" we mean the ability to secure outcomes one cannot secure without the instruments of power, then the paradoxical nature of our example suggests that we ought to be careful when measuring power or when implementing or amending, say, a political constitution in order to make some office more or less powerful. The preceding example is not, in fact, a mere anomaly. Recall the seemingly paradoxical character of our example of American football. Once again, we need to emphasize that such examples seem paradoxical only because we are too easily seduced by simple decision theoretic thinking. We too easily forget sometimes that other players in the game of politics adapt and react to any change in the status of one of the players and that deducing the ultimate consequences of any change requires a game theoretic analysis in which we attempt to assess the ultimate equilibrium of all adjustments. As we argue in future chapters, this is a lesson that must be learned when, in particular, national political constitutions are being drafted, lest they yield unanticipated and even undesirable consequences.

The Relevance of Irrelevant Things: Before proceeding to explicitly political matters, we want to offer a note of caution about what we observe and don't

	A	B
a	3, 1	0, 0
b	0, 0	1, 3

Figure 3.15 A simple coordination game

	A, A	A, B	B, A	B, B
T, a	2, 1	2, 1	−1, 0	−1, 0
T, b	−1, 0	−1, 0	0, 3	0, 3
NT, a	3, 1	0, 0	3, 1	0, 0
NT, b	0, 0	1, 3	0, 0	1, 3

Figure 3.16 Extended version of Figure 3.15

observe in the real world. Consider the 2 × 2 game in Figure 3.15, with payoffs in dollars. This game has two pure strategy Nash equilibria, (*a*, *A*) and (*b*, *B*) as well as a mixed strategy equilibrium. But notice that there is no guarantee that non-cooperative play results in any of these equilibria—the players confront a problem of coordination because in fact Figure 3.15 mimics the game in Figure 3.8, though perhaps not in such a dramatic fashion. If row chooser reasons that *a* is the best choice because it yields the greatest potential payoff, he must worry that column chooser thinks the same way, in which case he best settle for a payoff of 1 by choosing *b*. But if he reasons further that column chooser might mimic his thoughts, he is led back to choosing *a*, in which case the he-thinks regress begins anew. However, prior to playing this game, suppose player 1 (row chooser) is given the option of throwing (T) or not throwing (NT) away a dollar and that player 2 observes this decision prior to choosing a strategy.

Our instinct most likely is to assume that since 1's decision to throw away a dollar is made before the play of the game and since it only affects 1's welfare, we can ignore it. In modeling the full situation, though, notice that player 1 now has four strategies—(*T*, *a*), (*T*, *b*), (*NT*, *a*), and (*NT*, *b*) and player 2 has four strategies—(*A*, *A*), (*A*, *B*), (*B*, *A*), and (*B*, *B*)—which specify 2's choice, contingent on 1's prior selection of *T* or *NT*, respectively (for example, (*A*, *B*) reads "if 1 chooses T, then choose *A*; but if 1 chooses *NT*, then choose *B*"). Figure 3.16 shows this extended situation's strategic form.

Now consider what happens when the (weakly) dominated strategies are sequentially eliminated:

1. Delete (*T*, *b*) because it is dominated by (*NT*, *a*).
2. With (*T*, *b*) deleted, eliminate (*B*, *A*) and (*B*, *B*) because they are dominated by (*A*, *A*) and (*A*, *B*) respectively.

3. With (*T*, *b*), (*B*, *A*), and (*B*, *B*) eliminated, (*T*, *a*) then dominates (*NT*, *b*).
4. With (*T*, *b*), (*B*, *A*), (*B*, *B*), and (*NT*, *b*) eliminated, (*A*, *A*) dominates (*A*, *B*).
5. With all of 2's strategies but (*A*, *A*) eliminated, (*NT*, *a*) dominates (*T*, *a*), which yields the unique prediction (3, 1).

That 1's ability to throw away money affects our prediction might seem paradoxical, especially since player 1 does not choose *T* in the equilibrium arrived at by sequential elimination. However, the relevance of a seemingly irrelevant choice can be rationalized thus: Player 2 can reason that if 1 were to choose *T*, then it must be that player 1 intends to play *a*, since only the outcome generated by (*a*, *A*)—namely, (2, 1)—is preferred to the worst outcome that prevails if 1 chooses NT. And since 2 can assume that 1 can anticipate this reasoning, 2 must believe that 1 can anticipate 2's choice of the best response to *a*, *A*, in the event that 1 chooses *T*. So 2 concludes that it is common knowledge that *T* leads to the outcome (2, 1). But if 1 chooses *NT* instead, it must be that 1 intends to get a higher payoff. Since only the choice pair (*a*, *A*) yields 1 a more preferred outcome than (2, 1), the choice of *NT* must mean that 1 plans to choose *a*. Thus, 2 should choose *A*.

There is a theoretical note of caution that needs to be applied here. Notice that after (*T*, *b*) is eliminated, we eliminate (*B*, *A*) and (*B*, *B*) because (*A*, *A*) and (*A*, *B*) are superior in the event that player 1 chooses (*T*, *a*). But later we conclude that 1 will not choose (*T*, *a*), which appears to negate our original reason for eliminating (*B*, *A*) and (*B*, *B*). It is this type of contradiction that can arise with sequential elimination of weakly dominated strategies. This is not to say that such sequential elimination ought to be avoided, or even that it is incorrect in the present example. Rather, we are merely suggesting that it, as well as any other solution hypothesis, ought to be applied carefully, with an eye to possible problems and conceptual traps. Our example, however, is profoundly important, though, for a more substantive reason. If we accept the reasoning behind the conclusion that the outcome (3,1) will ultimately prevail we must also conclude that we are likely to overlook the potential relevance of seemingly irrelevant choices in other contexts, especially those in which coordination is required to achieve some equilibrium outcome. For example, consider the relationship that existed between President Reagan and Premier Gorbachev in the 1980s. Following the retirement of both from public office, any number of writers familiar with one or the other of these world leaders reported that both men questioned whether they'd be able to "push the big red button" in the event they were attacked by the other side. Suppose we assume that each of them, via perhaps their spy networks, understood that the other could not bring himself to order a retaliatory nuclear strike against the other. Under a naïve model of nuclear deterrence, deterrence should not then have worked. Yet it did and led even to a variety of substantial disarmament agreements that required long-term cooperative and coordinative actions to implement. Usually, of course, those agreements are credited to the relationship that developed between Reagan and Gorbachev and their mutual abhorrence of nuclear weapons. We are not ones to dispute this view, but our example opens the door to another possibility; namely, both sides were coordinated to a mutually agreeable outcome not because there was a chance the other might, with some small probability, use their "big red button" but rather simply because each side knew the other had such a button but with no intention of using it. Admittedly, such an interpretation of things might be a stretch, but the preceding example (and game theory generally) should make us aware of the possibility that our conclusions about political processes are sensitive to the finer details of how we conceptualize those processes, including the availability of actions no one would choose.

3.6 Finding Mixed Strategy Equilibria

By way now of illustrating further the character of mixed strategies and mixed strategy equilibria, we note that it is one thing to assert the existence of such a thing, but quite another to assert it can be readily computed or that people act in accordance with it. One issue that persists with respect to equilibria, regardless of type, concerns the hypothesis that they necessarily provide reasonable predictions about choice. We have already seen, for example, how non-uniqueness occasions problems, which is an issue we must address throughout the remainder of this volume. However, now is as good a time as any to pause, reflect on what we have discussed thus far, and see in particular whether we can avoid such problems for any special class of situations.

Let us begin with a simple scheme for calculating mixed strategy equilibria. Our first step in approaching any strategic form game of more than trivial complexity should be to eliminate all dominated strategies, iterating the process as many times as is possible. If we are lucky and are left with a 2×2 game, then we can solve for the mixed strategy equilibrium by repeating the steps we took to analyze the game in Figure 3.9. Specifically, after writing the expressions for the expected values of a_1 and a_2 in terms of q (2's mixed strategy), we solve for q by equating these two values. Similarly, we solve for 1's mixed strategy by equating the expected values of b_1 and b_2. But suppose we are left with an $m \times 2$ game, where $m > 2$. Fortunately, the same principles apply as before. Letting the player with two strategies, 2, use the mixture $(q, 1 - q)$, letting $u_1(a_i)$ denote the expected value to player 1 of strategy a_i, $u_1(a_i) = qu_1(a_i, b_1) + (1 - q)\, u_1(a_i, b_2)$, and letting $(p_1, p_2, \ldots, p_m)$ be 1's mixed equilibrium strategy, then if p_i and p_j are both greater than zero, it must be the case that $u_1(a_i) = u_1(a_j)$. And if $p_k = 0$, then it must be the case that $u_1(a_k) < u_1(a_i)$. Of course, if $0 < q < 1$, it must also be the case that $u_2(b_1) = u_2(b_2)$.

Example: Consider the 4×2 game in Figure 3.17. Assuming that person 2 abides by the mixed strategy $(q, 1 - q)$, and writing the expressions for $u_1(a_1)$ through $u_1(a_4)$, we get

$$u_1(a_1) = 12q + 2,$$
$$u_1(a_2) = 2q + 9,$$
$$u_1(a_3) = 11 - 6q,$$
$$u_1(a_4) = 13 - 11q.$$

	b_1	b_2
a_1	14, 3	2, 5
a_2	11, 10	9, 6
a_3	5, 2	11, 8
a_4	2, 1	13, 0

Figure 3.17 A 4 × 2 game

Plotting $u_1(a_i)$ against q (see Figure 3.18) shows that there are several instances in which two of these values are equal. However, there are only two instances in which equality applies and where player 1 would not, for the associated value of q, have an incentive to shift to some pure strategy—specifically, where $u_1(a_1) = u_1(a_2)$ and where $u_1(a_2) = u_1(a_4)$. There are, of course, other pairs of lines that intersect, but to see that they cannot correspond to an equilibrium, consider the intersection of lines a_3 and a_4. The value of q for which this occurs is 2/5 (as computed by solving $5q + 11(1 - q) = 2q + 13(1 - q)$). Such a value, though, cannot be part of any equilibrium because, if column chooser abides by this probability, row chooser can increase his expected payoff from 8 3/5 to 9 4/5 by adopting a_2 with certainty, in which case column chooser would then shift from $q = 2/5$ to choosing b_1 with certainty (thus increasing its expected payoff from 7 3/5 to 10), and so on. Thus, a mixture between a_1 and a_2 and between a_2 and a_4 are the sole candidates for a mixed strategy equilibrium. Focusing on the first of these possibilities, we can eliminate all strategies but a_1 and a_2 and solve for 1's mixed strategy—in this case we get ((2/3, 1/3, 0, 0), (7/10, 3/10)). There is, though, a second solution when the lines for a_2 and a_4 intersect, so equilibria are not unique but this second equilibrium can be solved for in an identical fashion.

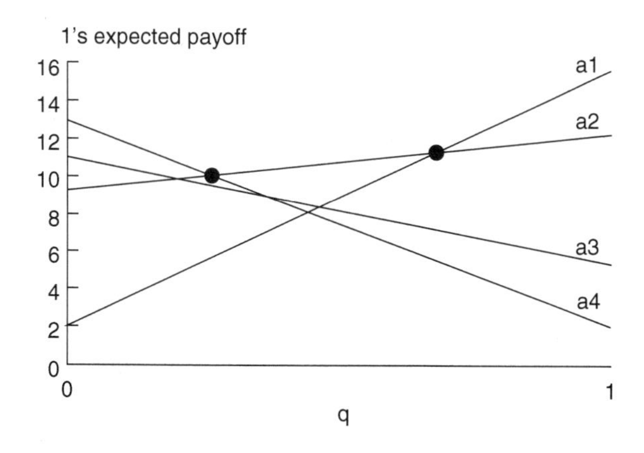

Figure 3.18 Finding the mixed strategy equilibrium

Solving larger games, including those with more than two decision makers, is more complicated. We may have to guess about which strategies have nonzero probabilities associated with them, and the equations we must solve will involve many more probabilities. But without minimizing the computation complexity involved, the procedures for solving larger games are essentially refinements and generalizations of the method we have just illustrated—they make use of the fact that any two strategies, say a and a', with nonzero probability associated with them must yield equal expected returns against the opponents' strategies. And if a'' is any strategy with zero probability assigned to it, then its expected return against the opponent cannot be greater than what a and a' yield. Some games can be solved, moreover, using various tricks. For instance, suppose a two-person game is symmetric in the following way:

> ***Symmetric two-person game:*** *A two-person game in strategic form is symmetric if both persons have identical feasible strategy sets, and if, when they switch strategies, they switch payoffs.*

Our previous examples of two-candidate elections illustrate symmetric games, because we did not suppose that either candidate held any advantage over the other and because both candidates could choose from an identical set of strategies. Suppose then that the players in a two-person game are indistinguishable in the way implied by symmetry. In this instance, if (a^*, b^*) is an equilibrium, then (b^*, a^*) must be an equilibrium as well, because the labeling of one player as 1 and the other as 2 is arbitrary.

This fact can frequently be used to facilitate a game's analysis, especially if it applies to constant-sum games. Briefly,

> ***Constant-sum game:*** *A game is constant sum if the sum of the payoffs across players is a constant, regardless of the strategies chosen by the players. A game is* **zero sum** *if this constant is zero.*

If candidates maximize plurality, then the corresponding election game is zero sum since the sum of pluralities is necessarily zero. Alternatively, if candidates maximize their probabilities of winning, then the game is constant sum because probabilities must sum to a constant, 1. Of course, any constant-sum game is equivalent to a zero-sum game, because we can always subtract this constant from one player's utility function without affecting the information that this function conveys about a person's ordinal preferences or attitudes toward risk.

The reason we isolate zero- and constant-sum games from the rest is that their two-person counterparts (but not such games with more than two players) have special properties that make them especially interesting and amenable to analysis. Consider the game in Figure 3.19, in which only the payoff in the (a_1, b_1) cell is specified. Suppose we are also told that both (a_1, b_1) and (a_3, b_3) are equilibrium strategy pairs. What we want to show now is that (a_1, b_3) and (a_3, b_1) are equilibria as well and that neither player cares which equilibrium prevails. We begin by observing that since (a_1, b_1) is an equilibrium, it must be true that neither row or column chooser has a positive incentive to switch to a_3 and b_3 unilaterally, in which case

$$1 \geq y \text{ and } -1 \geq -x.$$

And since (a_3, b_3) is an equilibrium, we must have

$$z \geq x \text{ and } -z \geq -y.$$

Eliminating minus signs and putting all inequalities together yields

$$1 \geq y \geq z \geq x \geq 1,$$

which is satisfied only if strict equality holds here and, thus, if the entries in all four corner cells are $(1, -1)$. This fact, in turn, implies that all four corner cells are equilibria. For example, since (a_1, b_1) is an equilibrium, player 1 (row chooser) has no positive incentive to switch unilaterally to a_2 from a_1 if 2 chooses b_1. But then it must be the case that 1 would not switch from a_3 to a_2 either. Repetition of this argument proves our assertion.

	b_1	b_2	b_3
a_1	1, −1		$x, -x$
a_2			
a_3	$y, -y$		$z, -z$

Figure 3.19 Interchangeability and equivalence in two-person zero-sum games

That (a_1, b_1) and (a_3, b_3) being equilibria implies that (a_1, b_3) and (a_3, b_1) are equilibria as well is a property we call *interchangeability*, and if people are indifferent as to which equilibrium ultimately prevails, we say that the equilibria are *equivalent*. That all equilibria for two-person constant-sum games satisfy these two properties tells us that *coordination problems cannot arise in two-person constant-sum games. If each player chooses a strategy that is involved in some equilibrium, then the players necessarily arrive at an equilibrium. Moreover, they have no preference as to which equilibrium ultimately prevails.*

If we combine this fact with what we already know about symmetric, two-person games, we can conclude that for any two-person, constant-sum game, *if (a^*, b^*) is an equilibrium, then symmetry implies that (b^*, a^*) is an equilibrium and, invoking interchangeability, that (a^*, a^*) and (b^*, b^*) are equilibria as well. Symmetry, however, also requires that both persons receive the same payoff at (a^*, a^*), so if the game is zero sum, then each must receive an expected payoff of zero in equilibrium. Hence, equivalence implies that both persons in a* *symmetric two-person zero-sum game must receive an expected payoff of zero from any Nash equilibrium.* Thus, if the candidates in a two-candidate election have wholly symmetric opportunities, then regardless of what we assume about the electorate or procedures, if Nash equilibria exist, there is one such equilibrium in which the two candidates adopt identical strategies. Moreover, regardless of which equilibrium prevails, the expected outcome is a tie. This fact may appear obvious, and it may hardly qualify as one that would be debated extensively by those who are concerned primarily with the substantive domain of politics. Nevertheless, it is comforting to know we can set this fact aside as an established rule about an ideal form of two-candidate, democratic elections.

We offer this discussion of symmetric, two-person, zero-sum games not merely because they model an interesting class of situations (those in which two persons are locked in a wholly conflictual situation), but also to show the kinds of abstract conclusions our approach sometimes allows. Our conclusion about equilibria in two-candidate elections employs a laundry list of assumptions that are almost certainly not satisfied in reality (notably, the assumption that citizens do not begin the campaign with inherent biases toward one candidate or the other, as well as the assumption that the candidates are afforded equal access to all strategies, which almost certainly is violated if one candidate is an incumbent). Nevertheless, once we accept those assumptions as a starting point for modeling elections, we reach a conclusion that allows considerable freedom in accommodating other aspects of an election. This is an enterprise we address in the next chapter.

3.7 Manipulation and Incentive Compatibility

It is tempting to suppose, owing to the use of mathematical notation and the presentation of theoretically derived results, that game theory and its applications are something new to the social sciences. It may be true that a mathematical formalization of ideas is new (at least if we label as "new" ideas that are 50 years old or less). It's far less clear, though, that explicit game theoretic reasoning is a 20th-century invention. Recall that our brief discussion of agendas (which we will elaborate on in subsequent chapters) shows that revealed preferences do not always correspond to sincere preferences—to the preferences people actually hold. Of course, the possibility that one might have an incentive to misrepresent one's true preferences hardly comes as a great surprise. Nevertheless, such examples reveal an important avenue of research—studying how political institutions affect people's incentives to reveal one type of preference as against another. Assessing the incentives for the strategic misrepresentation of preferences as revealed by the choices people make is in fact a part of a more general enterprise, that of *institutional design*, wherein we attempt to channel the revelation of preferences in ways that yield desirable social outcomes. And to underscore the fact that at least the informal components of game theoretic reasoning are not something new under the sun, we can say that the Framers of the U.S. Constitution understood the nature of the problem we are about to address in immediate practical terms. For example, in his notes on the Constitutional Convention, James Madison offers this summary of Benjamin Franklin's comments during the debate on how to best select federal judges (E. H. Scott, ed., *Journal of the Federal Convention*, Chicago: Scott, Foresman and Co., 1989: 108–9):

> *Doctor Franklin observed, that the two modes of choosing the Judges had been mentioned, to wit, by the Legislature, and by the Executive. He wished such other modes to be suggested as might occur to other gentlemen; it being a point of great moment. He would mention one which he had understood was practised in Scotland. He then, in a brief and entertaining manner, related a Scotch mode, in which the nomination proceeded from the lawyers, who always selected the ablest of the profession, in order to get rid of him, and share his practice among themselves. It was here, he said, the interest of the electors to make the best choice, which should always be made the case if possible.*

At first glance, it might seem that Franklin was merely inserting a bit of levity into a weighty debate, perhaps to take the edge off a number of sharp personal conflicts. But Franklin was no stand up comedian, and he said little that was without pointed purpose. In this instance, he was reminding the delegates of the theoretical underpinnings of their enterprise. Though directed at the specific issue of finding a way to select able judges, the general task of the delegates was the construction of an institution that was *incentive compatible* in the sense that equilibrium behavior—behavior directed by narrow, myopic and even supremely selfish motives—would nevertheless yield the desired social outcome as a (perhaps unintended) consequence.

In more contemporary terms, the problem of institutional design and incentive compatibility confronts legislators when empowering an executive agency to act. Will the interests of decision makers within the agency lead them to act so as to achieve the outcomes the legislature desires, and how can the legislature design an agency and mechanisms for monitoring agency performance so as to shape those interests in a particular way? To render such questions a tad more manageable, suppose our interest is merely finding a way to get people to reveal their preferences truthfully. Suppose, for instance, we want to tax property at its "true" value. The standard method, of course, is to develop a cadre of professional assessors who, on the basis of a property's general characteristics (i.e., location, size, etc.) form an estimate by comparing it to other recently sold properties. But assessors can make mistakes and even be corrupted, so suppose we want the owners of the property themselves to tell us what they think it's worth. The problem here is that since owners know the property will be taxed on the basis of what they report, it's in their interest to report as low a valuation as possible. But here, the Chinese leader Sun Yet-Sen suggested an interesting twist on self-reporting; namely, the owners must also be willing to sell the property at the price they reveal (see Niou and G. Tan's 1994 article, "An Analysis of Dr. Sun Yet-Sen's Self-Assessment Scheme for Land Policy," *Public Choice* 78,1, for a description of this scheme). Clearly this is much more likely to engender an evaluation closer to what owners truly believe their property is worth. Alternatively, consider a country club with a golf course where its members must choose a fee to charge themselves for the course's use. The club could take all suggestions and average them to set a fee, but then those who have golf as a passion have an interest in suggesting a lower fee whereas those who infrequently or even never golf would suggest higher numbers. So suppose instead that the club takes the median proposal. While perhaps not a perfect mechanism, this method might be more likely to induce the truthful revelation of what its members think is a fair fee.

To see how the requisite analysis might proceed, let us turn to a simple yet illuminating example of the opportunities for invention in this area. Suppose a social planner is interested in determining whether some costly public works project (such as increased expenditures on education or highways) is in "the public interest." Realizing the difficulty in defining this interest, suppose the planner settles on a simple scheme: People will be asked how much they value the expenditure in monetary terms, and if the total reported value exceeds the project's cost, then it will be deemed worthwhile and people will be taxed accordingly. The difficulty here is that those who anticipate paying little in taxes or who highly value the project will have an incentive to overstate their evaluations whereas this value will be understated by those who sincerely believe that the project's cost to them exceeds its benefits. This problem is commonly manifested, for example, in the public utterances regarding the benefits to be associated with local public works projects, such as the construction of highways, the development of sports complexes, and the building of mass transit systems. Contractors who might profit from being involved in the design and construction of highways would, quite naturally, want to overstate value, sports enthusiasts would love to have others pay for the facilities only they might enjoy, and local businesses who might benefit from any increased economic activity resulting from mass transit systems would most likely ignore the downside of proposals to invest public monies in such things. And since large-scale public works projects are typically funded by bond issues whereby costs are passed on to future generations, nearly everyone involved with making an assessment of things here has an interest to downplay those costs.

The specification of a system that elicits such preferences is, then, an example of the more general problem of institutional design, and to illustrate some possibilities, consider the following scenario.

Example: Suppose five people must choose between two projects, A and B, and suppose the respective benefits and costs of these projects are as shown in Table 3.1. With respect to these preferences, notice that in a majority vote, project A defeats B four to one whereas project B produces the greater net benefit. The greater efficiency of B (as measured by summed valuations), then, might lead these voters to hope that their elected representatives and government planners will ignore the majority preference and choose B. But suppose that the information in Table 3.1 is not common knowledge. Thus, if asked to state their valuations, persons 1, 2, 3, and 5 have an incentive to overstate the value of A to them whereas 4 has an incentive to overstate the value of B. Rather than abide by the results of a simple poll of voters in this example, suppose as an alternative we implement the following taxation scheme: (1) Each person reports a net valuation for each project; (2) the reported net valuations are summed, and the project with the highest sum is chosen; (3) each voter's evaluations are deleted and replaced, one at a time, to ascertain if that person's report materially affects the group's decision; and (4) if a different project would be chosen had that person been absent from the poll, then that person pays an incremental tax equal to the difference in valuations between the two projects without that person's preferences being taken into account. With respect to our example, if everyone tells the truth, the numbers in Table 3.2 summarize the total valuations for the two projects after we exclude one voter's report. Notice that only voter 4 is decisive for project B, so in the event of sincere revelation of preferences, only voter 4 pays an incremental tax, which in this case equals \$45 (i.e., (90 – 35) – (100 – 90) = 45).

Table 3.1 Evaluations for Two Projects

Voter	*Project A*			*Project B*		
	benefit	*tax cost*	*net value*	*benefit*	*tax cost*	*net value*
1	60	40	20	65	50	15
2	40	40	0	30	50	−20
3	25	40	−15	25	50	−25
4	75	40	35	140	50	90
5	90	40	50	90	50	40
sum	290	200	90	350	250	100

Table 3.2 Computation of Incremental Tax

Excluded voter	*Summed net valuations*		*Incremental tax*
	A	*B*	
1	70	85	0
2	90	120	0
3	105	125	0
4	55	10	45
5	40	60	0

The remarkable feature of the tax scheme just outlined is that telling the truth is not only an equilibrium strategy, it is dominant as well. To see that telling the truth is dominant in general under the proposed tax scheme, suppose that, excluding the voter in question, the summed valuations for A and B are u_A and u_B (which may or may not be based on sincere preferences); suppose that $u_A > u_B$ (so A is chosen if the voter in question is ignored); and suppose that the voter's actual net utilities from A and B are u^*_A and u^*_B. What we want to establish is that telling the truth is dominant regardless of the valuations that others report. An appeal to symmetry then establishes that truth-telling corresponds to the unique Nash equilibrium. To see this, notice that we have two possible contingencies: Telling the truth renders the person in question decisive; and telling the truth does not render that person decisive. With respect to the first possibility, if telling the truth changes the outcome, the truth costs our voter $u_A - u_B$ in additional taxes. However, since the truth is decisive, it must be the case that when we add this voter's valuations, B is chosen. Hence, $u_A + u^*_A < u_B + u^*_B$. But this inequality implies that $u_A - u_B < u^*_B - u^*_A$, which is to say that the tax, $u_A - u_B$, is less than the difference in value between the projects, so our voter prefers to pay the tax rather than have the less preferred project chosen. Now suppose that telling the truth does not render the voter decisive, in which case $u_A + u^*_A > u_B + u^*_B$, and telling the truth entails no additional tax. But if our voter misleadingly inflates the reported value of B so as to reverse this inequality in order to secure the more preferred project, then the voter must pay a tax of $u_A - u_B$, which exceeds $u^*_B - u^*_A$. Hence, the incremental tax exceeds whatever benefit the voter associates with project B as against A.

We can imagine applying such a scheme to a variety of decisions in addition to those pertaining to public policy and public works projects, such as when university faculty must choose between candidates for some vacancy. Although debating the issue may make people aware of the advantages and disadvantages of alternative candidates, it has been our experience that colleagues typically overstate the qualifications and research potential of those they sincerely wish to hire (typically those who fit their private research agendas) while understating the character of competitors. So rather than choose whom to hire with a majority vote, faculty could be asked how much they are willing to pay for each candidate as a way to reveal more honest valuations and judgments.

There are, nevertheless, some practical difficulties with this scheme, regardless of the context to which it is applied. First, although we have shown that telling the truth is a dominant strategy, we have not shown what might happen if people collude to form coalitions. Second, there is the issue of what to do with the incremental taxes that are collected. Unfortunately, we cannot redistribute the money among the participants without, in theory at least, destroying the incentives for truth-telling. In particular, if people know that part of their incremental tax will be returned to them, they may have incentives to "shade" the truth. To maintain incentives the money must be destroyed. Destroying money, though, is not Pareto efficient and we might hope that the amount burned is less than the gains associated with choosing the correct project or candidate. Alternatively, we might try to design a mechanism with different properties. Indeed, several tax schemes have been devised that eliminate this type of inefficiency. Unfortunately, they require abandoning the possibility that truth-telling is dominant and settle instead on establishing that truth-telling is merely a

Nash equilibrium. The disadvantage here, of course, is that in practice we may require several iterations before responses converge to an equilibrium. Finally, there is the possibility of "income effects" wherein some people are unable to pay an incremental tax and are thereby forced to understate their preferences or to not participate in the decision at all.

There is, in fact, one additional problem with this scheme that is more evident if we suppose that more than two projects are under consideration. Specifically, by representing the value of each project in terms of dollars, and by letting people be compensated for the selection of one project as against another by a monetary transfer, we have assumed that utility and money are equivalent and that a person's utility for money does not change as a function of the project selected. Indeed, this assumption permits us to circumvent the consequences of a profoundly important theorem about social decision processes. Known as *Gibbard and Satterthwaite's Manipulability Result*, the theorem states that

> *If social processes concern more than two alternatives and if they satisfy all of Arrow's axioms except the requirement of transitive social orders (including the axiom that all preferences over the alternatives are feasible, which is invalidated by the assumption that utility and money are equivalent—the assumption that we can write $u(x) = ax + b$ for any person's utility for the amount of money x, where a and b are constants), then we cannot preclude the possibility that, regardless of what others do, one or more persons will have a strategic incentive to disguise their true preferences.*

In particular, we cannot preclude the possibility that for any social decision process, preferences will be configured in such a way that "telling the truth" is not a Nash equilibrium—indeed, that there will be no equilibrium, truthful or otherwise.

We have, of course, already seen the incentives for strategic maneuver in agendas and thus we are able to identify those circumstances in which the votes of legislators on bills and amendments may not correspond to their sincere preferences. Gibbard and Satterthwaite's result tells us that for any agenda with more than two alternatives and two voters, there exists at least one set of preferences such that one or more voters will strategically cast a ballot that does not correspond to their sincere preference. We can also conclude that under plurality voting, if there are three or more candidates, then there are voter preferences such that some subset of the electorate will vote for their second choice.

To illustrate matters in a different context, suppose Congress is designing a bill and that we can represent the relevant alternative policies as positions on a line. Suppose the status quo is the point 0, with the president, who must sign the bill, preferring the position $P > 0$ while the majority of the legislature prefers policies greater than P. Assuming that everyone prefers policies as close as possible to their ideal, the question is whether members of the legislature will reveal their preferences sincerely. Suppose the legislature's median preference is $L_m > P$. If the median prevails under majority rule and if L_m is closer to P than is the status quo, 0, the president will, of course, sign the bill. In this event, the decisive median legislator has no incentive to report anything other than his sincere preference. On the other hand, suppose L_m is further from P than P is from 0, in which case, if L_m prevails in the legislature, the president will veto the measure in favor of the status quo since it is closer to the president's ideal than is L_m. In this event, then, at least a majority of the legislature (all those with preferences at and to the right of L_m) have an incentive to report a bill closer to P than is L_m in order to avoid the veto. This legislative outcome is little more than the implicit bargaining that frequently occurs between executive and legislative branches. Our example also hints at the incentives a president possesses for trying to convince the public and the legislature beforehand that he prefers policies closer to the status quo than is in fact the case. Hence, although our scenario is simple, it nevertheless reveals how strategic maneuver is an inherent part of politics.

This result is important in the study of politics, then, because it reveals that strategy is a pervasive feature of our subject and that we can never be certain (barring some strong assumptions about preference) that revealed preferences are sincere. Our only hope for understanding the relation between preferences and choice, then, is to model social processes carefully so as to learn the incentives for strategic maneuver and deception.

3.8 Key Ideas and Concepts

simultaneous choice
Nash equilibrium
unilateral moves
Median Voter Theorem
coordination
dominant choice
pure strategy
he-thinks regress
mixed strategy
Colonel Blotto game
fixed point
paradox of power
incentive compatibility
constant sum
zero sum
symmetric game
equivalent strategies
interchangeable strategies

Exercises for Chapter 3

1. Going into the home stretch of a research and development project, you are six months ahead of the competition. To bring the project to completion

requires finishing the development stage, and you have two strategies: Risky and Safe. Safe takes two years but is guaranteed to work. Risky takes only a year, but there is a 50% chance it will get you nowhere, in which case you will have to return to the safe strategy and take an additional two years. In six months, your competitor (who cannot observe your decision) will confront an identical choice between Safe and Risky strategies with the same properties as your alternatives, and they too must switch to Safe if their Risky choice fails. Only the first to complete is awarded the patent, and due to limited resources it is not possible to pursue both strategies simultaneously.

a. Which strategy should you pursue to maximize your chance of winning?
b. Do you want to keep your move hidden?

2. Draw the extensive form corresponding to the following description and solve for equilibrium strategies: "There are two dark boxes. Player 1 hides a pearl in one of them; then player 2, not knowing which box contains the pearl, peeks into one of them. If the pearl is in box 1 and she looks there, she sees it with a one-half probability. If it is in box 2 and she looks there, she sees it with a one-half probability. If she looks into the wrong box, she sees nothing (and is not even told that the box is empty)." The payoff is 5 to player 2 and –5 to player 1 if player 2 finds the pearl; otherwise, there is no payment.
3. Consider the following scenario: With one opportunity to bet remaining in the game show, player A has $7,200, player B has $5,000, and player C has $3,601. Assume that there is no benefit to being second versus third, and that the player with the most money wins that amount of money in cash. Each player must decide whether to bet *All* or *Nothing*. Prior to betting (which they must do simultaneously), nature tosses a fair coin to determine which "state of the world" will pertain: in State 1, players A and B win their bets, but C loses; in State 2, players A and B lose their bets, but C wins. Assume that if a player bets "all" and wins, his wealth is doubled. If he loses, his wealth is zero. If a player bets nothing, his wealth does not change. The player with the most money at the end of the game wins. Draw this situation's extensive form and show what each player does in equilibrium.
4. "Cat" and "mouse" each start at opposite corners of the simple maze shown below. It takes five seconds for both animals to traverse one segment, but the passages are sufficiently narrow that neither animal can turn about in the maze. If the cat eats the mouse, its payoff is +1 and the mouse's is −1; otherwise, these payoffs are reversed. After twenty seconds, the mouse, if available, will be rescued from the maze.

   a. Letting a strategy be a complete plan as to which way to turn (left, right, straight ahead) at each juncture in the maze, does this game possess an equilibrium in pure strategies, assuming that neither the cat nor the mouse can observe the other as it traverses the maze (until of course, it is "too late")?

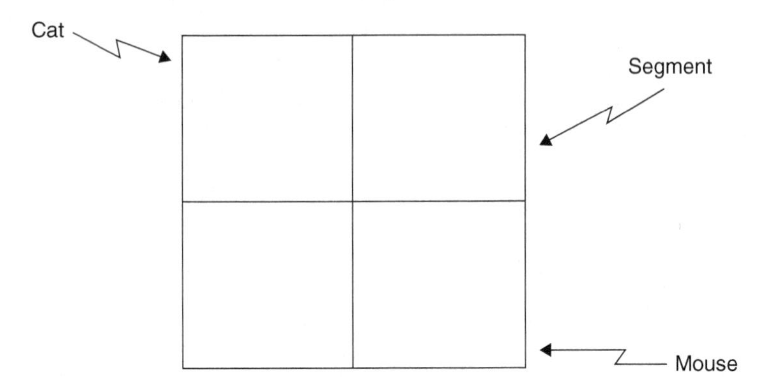

Figure 3.20 Cat and mouse maze

b. Does your answer change if the mouse is rescued only after thirty seconds?

5. Prior to playing a game, a fair coin is tossed to determine which of the following two 2 × 2 games will actually be played (the first number in each cell denotes your payoff—row chooser—whereas the second number denotes your opponent's):

	b_1	b_2
a_1	9, 5	1, −3
a_2	3, 7	4, 6

b_1	b_2
3, 0	9, −2
9, 9	3, 8

a. Assuming that neither person observes the outcome of the coin toss and that both of you must choose simultaneously, portray the situation's extensive form.

b. Portray the extensive form, assuming that you can secretly pay $3 to learn the outcome of the coin toss.

c. If neither player observes the outcome of the coin toss, how much would you pay to learn the outcome of the toss (assume that all payoffs are in terms of dollars and that utility and money are equivalent) and for what information cost is 2 indifferent between learning and not learning the outcome of the coin toss?

6. You are a member of a committee that must decide between A and B. As things stand now, B defeats A, but you hate B, so you've introduced alternative C, which will be put to a vote against B; then the winner will be pitted against A. You are hoping that C will beat B, but lose to A. The outcome, though, depends on your argument for C and the argument of your opponent. The social ordering will be determined by those arguments. Your preferences are A > C > B, and your opponent's preferences are B > C > A. You (row chooser) and your opponent (column chooser) each have two alternative arguments, and they yield the following social preference orders under majority rule:

	Argument 1	Argument 2
Argument 1	B > C > A (transitive)	C > A > B > C (a cycle)
Argument 2	C > B > A > C (a cycle)	C > B > A (transitive)

If you must each choose an argument without knowing your opponent's choice, and if everyone is sophisticated, what outcome prevails?

7. Let two political candidates each have two strategies, as indicated in the table below, and let the probability that candidate 1 receives specific pluralities from the four possible joint strategy choices also be as indicated. If the objective of both candidates is to maximize their probability of winning the election, what, if any, is the equilibrium to this game? How does this equilibrium change if both candidates seek to maximize their expected plurality? Note that $p(x)$ is the probability of winning a plurality of x votes.

	b1	*b2*
a1	$p(0) = 5/8$ $p(400) = 1/8$ $p(800) = 2/8$	$p(8) = 1/2$ $p(0) = 1/8$ $p(-80) = 3/8$
a2	$p(-400) = 1/2$ $p(400) = 1/2$	$p(0) = 3/4$ $p(-75) = 1/8$ $p(-125) = 1/8$

8. Players 1 and 2, acting simultaneously, must each first choose whether to play L or R. If they both play L, they then play game A below (all payoffs are in dollars); otherwise, the toss of a fair coin determines whether game B or C is played. (The outcome of the toss is revealed immediately.) Prior to their initial choices, though, player 1 (row chooser) can offer 2 any amount of money for the right to make 2's choice of R or L. Determine what amount, if any, player 1 ought to offer 2, as well as the value of the scenario to players 1 and 2.

A

	X	Y
X	0, 4	0, 2
Y	4, 2	1,1

B

	X	Y
X	1, 3	2, 1
Y	4, 3	3, 2

C

	X	Y
X	1, 4	6, 1
Y	5, 1	0, 4

9. Recall the game from elementary school called "one, two, three, shoot!" One of the players chooses "even," and the other player gets "odd." On the count of three, each of the two players simultaneously casts out either one or two fingers. If the total number of fingers is even, then the "even" player wins, while if the sum is odd, then the "odd" player wins. Suppose the payoff is 1 for the winner and –1 for the loser.

   a. Show the extensive form for this game.
   b. Show that this game has no pure strategy equilibria.

c. Prove that "acting randomly" is the only equilibrium.
d. Suppose the rules of the game are changed so that at the count of three a person can hesitate. If both hesitate, each receives a payoff of 2 whereas if only one hesitates, that player loses 1 and the opponent gets 0. What is the outcome now?

10. Consider the following two games:

Game 1

		Player 2: A	Player 2: B
Player 1	A	10, 4	4, 0
	B	8, 2	8, 6

Game 2

		Player 2: A	Player 2: B
Player 1	A	14, 14	4, 16
	B	10, 2	0, 20

a. If a coin is flipped to determine which game will be played, what is the Nash equilibrium of the situation in which neither player knows the outcome of the coin flip and the players choose their strategies simultaneously?
b. Model part (a) in extensive form.
c. If, in part (a), one of the players can pay the coin flipper, who is not one of the players, to reveal the outcome of the coin toss to both players before they choose their strategies, which player (row or column) would be willing to pay more for this information?

11. In a two-player game, the following pairs of actions lead to the following outcomes:

(Player 1's Action, Player 2's Action) →	*(1's Utility, 2's Utility)*
(L, L)	*(3, 6)*
(R, R)	*(1, 5)*
(C, C)	*(6, 8)*
(L, R)	*(9, 6)*
(R, L)	*(9, 9)*
(L, C)	*(6, 1)*
(C, L)	*(2, 2)*
(R, C)	*(4, 4)*
(C, R)	*(4, 3)*

a. What is the outcome of this game if 1 goes first and 2 observes 1's action and then chooses his own action?
b. What is the outcome if 2 goes first and 1 observes 2's action and then chooses her own action?
c. What is the outcome if each player has to choose an action before he or she knows the other player's action?
d. A third player enters the game. If the third player chooses L, he gets the average of the payoff to players 1 and 2 for any given outcome. If the third player chooses R, he gets a payoff equal to whichever of player

1's payoff or player 2's payoff is higher. What is the outcome of the game where 1 goes first, 2 observes 1's action and chooses, and then 3 observes 1 and 2's actions and chooses?

e. What is the outcome of the three-player game if all players must choose their actions at the same time?

		Player 3 chooses L			Player 3 chooses R		
		Player 2			Player 2		
		L	C	R	L	C	R
Player 1	L	(3, 6, 4.5)	(6, 1, 3.5)	(9, 6, 7.5)	(3, 6, 6)	(6, 1, 6)	(9, 6, 9)
	C	(2, 2, 2)	(6, 8, 7)	(4, 3, 3.5)	(2, 2, 2)	(6, 8, 8)	(4, 3, 4)
	R	(9, 9, 9)	(4, 4, 4)	(1, 5, 3)	(9, 9, 9)	(4, 4, 4)	(1, 5, 5)

12. Persons 1, 2, and 3 each have a budget of B dollars, which must be allocated between goods x and y at unit prices (i.e., $p_x, p_y = 1$). Person i's utility is $u_i = x^{a_i}y$, where $a_1 > a_2 > a_3$. There are the following ways of making final decisions about how much of x and y each person will consume:

    a. People independently make decisions in the open market.
    b. With the government expropriating all wealth, two candidates for "dictator" each announce an election platform that corresponds to an identical (and feasible) commodity bundle for each person; the three persons then vote for one candidate or the other, and the victorious candidate implements his or her platform.
    c. A "dictator" is selected thus: Person 1 vetoes one person from this set, and person 2 vetoes one of the remaining two persons. The survivor, using the government's budget, then selects a representative feasible commodity bundle that each person must consume.

    i. Describe each person's preferences over his or her budget constraint.
    ii. Describe the outcomes achieved under each method.

13. Let four individuals have the following net valuations over the three alternatives, A, B, and C:

Person	A	B	C
1	30	0	50
2	45	65	0
3	10	20	45
4	50	35	0

Assume that each person must report a valuation for each alternative and that the alternative chosen is the one with highest summed valuation. Assume also that taxes are collected as described in Section 3.7.

a. How much incremental tax will be paid by each person?

person	A	B	C
1	30	0	50
2	45	65	0
3	10	20	45
4	50	35	0
Total	135	120	95

	Summed net valuations			
Excluded voter	A	B	C	**Incremental tax**
1	105	120	45	15
2	90	55	95	5
3	125	100	50	0
4	85	85	95	10

b. Suppose persons 2 and 3 can hire an agent who will coordinate their responses (reported evaluations), including lies. Should they hire such a person if the fee is not too great?

14. Congressman Pork represents a district with a significant electronics industry that is in an economic slump because of foreign competition. Accordingly, he introduces a bill for a $100 billion federal subsidy for consumer electronics, but he knows that several committee members led by Congresswoman Pam Sonic are opposed to his bill. The bill will be debated next week and Pork can offer one of two arguments on behalf of it: (1) The United States needs the technology of consumer electronics for possible conversion to military use, in which case $50 billion would be appropriated. However, if Sonic anticipates this argument she can seek out expert witnesses to testify that such technology is irrelevant to defense needs, destroying Pork's argument and defeating the bill. (2) Pork could emphasize the jobs that the federal subsidy would provide for constituents of the majority of the committee members. In this case, he could get $40 billion. Sonic's counterargument mentioned above would have no effect on this approach, but Sonic could organize a letter-writing campaign of consumers against the bill. In this case, the committee would compromise at $25 billion. However, the letter-writing approach would have no effect if Pork's military security argument goes uncontested, since the committee feels that money is no object when it comes to national security. Portray this situation in strategic form by using a reasonable attribution of utility to the outcomes for both Pork and Sonic.

4 Zero-Sum Games with Spatial Preferences

4.1 Introduction

The quest for the American Presidency within the Republican party in 2012 began with Mitt Romney as only one of a multitude of suitors. Romney's candidacy, moreover, was met with anything but enthusiasm among Republican activists. Eschewing much of the far right rhetoric that appealed to the more conservative wing of the party, he simply seemed "not conservative enough" and, as a consequence, his candidacy, rather than being embraced, was met by a long series of fiery challengers that yielded names like Santorum, Gingrich, Paul, Perry, Bachman and Cain. Yet each in turn, while winning a statewide primary here or there, fell by the wayside until Romney clinched the nomination. This, of course, is not how it has to be. In 1964 the Republicans nominated the conservative Barry Goldwater to run against Lyndon Johnson and in 1972 the Democrats swung to the far left with the nomination of George McGovern as its Presidential candidate against Richard Nixon. The fact that both Goldwater and McGovern suffered, by American standards, defeats of historic proportions (61% of the vote vs. 38.5% and 60.7% vs. 37.5%, respectively) seemed not to bother the supporters of Romney's intra-party challengers. It was almost as if the Republican party, against its wishes and after having tried all the alternatives, was being dragged, kicking and screaming, in 2012 to the ideological center of the American electorate. And while Romney lost the election, he did so by a far smaller margin than did either Goldwater or McGovern (51.1% vs. 47.2% so that while Johnson won 486 electoral votes and Nixon won 520, Obama won but 332).

Those familiar with two-person, zero-sum games, their application to the study of elections and the Median Voter Theorem are unsurprised at this result. It is, in fact, the elections of 1964 and 1972 that are deemed anomalous. Anomalies, however, require explanation for it is there that we learn where our theories and models require elaboration and further development. Before we do so, however, we should first look more closely at the class of games with which we ended the previous chapter—zero- or constant-sum games, especially their two-person variant—and which, in fact, form the basis for much of the subsequent theorizing about elections.

Briefly, we already know that in the two-player case, if the game is zero or constant sum, there is no need for coordination in order to assure that an equilibrium is achieved. Such games, moreover, share a property that, although it does not necessarily rationalize mixed strategy equilibria, gives us some confidence that a pure strategy equilibrium prevails as the final outcome whenever such a thing exists and also gives us an easy way to identify games of that type that do not have such equilibria. Keep in mind that two-person, zero-sum games are games of pure conflict—whatever one player gains, the other loses. Thus, it is not unreasonable to suppose that people engage in such games with a degree of pessimism—that their opponent can anticipate whatever strategic ploy they might attempt. So suppose each person evaluates each of his or her strategies on the basis of the worst outcome that can prevail if it is chosen. That is, suppose each person reasons as follows: "Regardless of what I decide to do, my opponent can anticipate my thoughts and select a best response, which, owing to the zero-sum character of the situation, does me the greatest harm. Thus, I should choose a strategy that maximizes my minimum payoff (or equivalently, minimizes my maximum loss)." We would not want to extend this reasoning to other types of games, especially if there are mutual gains to be realized from coordination, but it is useful to know that

> *If a two-person, zero-sum game has a pure strategy Nash equilibrium, and if both persons choose strategies that maximize their minimum payoff (a* ***maxmin strategy****), a Nash equilibrium is realized.*

Because the proof of this proposition is simple and illustrates one property of equilibria for this special class of games, we review it here. Briefly, if (a^*, b^*) is an equilibrium that yields the utility outcome $(c, -c)$, then c is the minimum payoff that player 1 should associate with a^*—if there was some payoff lower than c associated with a^*, the corresponding outcome must give 2 more than $-c$, in which case player 2 would unilaterally switch from the presumed equilibrium strategy of b^* in order to receive that payoff. Similarly, since a^* is an equilibrium strategy for 1, it must be that a unilateral move cannot raise 1's payoff and may even decrease it. Hence, the minimum payoff 1 associates with any other strategy cannot exceed c. So a^* is player 1's maxmin strategy, and a parallel argument establishes the same thing for player 2. And although this argument is made with reference only to pure strategies, we can readily give a^* and b^* a broader interpretation to see that it applies to mixed strategies as well.

We might question whether such pessimism is warranted in all circumstances—even zero-sum ones. But if we believe that a zero-sum game models a situation with "sufficient" accuracy, choosing maxmin strategies seems only reasonable. Moreover, the notion of maxmin and the preceding analysis provides a convenient way for determining whether a two-person, zero-sum game has an equilibrium in pure strategies or whether we must resort to mixed strategies in order to ascertain a game's equilibria. Specifically, if a game is zero sum, it must be the case in any equilibrium that the payoff to one person is the negative of the payoff to the other. If, then, c is the maxmin value to player 1, then $-c$ must be the maxmin value to player 2. It follows from this fact that:

> *If player 1's maxmin value in a two-person game, computed on the basis of looking only at pure strategies, is not the negative of player 2's maxmin value, then the game has a solution only in mixed strategies.*

Of course, such properties are hardly a compelling theoretical reason for giving the quite special case of zero sum, especially those limited to two players, much attention or prominence over other classes of games. There is, though, a good substantive reason for focusing on this limited class. Specifically, they model a great many political situations, where the most evident is a two-candidate competitive election wherein the Median Voter Theorem establishes the ideal preference of the median voter in a one-issue election as the Nash equilibrium to the game "played" by the two competing candidates. As a model of actual political processes, however, that theorem leaves a lot to be desired, since the assumptions employed by it are indeed heroic. For example:

> **Election Rules:** In the 2000 U.S. Presidential contest, George Bush defeated Al Gore despite the fact that nationally, Gore won a positive plurality of the popular vote. Intervening between that vote and the determination of a winner is a peculiar American institution, the Electoral College, wherein the winner is determined only indirectly by the popular vote. A reversal of this type, however, was not unique to the 2000 contest—reversals also occurred in 1960, 1888, 1876 and, arguably, 1824 as well. Thus, the Median Voter Theorem seems irrelevant to one of the most important democratically filled posts in the world. The Electoral College, however, is only one of a wide range of institutional devices employed to muddle the relationship between votes cast and the determination of winners. What, for example, of such things as proportional representation, elections with runoff provisions, elections that are deemed valid only if a certain level of turnout prevails, and so on? Indeed, it seems at times as if the imagination and inventiveness of the human species is never so clearly displayed as when an electoral system is being designed.
>
> **Third Parties:** The 2000 U.S. Presidential contest was also characterized by a considerable dispute over the counting of votes in the state of Florida—a dispute that concerned a margin of victory for Bush there of under 1000 votes out of over six million cast. Florida's vote was pivotal, and if the state had been declared for Gore, Gore would have won the presidency. However, there would arguably have been no dispute at all had a third candidate, Ralph Nader, not been on the ballot. Most pundits agree that the majority of Nader's 97,000 + votes would have gone to Gore had Nader not been on the ballot, thereby giving Gore an indisputable plurality in the state. Indeed, the Gore campaign knew beforehand the problems a Nader candidacy would cause them, and they surely encouraged Nader

to forgo the ego trip of running for the presidency. Once again, then, the Median Voter Theorem, which is limited to two-candidate contests, seems irrelevant.

Voter Ignorance: The Median Voter Theorem also makes heroic assumptions about the information of both voters and the candidates, along with the nature of the candidate's strategies. First, those strategies are presumed to be precisely defined points within the issue space, with no room for ambiguity or uncertainty. It appears, therefore, difficult to apply the theorem to, say, the 1968 U.S. Presidential contest whereby Nixon's asserted policy with respect to American's involvement in Viet Nam was little more than "I have a plan" or Eisenhower's campaign promise in 1952 in the context of the Korean War that "I'll go to Korea." No more precise was Obama's 2008 campaign slogan "Hope and Change" or his campaign theme in 2012 of "Forward" wherein Obama gave little indication of his plans for a second term and instead directed his campaign principally at simply denigrating his opponent. Indeed, with the microphone accidentally on, his whispered comment to the then-President of Russia, Dimitry Medvedev, that he'd have more flexibility after the election surely suggested that he was being less than forthright with the American electorate, at least in the realm of foreign policy. Nixon, Eisenhower and Obama, however, were hardly the first presidential candidates to obscure positions on salient campaign issues—witness Lincoln's somewhat vague stand on slavery in the 1860 contest. But the Median Voter Theorem's informational assumptions go far beyond any assumption about the nature of strategies. Those assumptions also require that the candidates know with certainty the distribution of the electorate's preferences on the issue in question, that voters know their own preferences on the issue, and, perhaps most daunting of all, that each and every voter knows precisely what the policy positions of the two candidates are. It is this last assumption, of course, that defies the findings of essentially every public opinion poll in existence that attempts to tap an electorate's knowledge of a campaign.

Turnout: One of the cardinal rules of campaigning for virtually every office filled by direct election anywhere, is that candidates or parties must make every effort at getting out the vote of their base supporters. The presumption here is that a candidate's ideological predisposition, sex, ethnicity, tribe, religion, residence and so on prejudices members of the electorate so that for at least a subset of the eligible electorate, specific issue positions on the salient policy issues of the day are either unimportant or have, as best, a marginal impact. This rule of thumb, though, merely corresponds to the obvious fact that only in the corrupt "elections" of the former Soviet Union or in today's Russia does turnout even approach 100% and that unless one maximizes the turnout of one's "natural" vote, it generally matters little what influence a campaign might have on those voters who are attuned solely to salient policy issues. Nevertheless, the Median Voter Theorem assumes that all voters vote.

Candidate Objectives: The assumption of 100% turnout in combination with the absence of any accommodation of uncertainty and flawed or imperfect information closes the door on another possibility ignored by the Median Voter Theorem: Alternative formulations of each candidate's objective function. In the 1964 U.S. Presidential contest, for example, it seems that Lyndon Johnson's objective was not simply to win, but to do so with the largest plurality possible. Appreciating the fact that he did not have his predecessor's media-cultivated charisma, Johnson believed that to be effective and to keep Kennedy's retinue from abandoning or attempting to undermine his administration, he needed an overwhelming electoral mandate. In 2000, in contrast, it is almost certainly the case that, knowing the contest would be close, both Bush and Gore were satisfied with adopting strategies that gave them the greatest likelihood of a positive plurality (or plurality in the Electoral College). For the imposed assumptions of the Median Voter Theorem, however, all objectives such as maximizing one's vote, one's share of the vote, and one's probability of winning are wholly equivalent. But if there is some uncertainty about things, we must then ask whether it is better to assume that candidates maximize, say, their expected plurality or their probability of winning—or, to state things differently, does the nature of equilibria depend on what assumption we impose about objectives?

Multiple Issues: Chapter 3 also underscored another critical assumption of the Median Voter Theorem; namely, that the election concerns but a single issue. It may be that a considerable share of electoral contests around the world are disputes over ideology, which can be represented by a single dimension from Left to Right, from Liberal to Conservative, from pro-Business to pro-Labor, and so on. But a more realistic and general analysis allows for subsets of the electorate to care about different things—for farmers to care about farm subsidies, the elderly to care about retirement programs, women to care about policies on abortion, environmentalists to care about government regulation of various industries, students and parents to care about government funding of higher education, motorists to care about gasoline prices, newlyweds to care about housing prices, racial or ethnic minorities to care about civil rights, and so on and so forth. But now notice that people can be members of several such categories simultaneously—one can be a woman, a newlywed and a motorist just as one can also be a farmer, an environmentalist and an ethnic minority simultaneously. It may be, of course, that preferences on the various things that concern people correlate and thereby yield what we take to be ideology, but there is no reason to suppose that this is necessarily the case. To do so, moreover, precludes consideration of an important strategy candidates often employ in a campaign, challengers to an incumbent especially. Specifically, suppose an electorate is initially concerned with a single issue and that the incumbent, having read Chapter 3 of this book, has skillfully used his or her term of office to position themselves at the electorate's median preference on that issue. The challenger, then, is confronted with a dilemma: If he adopts a different position, he loses; but if he matches the incumbent's position, the electorate can simply reason, "Why change and take a chance that the challenger is being untruthful in what he or she is saying?"

This, of course, is the problem that confronts any new political party attempting to form and enter the electoral game, and here we have in mind America's Republican party of the late 1850s that was soon to nominate Abraham Lincoln as its standard bearer. The potential solution to this problem is suggested by our analysis of spatial games in Chapter 3—make the election multi-dimensional. In the case of Republicans in the 1850s, the strategy was to put emphasis on an issue that the two established parties, Whigs and Democrats, had thus far attempted to finesse via a series of compromises: slavery. Despite the fact that those favoring or arguing for the universal abolition of slavery were a decided minority both in the North and South, the two established parties knew this issue would splinter each if its salience were to increase, with Democrats in particular split between Northern and Southern wings of the party. Thus, both parties attempted every compromise to keep the issue from bubbling up to disrupt both the country and their political survival. Lincoln's Republicans, however, had little incentive to keep either established party from splintering and, unable to secure any advantage on the traditional economic issues that divided Whigs and Democrats, every incentive to add this new dimension as a criterion of choice for voters in the election of 1860.

Redistributive Politics: In addition to all of the other limiting assumptions made in establishing the Median Voter Theorem, there is one more—the presumption of spatial preferences themselves. There is, as a consequence, one class of election games not covered—or rather, covered to little effect—by a spatial representation: Redistributive elections wherein the election concerns only the redistribution of wealth and candidate strategies are proposals to distribute some fixed amount, W, of wealth among the electorate. This is not to say that we cannot represent preferences and the policy space spatially. For instance, if we assume again that there are three voters, we can represent the issue space—all possible three-way divisions of W—as a triangle where vertex i represents the outcome "voter i gets W and the remaining two voters get nothing," where the midpoint of the opposite side of the triangle represents "voter i gets nothing and the remaining two voters evenly divide W," and where i's indifference contours are straight lines parallel to the side of the triangle's opposite vertex i such that i's utility declines as we move away from this vertex. The reader, in fact, should draw this figure to convince themselves of two things about redistributive politics: First, regardless of the point we choose, the sum of the payoffs to all three voters is *W*, and any point in or on the triangle can be defeated in a majority vote by any number of other points. In this case, however, a spatial representation gains us little over simple intuition or common sense. For example, suppose the candidates must propose alternative divisions of $90 among our three voters. If one candidate proposes ($45, $45, $0), the second can win with ($0, $50, $40), which is in turn defeated by ($45, $0, $45), which is defeated by ($60, $30, $0), and so on. Thus, even without the spatial representation of preferences required by a redistributive electoral conflict, we see quite directly that the candidates can cycle endlessly through alternative distributions. Or, to state matters differently, if an incumbent must choose a specific redistribution during his or her tenure in office, a challenger in the next election should be able to propose an alternative that defeats the incumbent. Thus, the second thing we learn here about redistributive politics is that candidates, and incumbents especially, will prefer to speak in generalities when addressing policies that specifically refer to the redistribution of wealth. Thus, in American politics at least, instead of explicitly identifying who will and who will not benefit from one's proposed policies, politicians instead make reference instead to such vague promises as "making sure everyone pays their fair share" or labeling their policies as beneficial to "the middle class," whomever that might include.

4.2 Plott, McKelvey and the Core Results of Spatial Theory

At best, then, the Median Voter Theorem is but a first step in the application of zero-sum games to the study of elections. But before we see what next steps might be taken, we need to consider several more fundamental theoretical results. The first, which we can attribute to Charles Plott (see the 1967 article, "A Notion of Equilibrium and its Possibility Under Majority Rule," *American Economic Review* 57), directly confronts the issue of the existence of equilibria in multidimensional choice spaces. To introduce the results of research here, let us look back to the example in Chapter 1 of the 3-voter spatial scenario portrayed in Figure 1.6 with two dimensions or issues. There is little need to reproduce that figure, which we introduced earlier to illustrate a situation in which there is no Condorcet winner. But what needs to be noted is that the lines connecting pairs of voters have a specific property; namely, because indifference contours are represented as concentric circles about each voter's ideal, the line between any pair of ideals describes the outcomes that are *Pareto Optimal* for the two voters in question. Specifically, when at a policy on such a line—which is called a *contract curve* in economics and corresponds to the tangencies of both voters' indifference curves—we cannot move along or off it without moving to outcomes that are less preferred by one or both persons. And from any point off that line, we can drop a perpendicular to the line (or its extension past either ideal point) and move along the perpendicular to make both persons better off. Thus, if an outcome is a Condorcet winner in this three voter game and, thus, a Nash equilibrium to the corresponding two-person, majority rule election game, it must lie simultaneously on the lines connecting each pair of voters. Stated differently, if a candidate's campaign policy (position) is not Pareto optimal for each pair of voters (for each majority coalition), the opponent can defeat the candidate in a majority vote with a different policy that will be preferred by two voters. But in our example, there is no simultaneous intersection of all three lines connecting ideal points. Thus, regardless of the position a candidate adopts, the challenger can drop a perpendicular to the line connecting some pair of voters and find a position that is preferred by a majority. It follows that, as we already know, there cannot be a pure strategy Nash equilibrium to the two-candidate election game of our 3-voter example.

There is, though, nothing special about our example aside from the fact that a majority consists of but two voters and that the Pareto optimal policies for a winning coalition consist simply of a straight line. If, instead, we expand the electorate to consist of n (odd) voters, all with circular indifference curves, the reader should be able to verify that the set of Pareto optimal policies for any majority of $(n + 1)/2$ or more voters is described by the convex hull that encompasses or is bounded by the ideals of that majority. That is, if we connect the ideal points of the members of that majority much like a child might draw a "connect the dots" picture with straight lines in such a way that all ideal points lie on or inside the figure formed by those lines, the policies on or inside that figure are Pareto optimal for that set of voters taken as a whole. A Condorcet winner and thus a pure strategy Nash equilibrium to the corresponding two-candidate, n-voter game will exist if and only if the Pareto optimal sets for ALL majority coalitions have a common intersection.

It might seem, of course, that this construction lends itself to no immediate application or geometric intuition as to what might be required for such an equilibrium to exist, but Plott observed that if, for example, we limit things to two dimensions and circular indifference contours and if at most a single voter occupies the presumed Nash equilibrium (Condorcet winning) policy, then this requirement about intersecting Pareto optimal sets is satisfied if and only if we can pair the remaining $n - 1$ voters in such a way that the lines connecting their ideals all intersect at the ideal of the nth voter. Put differently, if there is a voter at the point x, then a Condorcet winner exists only if we can draw a line from x to and through the ideal of the nth voter (whose ideal is presumed to be the Condorcet winner) and find a second voter whose ideal lies on that line to the opposite side of the nth voter's ideal.

Plott, of course, understood that two dimensions and circular indifference contours were special. But he was able to extend this result to establish that if indifference contours merely satisfied some rather weak and seemingly reasonable conditions (e.g., that they formed convex sets and are continuous), his result could be restated to require roughly the following: *If x^* is the presumed Condorcet winner and if at most one voter holds x^* as their ideal preference, then*

x^* is a Condorcet winner ***if and only if*** *we can pair all remaining voters such that the contract curve (the curve describing the tangencies of the indifference curves of the two voters in question) for each pair passes through x^*.*

Clearly, this is a strong condition, and is especially problematical when one considers the fact that subject to the condition that *at most one person prefers the presumed Condorcet point*, it is both necessary and sufficient. Moreover, if we extend things, as we sometimes must do in the world of statistical uncertainty, and speak of continuous probability distributions of ideal points, then as long as there are no point masses to that distribution (which is essentially the requirement that it be continuous and that masses of voters do not have their ideal points piled up at specific points), then Plott's condition is necessary and sufficient without further qualification. (To see that Plott's result establishes a general sufficient but not necessary condition for there to exist a Condorcet winner, suppose the idea points of three voters form a triangle and that two voters share the same ideal in the interior of that triangle. That shared ideal, then, is a Condorcet winner, but ideals cannot be paired so as to yield a common intersection of all contract curves.) Geometrically, of course, Plott's condition requires a degree of symmetry (termed radial symmetry in mathematics) to distributions of preference where that symmetry must be precise for Condorcet winners to exist. Plott's result, then, establishes that Condorcet winners are, mathematically at least, unlikely to exist when the domain of outcomes is multi-dimensional.

Things would seem to grow even worse, moreover, from the perspective of making predictions about the strategies of candidates to two-candidate spatial election games when we consider the next result under the assumption of spatial preferences. Richard McKelvey asked how serious the intransitivity might be when Condorcet winners are absent in a spatial context (in the 1976 article, "Intransitivities in Multidimensional Voting Models and Some Implications for Agenda Control," *Journal of Economic Theory* 12). After all, what if, for instance, three alternatives were ordered intransitively but all three defeated the thousands of substantively practical alternatives? Might not being able to limit our predictions to these three alternatives (called a *top cycle set*) be good enough? McKelvey, however, crushed this possibility. Under assumptions about preferences essentially equivalent to those imposed by Plott, he showed that if there is no Condorcet winner and if x and y are ANY two policies in a multidimensional space where x defeats y in a majority vote, one can find a sequence of policies $(z_1, z_2, \ldots, z_m)$ such that z_1 defeats x, z_2 defeats $z_1, \ldots, z_m$ defeats z_{m-1} and y defeats z_m. *Thus, in the usual spatial context, if there is no Condorcet winner, then the majority preference relation is intransitive over the entire policy space.*

Initially, academics who had not read his writing on the subject carefully interpreted McKelvey's result to imply that the simple democratic system of two-candidate majority rule elections was necessarily chaotic—that anything was possible. Some even took to calling the result McKelvey's "Chaos Theorem." But there were also those, McKelvey included, who deemed chaos counterintuitive. Specifically, consider the following circumstance: Imagine seven concentric circles (not indifference contours, but simply circles) and suppose one million voters have their ideal points uniformly distributed on each circle and that the ideal point of the 7,000,001th voter, x^*, is at the precise center of these seven circles. Given, then, the perfect symmetry of ideals, it follows that x^* is the Condorcet winner and corresponds to the Nash equilibrium of the corresponding two-candidate election game. Now, however, suppose we take one voter's ideal on any of the circles and move it ever so slightly to one side or the other. The distribution of ideals is no longer perfectly symmetric and the contract lines between pairs of voters will not all intersect at x^*. Mathematically, then, there is no Condorcet winner, and McKelvey's result kicks in to establish that the majority preference relation over the entire policy space is wholly intransitive. Such is the sensitivity of Plott's necessary and sufficient condition. However, imagine that we have made this situation a parlor game in which, prior to moving anyone's ideal, we ask participants to pick a policy after telling each of them that if anyone chooses a policy that defeats them, they lose the game and must pay some manner of fine. In this instance, we suspect that most if not all players would choose x^* (especially after reading this book). Now suppose we repeat the game after moving that one ideal slightly on a circle. Suppose, in fact, that we've drawn our circles on a standard 11×8 sheet of paper and that we move that one ideal 0.00000001 inches to the side. Knowing that there is no longer a Condorcet winner, players might no longer choose x^* as their strategy. But we suspect they will not choose a point very far from it. They might shift their choice ever so slightly one way or another in the direction of the sole perturbed ideal, but we doubt whether they would take McKelvey's "Chaos Theorem," even if they had absorbed the dense mathematics within it, to mean they should choose something radically different from x^*.

The question now, however, is whether we can formalize this intuition in a way that lends itself to a precise, general and intuitively meaningful specification. And this is precisely what McKelvey did (in the 1979 article, "General Conditions for Global Intransitivities in Formal Voting Models," *Econometrica* 47 and in the 1986 article, "Covering, Dominance and the Institution Free Properties of Social Choice," *American Journal of Political Science* 30,2). Admittedly, his result, fully presented, requires mathematics far beyond what we can or should offer in this volume; we should also emphasize that his analysis and conclusions are not limited to the special case discussed here of two issues and circular indifference curves. But for ease of presentation, we return again to that special case and suppose that with n (odd) voters, we draw all *median lines* on the page—all lines such that half or more voter ideal points lie on or to one side of the line and half or more ideal points lie on or to the other side of the line. Necessarily, if Plott's condition is satisfied, all such lines (of which there are an uncountable infinity) will intersect at the presumed Condorcet winner. However, if that condition is not satisfied, then there is no common intersection. But now comes the interesting and unanticipated result: Suppose, after drawing all such lines (impractical since there are an infinity of them), we draw the smallest circle possible such that every median line touches or passes through that circle, and suppose that circle has a radius of r. Now draw a larger circle of radius $4r$ with the same center as the first circle. McKelvey's result is that for any point x outside of the $4r$ ball, there exists at least one point inside of it that defeats x in a majority vote. In fact, we can even say something stronger: Not only is there a point inside of the $4r$ ball that defeats x, but at least one such point defeats everything that x defeats. In other words, all points outside of the $4r$ ball are dominated by something inside the ball.

There is no reason to suppose that this result is intuitive in any way to any person, and indeed its mathematical proof is beyond the limits of formalism we have set for this volume. We can, though, give some intuition behind it. Suppose the point z lies outside of the ball of radius r. Now draw a line l from z into the ball. From the definition of the ball, there must exist a median line perpendicular to l that intersects the ball (and in general cuts through it). Pick a point x, now, that lies on l and either on or arbitrarily close to that median. Since x is closer to the median line than is z, it follows that x necessarily beats z in a majority vote. This doesn't necessarily mean, of course, that x dominates z, but the deeper mathematics of McKelvey's proof is to show that if z is at least $4r$ from the center of the ball, then x will dominate z. It might seem, now, that such a result cannot tell us much about politics. But here at least first impressions are deceiving, especially if we allow ourselves qualitative as opposed to quantitative conclusions. First, let us return to the example of seven concentric circles and 7,000,001 voters. Prior to perturbing anyone's ideal preference, all median lines will pass through the center of those circles, x^*, and thus McKelvey's minimal ball will have a radius of 0. And four times zero is still zero, which is what must be the case with a Condorcet winner. Referring now to Figure 4.1a, suppose we perturb the ideal point x_i to x_i'. Now not all median lines will intersect at a common point, but those that are perturbed will only be perturbed slightly, in which case a ball that touches all medians will be quite small and McKelvey's $4r$ ball will still be confined to the near center of the distribution. In this way, then, we can rationalize the choice of positions close to x^* with the admonition "choose a position that is not dominated." In other words, if the distribution of the electorate's preferences is *nearly* symmetric, we can infer that the candidates of a two-candidate election will not choose a policy that is far from the "near center" of the preference distribution.

We can say something even more meaningful to the study of politics. In addition to maintaining the assumption of circular indifference contours, suppose the coordinates of the issue space have substantive meaning—that they correspond to specific identifiable issues under debate during an election campaign. If the overall distribution of preferences is not symmetric, then we know that the combination of the median preference on each issue cannot correspond to a Condorcet winner for the simple reason that there is no Condorcet winner and, thus, no pure strategy Nash equilibrium to the two-candidate election

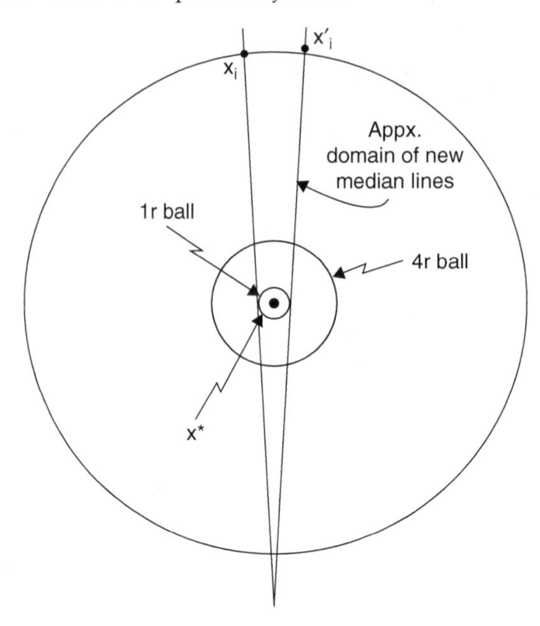

Figure 4.1a McKelvey's 4*r* ball with a small perturbation in one preference

game. However—and here the reader should refer to Figure 4.1b—since indifference contours are by assumption circles, if we project ideal points onto, say, the horizontal axis, identify the median ideal there and perpendicularly project a median line up from that issue-specific median preference, that median line must by definition pass through McKelvey's $4r$ ball. It follows that if $4r$ is "small," then even in the absence of a Condorcet winner, candidates that seek to avoid dominated policy positions will not adopt a position on that issue far from the preference of the median voter. In other words, despite the absence of a pure strategy equilibrium, McKelvey's analysis gives us a reason to reassert something like the Median Voter Theorem in multi-dimensional contexts.

For this assertion to be meaningful requires that our $4r$ ball not be "too big," and we do not want to leave the impression that r is always necessarily small. Consider, then, our original 3-voter example whereby the ideal points of those voters form a triangle. A simple paper-and-pencil exercise should convince the reader that McKelvey's initial circle of radius r will be the circle inscribed inside the triangle formed by the ideal points and that touches all three sides of the triangle. McKelvey's $4r$ ball, then, will wholly dwarf that triangle and, thereby, tell us very little about the policies two candidates might choose as their election strategies. Notice, moreover, that it matters not at all whether there is but one voter at each identified ideal point or one million, or two million, or whatever. And indeed, if there are several million at each vertex, then we might say we are modeling a sharply divided electorate wherein preferences cluster at one of

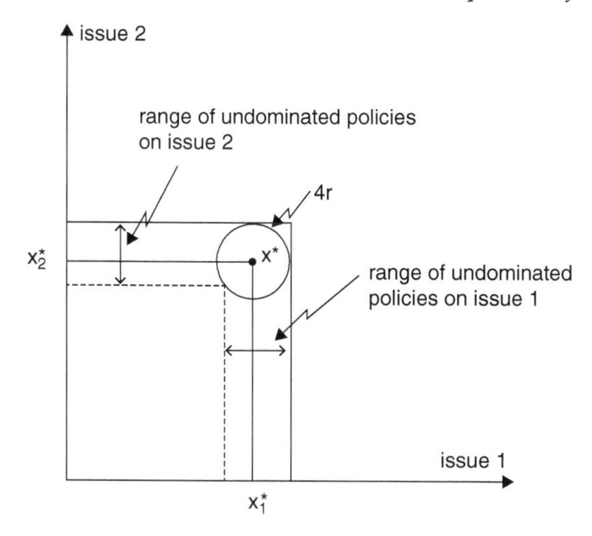

Figure 4.1b Domain of undominated policies with 4*r* ball projection

three alternative programs. And in this case, all we can predict is the possibility of wide swings in policy as the candidates jockey for position.

Substantively, then, McKelvey's result tells us that absent any divisive polarization of the electorate that might render the distribution of electoral preferences highly asymmetric as in Figure 1.6, the policy differences between candidates will be relatively minor and that the candidates themselves might focus more on personalities than on policy disputes. On the other hand, with a polarization that divides the electorate into multiple and sharply conflicting interests, it will be conflict over the issues themselves that is likely to take center stage and that regardless of the outcome, a significant share of the electorate is likely to be dissatisfied with the outcome.

Of course, at this point the reader might wonder why, in the absence of a Condorcet winner and a pure strategy Nash equilibrium, we have not reverted to the "solution" offered in the previous chapter of mixed strategies. There are three reasons for this, two technical and one substantive. One technical reason is that we are uncertain as to whether a mixed strategy equilibrium exists. The existence theorem for Nash equilibria offered in the previous chapter assumes that the number of pure strategies available to the players is finite, whereas here, if we conceptualize the policy space as a coordinate system, those strategies are infinite and uncountable. It is true that existence theorems exist for games with a continuum of strategies, but the most general impose a variety of assumptions on payoff functions, the most common being that those functions be continuous. Once again, absent introducing such things as probabilistic voting, the spatial election models discussed here have discontinuous payoffs (i.e., if one candidate passes the other on an issue, their payoffs exhibit a discontinuity whereby the candidates switch payoffs). We suspect, nevertheless, that equilibrium mixed strategies do exist for the most common spatial election games, but then we encounter a second technical matter, that of calculating those strategies.

Earlier, in Chapter 3, we discussed a zero-sum game of pure distribution or redistribution called the Colonel Blotto Game, noting that the allocation of time and resources in an American Presidential election can be thought of as such a game but with 50 unequally weighted battlefields. And in discussing that interpretation of Colonel Blotto, we noted that it is simply unbelievable to suppose that candidates in a real presidential election would or even should abide by some mixed strategy based on a static model of an election campaign. That same argument applies here. Our models here take an incredibly complex dynamic process and reduce it to a snapshot. Campaigns do not unfold in a day or even a month. They are typically ongoing events whereby the candidates try out different appeals to see which ones resonate best with the electorate, which tactics best facilitate securing campaign contributions, and what compromises have to be reached so as to mollify whatever disputes might have arisen within a candidate's party. Candidates are also busy adjusting their tactics to the actions of their opponents—deciding what to say and where to say it based on their opponent's tactics. Thus, reducing an election to a simple choice of policy in some n-dimensional policy space is a heroic simplification. We make such simplifications, though, in the hopes of eventually building better models as some fundamental processes and forces that we believe operate universally are uncovered. A spatial model of an election is no more a model of a real election than is the kinetic theory of gasses a model of what occurs inside of a balloon as it is blown up. All we do know is that as the balloon expands with each breath, whatever occurs within it is doing so in accordance with that theory and, hopefully, what we are discovering with our spatial models is uncovering some fundamental forces that operate in all elections and that minimally put boundaries on what can actually happen.

But there is also a substantive reason for supposing that the candidates need not compute mixed strategy solutions even if a spatial model does describe any real ongoing election campaign. Specifically, *one feature of a mixed strategy Nash equilibrium is that each of the mixed strategies involved in an equilibrium will give zero weight to any dominated pure strategy.* Were a mixed strategy to give positive weight to a dominated pure strategy, then the mixed strategy could be improved by shifting that weight to some undominated strategy. Thus, were a candidate, for some unknown reason, determined to abide by a mixed strategy, the first step should be to identify all undominated pure strategies. But this means that a candidate should limit the range of choices to those within McKelvey's $4r$ ball, since if a pure spatial strategy is undominated, it will lie inside that ball. Correspondingly, from the perspective of the analyst attempting to offer a prediction about a candidate's strategy, then that analysis can begin by predicting that the candidate in question—regardless of whether he is employing a mixed strategy or some other algorithm—will choose a policy inside of the $4r$ ball. And unless the electorate is sharply divided into virtually warring camps, that may be good enough.

4.3 Two-Candidate Elections and the Electoral College

Our focus on elections in the context of zero-sum games derives in part from the fact that no class of strategic environments lies more at the heart of political process in a democratic state than that of elections. Indeed, it seems that today even authoritarian states prefer to at least give the impression that their leaders are chosen by popular vote even if electoral fraud is self-evident, as is the case in places such as Russia, the Asiatic former components of the USSR and any number of African countries. The legitimacy of regimes today no longer resides in the divine right of kings, but in "the voice of the people," however imperfectly that voice is allowed to be expressed. The contribution of game theory to all of this lies, in part, with results such as the Median Voter Theorem, which, while dealing with a highly abstract representation of elections, moves us in the general direction of a theoretical understanding of things. That it does not accurately model any real election with which we are familiar is evident, but it can be used to address some of the questions that arise with respect to real elections. And here we can look again at the United States and its peculiar method of electing Presidents.

As previously noted, election to the presidency of the United States is complicated by the existence of a peculiar institution, the Electoral College. Implemented at a time (1787) when political parties as we know them were unknown, when the geographic expanse of the new nation exceeded that of most countries and, in particular, of Europe itself, and when the technologies of communication were at best rudimentary, the Framers of the U.S. Constitution assumed that voters would know little about politics and politicians except for what occurred immediately around them. It was not assumed that voters in, say, Massachusetts would know much if anything about the interests and politicians of far-away states such as South Carolina or even Pennsylvania. In addition, the Framers themselves, distrustful of voters and the passions that might control them and seeking various ways to blunt the powers of the national government in favor of those of the individual states, implemented America's peculiar institution whereby each state would be given a voting weight equal to the number of its representatives in the new national Congress plus two. The states, in turn, would, in accordance with rules established by their individual legislatures, choose how its electors to the Electoral College would be selected, where it would be those electors and not any mass electorate who would vote to fill the offices of President and Vice President.

Needless to say, the door was open to a great many ways to choose electors. One possibility was having state legislatures themselves make the choice, in which case when voting for state representatives, voters would in effect be voting for President. Another possibility was to allow electors to be voted on directly by the electorate (which back then meant white propertied males). But even in this case, there were several possibilities: Electors could be chosen by district, they could be allocated in proportion to the votes cast for the individual candidates, or they could be awarded as a block to the candidate (or, once political parties appeared on the scene, to the party) receiving the most votes. Individual states experimented with all of these alternatives during the early years of the republic wherein the method chosen typically was the one that best served the interests of the majority party of the state in question. For example, in the strongly contested election between Adams and Jefferson in 1800, Madison and Jefferson feared that some of Virginia's vote, if allocated by district, would go to the Federalist Adams. They thus succeeded in having Virginia's legislature implement the unit rule whereby all of that state's electoral votes would be cast for whoever secured a majority of the vote in that state, which they knew would be Jefferson. Massachusetts, in turn, dominated by Adams's party, the Federalists, wasn't about to leave its favored candidate in the lurch, whereupon its legislature, the General Court, also implemented the block vote scheme. In fact, by 1836 all states with but one exception had converted to this winner-take-all method whereby the Electoral College, in combination with the 12th Amendment to the U.S. Constitution that required electors to cast their votes for a slate pledged to a candidate for President and one for Vice President, took the form we see today.

A variety of issues arose with these changes that remain with us today, including (1) the possibility that someone can win the presidency with a negative popular vote plurality; (2) an assessment of the extent to which the Electoral College, as compared to a national direct popular vote, discourages the formation of third parties; (3) a measure of the advantage given to small versus large states; and (4) an assessment of the policy biases inherent in the Electoral College. Today, of course, with memories of the 2000 and 1960 contests still alive in some people's memories, it is the first issue that seems most salient. In the early years of the Republic, however, it was perhaps the fourth issue that was the greatest concern. Here we need to remember the Constitution's peculiar and unsavory compromise with slavery whereby in allocating representatives to the national Congress on the basis of population, a slave was counted as three-fifths of a person. Thus, while disallowed from voting in the South and treated essentially as property, the South held an advantage over those states with few slaves or states that had banned slavery from their territory. And since a state's Electoral College weight was dictated by the number of representatives it sent to Congress, the South and its policy interests gained an advantage in the selection of president.

We cannot address all of these issues, but we can begin to see some of the Electoral College's influences by, believe it or not, returning to the Median Voter Theorem. Actually, it is not that theorem specifically we will use here but a subsidiary result that holds true when the conditions of the theorem apply. So suppose two candidates must compete on a single policy dimension on which all voters have well defined (single-peaked) preferences. The Median Voter Theorem tells us, of course, that if the election were held by direct vote, without the intervention of the College, both candidates should converge to the electorate's median preference. But notice that if we apply this result to individual states, we see that there is an equilibrium policy position within each state—if preferences are single-peaked overall, then they are single-peaked among any subset of citizens. However, here the formal proof of the Median Voter Theorem can be extended to imply something stronger:

> *If preferences are single-peaked, not only does the median preference stand highest on the social preference order under majority rule (because it is the Condorcet winner), but the order is wholly transitive—if x, y and z are any three alternatives, if x is majority preferred to y and y majority preferred to z, then x is majority preferred to z.*

To see that this is true, consider three alternatives *Z*, *Y* and *W* that lie on the issue from left to right in that order, and suppose that *W* defeats *Y* and *Y* defeats *Z*. What remains to be shown is that *W* necessarily defeats *Z*. If preferences are single peaked, then any voter who prefers *W* to *Y* will prefer *W* to *Z*. But since *Y* is closer to *W* than is *Z*, there will be some voters who most prefer *Y* who will also prefer *W* to *Z*. Thus, if *W* defeats *Y* in a majority vote it must defeat *Z* in such a vote as well (and by even a greater plurality). This result suggests that if individual states employ a winner-take-all voting scheme, we can represent states as individual voters with transitive single peaked preferences wherein the peak preference is the ideal of the median voter within it. Thus, in those instances in which a single dominant issue occupies the stage, we can be anthropomorphic in our thinking and speak of a state as a voter with an ideal point at the median ideal point of the voters in that state, with a utility function that declines monotonically as we move away from the median in either direction, and with a voting weight equal to its vote in the Electoral College.

Using this argument to analyze the potential policy biases of the Electoral College, consider the following simple example:

> **Example**: Suppose there are three states with 7, 5, and 3 voters and three alternative policy platforms. Figure 4.2 gives a possible preference distribution in these states and shows that, although more than half of the voters most prefer policy 1 (eight voters versus five preferring policy 2 and two preferring policy 3), which is thereby the equilibrium in a two-candidate direct vote election, policy 2 is the median preference in states 2 and 3. So, regardless of whether electoral votes are awarded to states on the basis of their population, or whether each state receives two extra electoral votes, the electoral votes of states 2 and 3 are sufficient to ensure that policy 2 is the equilibrium platform under the Electoral College.

Our example shows, then, that the Electoral College can occasion a policy distortion whenever there is a strong asymmetry in the distribution of preferences within states. One might ask, then, as to the biases that exist today or existed in the past, and here research suggests that whatever biases existed in the past have

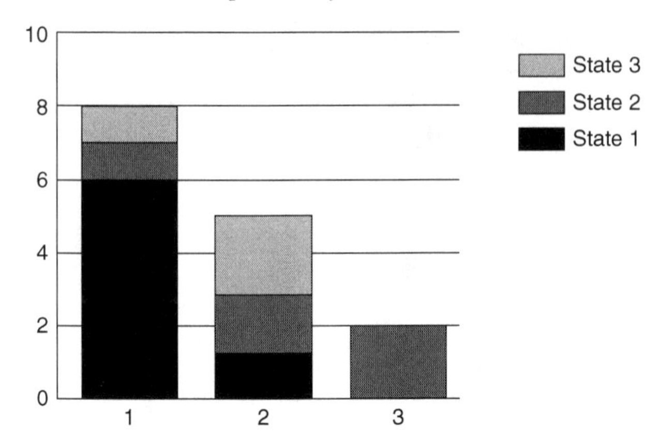

Figure 4.2 The potential for Electoral College bias

largely disappeared owing to the increasing homogeneity of the United States. Blacks, for example, are no longer concentrated in the South, nor is industry concentrated in the Northeast or in such states as Ohio and Pennsylvania. This isn't to say that bias has been wholly eliminated—only that it is likely to be less than what it had been in, say, the 19th century. There has emerged of late a sharp ideological divide, some say, between the East and West coasts versus the rest of the country—of "blue" states versus "red"—but demographically at least (and in the core economic interests that correlate with such variables) the country seems far more homogeneous and the opportunities for Electoral College bias far less pronounced today. The important lesson here, however, is not contained in any substantive conclusions we might draw about America's peculiar method of electing presidents, but rather in how we can use an abstract result, the Median Voter Theorem, to study in a theoretically meaningful way an electoral system that doesn't quite seem to match the assumptions imposed by that theorem.

4.4 Turnout and Responsible Political Parties

Among the other implications of the Median Voter Theorem is that in equilibrium, voters are not presented with meaningful choices—both candidates advocate the same policy. That theorem tells us, then, that under its admittedly severe assumptions, majority rule elections establish a strong incentive for the two major parties or their candidates to move their election platforms to the "center" of the electorate's preference distribution. This implication, though, flies in the face of those who prefer that the major parties offer distinct choices so as to give the electorate an opportunity to choose between, say, a liberal versus conservative policy. Thus, commentators on American politics bemoan the empirical manifestations of convergence to the median wherein

Democrats and Republicans often seem to not be worth "a dime's worth of difference" and where election campaigns for national office devolve into slick advertising and the denigration of an opponent's character or personality. At the same time, however, these same commentators lament the low voter turnout rates that characterize U.S. elections—generally, for Presidential contests, in the vicinity of 55%—as compared to the rates of other democracies where turnout commonly exceeds 70%. Some persons blame these rates on the indistinguishability of the candidates while others blame the indistinguishability on these rates—implying that, whatever the relationship, there is something amiss in American democracy.

To explore the issue of whether low turnout and candidates who fail to be distinct are symptomatic of "something wrong" with the way majority rule functions, suppose, contrary to the assumptions of the Median Voter Theorem, that not only are citizens willing to abstain from voting, but that their likelihood of voting is a function of the utility difference they perceive between the candidates. Thus, let us formulate a model in which citizens punish candidates (by staying home on election day) for failing to provide a distinct choice. For purposes of a numerical example, assume that if the two candidates converge identically to the median, then, regardless of policy preference, each citizen's probability of voting is some small number $0 < p_o << 1$. Assume further that the feasible policy space that serves as the candidates' alternative election strategies is represented by the interval $[0, 1]$ and that each person's utility for the policy x is given by the function

$$u_i(x) = 1 - |(x_i - x)|$$

where x_i denotes the citizen's ideal position on the issue and where $|.|$ denotes absolute value. Thus, utility declines linearly as we move away in either direction from a person's ideal. Finally, to model abstention, suppose person i's probability of voting is given by the function

$$p_i(x_A, x_B) = p_o + a|u_i(x_A) - u_i(x_B)|,$$

where a is some number that ensures that p_i doesn't exceed 1, and where x_A and x_B are the positions adopted by candidates A and B respectively. Thus, in accordance with the idea that voters ought to punish candidates for appearing similar and that they should reward a preferred candidate for offering a choice that is distinct from the opponent, the probability of voting increases as the utility difference between the candidates increases. With respect to specific probabilities of voting as a function of a voter's ideal,

1. A voter whose ideal lies precisely midway between x_A and x_B is indifferent between the candidates' policy positions. Thus, such a person votes with the minimum probability, p_o.

2. Assuming, without loss of generality, that $x_A < x_B$, a voter to the left of x_A votes with a probability

$$p_o + a|u_i(x_A) - u_i(x_B)| =$$
$$p_o + a|1 + (x_i - x_A) - 1 - (x_i - x_B)| =$$
$$p_o + a|x_A - x_B|,$$

which is a constant since this probability does not depend on x_i. Similarly, everyone with ideal points to the right of x_B votes with the same constant probability.

3. As we move from x_A to $(x_A + x_B)/2$, the probability of voting declines from $p_o + a|x_A - x_B|$ to p_o, and as we move from $(x_A + x_B)/2$ to x_B, it increases from p_o to $p_o + a|x_A - x_B|$. The dark line in Figure 4.3 illustrates this function for $p_o = 0$.

Now consider the remaining part of Figure 4.3, which, by way of a numerical example, shows a symmetric and bimodal distribution of preferences for an electorate of 67 voters. It might seem now that with this distribution of ideal preferences, we've given the candidates a strong incentive to adopt policies away from the median and that our assumptions about the likelihood of voting reinforces this incentive. However, to see that this is not the case—to see that the median remains the unique Nash equilibrium to the associated two-candidate majority rule game—suppose candidate B is at the median preference, policy 7, and that candidate A is at one of the two modes of the distribution, policy 3. Notice first that, because the probability of voting is a constant (0) through the electorate if A and B both choose the median, such convergence yields a tie in which neither candidate receives any votes. Candidate A's shift from the median increases turnout, but the question is whether this shift produces a relative gain for A. To see that the answer is "No," notice that by moving to 3, A increases the

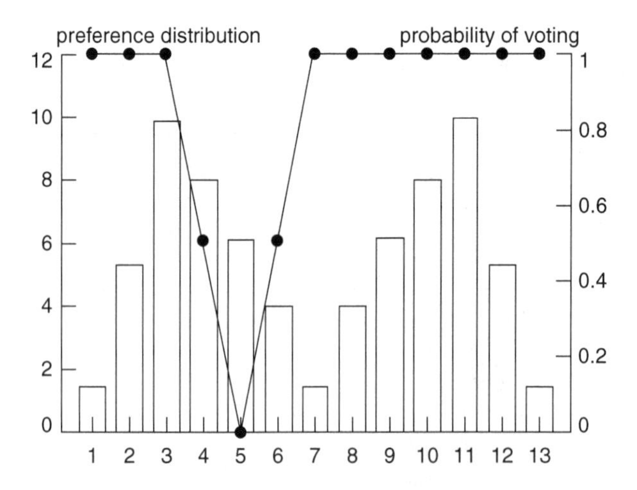

Figure 4.3 Abstention from indifference

probability of voting to 1 for all voters at and to the right of the median, and these voters prefer B. In addition, B gets additional expected votes from those with ideal points at 6, and thereby is the expected victor. In fact, this move by A yields B an expected vote of 36 and yields A an expected vote of 21, so A's expected plurality declines from zero to −12 if he shifts from the median to 3.

We appreciate the simplicity of this example, and a more realistic one should consider more general utility functions in multidimensional issue spaces, as well as other relationships between utility and probabilities of voting. One alternative, for example, is to assume that rather than be concerned with differences between the candidates, a potential voter's decision as to whether or not to vote depends on how satisfied they are with their most preferred candidate (such a model, assuming what we might call *abstention from alienation*, has been examined and the conclusion is the same as what we offer here provided, though, that the distribution of preferences is unimodal). Nevertheless, the implications of this example are clear. First, a low turnout rate does not indicate necessarily that there is something amiss in democracy. Indeed, it may indicate that democratic institutions are working as intended and that low turnout is the result of there not being rigidities in the system precluding one party or the other from nominating an effective moderate candidate. High turnout, on the other hand, might indicate that extremist activists have gained control of a party and secured the nomination of candidates that are removed from the mainstream of public opinion. We are reminded here of the fact that the election in Germany that resulted in Hitler being appointed Chancellor was, in fact, the highest turnout election of the Weimar regime. Minimally, our example suggests that the failure of the two parties in a two-party system to nominate candidates who offer distinct choices may be merely the logical consequence of majority rule electoral institutions, and that it is incumbent upon those who bemoan this fact to propose different institutions as well as good arguments why median policies, in the abstract, are inferior to policies that lie nearer one extreme or the other.

4.5 Multi-Candidate Elections

Perhaps the most glaring assumption in our discussion of elections that would seem to make our analyses of marginal interest to people living in countries other than the United States is that there are only two candidates or parties. This assumption makes our analysis especially simple for two reasons: (1) We can rely on the specialized results about two-person, zero-sum games as well as about symmetric versions of those games; and (2) citizens are presented with a relatively simple task of abstaining or voting for the candidate they most prefer without having to concern themselves with voting strategically for second or third choices. Once we admit the possibility of more than two candidates, however, a number of new questions arise:

1. If parties or candidates are free to enter an election, what is the equilibrium number of parties?

2. If parties or candidates are free to enter an otherwise two-party system, what strategies will the two established parties adopt in anticipation of new entrants or in the attempt to forestall entry?
3. If the number of candidates exceeds two, what spatial positions if any are in equilibrium?
4. If voters must choose among more than two candidates, how can we accommodate the fact that some voters may choose to vote for their second-ranked candidate whenever their most-preferred candidate has little chance of winning?
5. If, to accommodate the possibility that with more than two candidates, no candidate will secure an absolute majority of the vote, constitutional designers implement a system wherein a runoff election is held between the top two vote-getters if no one in fact wins a majority in the first balloting, what impact will such a constitutional provision have on party or candidate spatial positions?

Each of these questions is related to the others, and any general analysis must answer the fourth question, for example, before it can satisfactorily answer the first or the second. And at this point we must admit that we do not have general answers to all of these questions. Instead, we can answer some by assuming answers to the others. The opportunities for future research are made even more apparent, moreover, if we realize that winner-take-all plurality rule is but one voting procedure out of a vast array of possibilities. There are, of course, the various forms of proportional representation where it isn't candidates per se who compete, but rather the parties themselves. Plurality rule itself is but a special case of systems in which several candidates can be elected simultaneously from the same electorate. Even this system has multiple variants that include allowing voters a single vote that they can cast for one candidate versus allowing them to vote for more than one candidate. And in the event that elections are merely a preliminary step to the selection of a chief executive in the context of a parliamentary system, voters must be concerned as well with how their votes affect the subsequent process of coalition formation in parliament.

We cannot begin to address the vast array of possibilities here, nor can we survey the research, however limited, that already exists about such matters. Instead, by way of the more limited objective of illustrating the application of zero-sum game theory and the notion of a Nash equilibrium, let us focus in a simple way on the second question we list above—the strategies that two established candidates take in anticipation of the entry of a third opponent.

Third Candidate Entry Under Plurality Rule: Assume that an election concerns a single issue represented by the interval [0, 1], that all voters vote sincerely for the candidate closest to their ideal, and that voter ideal points are distributed *uniformly* on the interval [0, 1]. Even limiting ourselves to plurality rule, we now have two possibilities. Either a third candidate is already in the race or such a candidate is merely threatening to enter. Suppose first that there already are three candidates. In this case it's straightforward to see that there is no Nash equilibrium. If the candidates are positioned from left to right at A, B and C, then clearly both A and C have an incentive to squeeze B between them. But at some point in this process, the candidate at B will have an incentive to "jump out" from between its opponents and render either A or C the squeezed competitor.

So let us now consider the situation where a third candidate is merely threatening to enter and will do so only if it can win the election. If we stay with the assumption of a uniform distribution of preferences and if individual preferences are themselves symmetric in form, it is useful to note that if x and y ($x < y$) are any two positions in the interval [0,1], then the share of the vote between x and y is simply $y - x$. The first thing to ask now is what positions for the existing two candidates guarantees that a third entrant cannot win and thus will not enter. Let those positions be A and B again with A < B and to simplify things here we will short-circuit some algebra and simply assume that A and B are symmetric about the median, 1/2. Clearly, if entry is to occur it would not be to the left of A or right of B since that would only throw the election to B or A respectively as it would leave one of them with 1/2 the vote. So suppose that a third candidate, if it enters, would do so by choosing a position between A and B. If it does, the voters who would most prefer it lie in the interval $[(A + C)/2, (B + C)/2]$ so that its vote share equals $(B + C)/2 - (A + C)/2 = (B - A)/2$. However, if this third candidate enters only if it can win, its best strategy is to enter precisely at 1/2, since entering to the right or left of the median leaves one of its opponents with a positive plurality. Thus, the voters who prefer A with a competitor at $C = 1/2$ all lie in the interval $[0, (A + 1/2)/2]$, thereby giving A's share of the vote as $(A + 1/2)/2 - 0 = A/2 + 1/4$. The third candidate will not enter then as long as $A/2 + 1/4 > (B - A)/2$. But we already know that B is symmetrically opposite A relative to 1/2, or simply $B = 1 - A$. Thus, some simple algebra gives us $A > 1/6$. That is, if A locates at 1/6 or greater and B locates at 5/6 or less, a third candidate cannot win by entering the competition.

Notice, though, that $A > 1/6$ is merely a sufficient condition for keeping a third-party candidate from entering, which is not to say that (1/6, 5/6) is a Nash equilibrium since there is nothing to keep either of the competitors from moving toward the center. A strategy pair such as (0.25, 0.75) for example, also keeps a third competitor out of the contest, as does the policy pair $(1/2 - \delta, 1/2 + \delta)$ for any arbitrarily small δ. Thus, as long as A and B stay away from the extremes, a third-party candidate will not enter, but there is no Nash equilibrium for the two original candidates.

Third Candidate Entry Under Majority Rule with a Runoff: Almost certainly the most popular method for electing a chief executive by direct vote is not simple plurality rule but rather majority rule with a runoff whereby if no candidate receives 50% or more of the vote in the first round of balloting, then a runoff is conducted between the two top vote-getters. So consider this possibility and suppose a third party will not enter unless it can secure at least a second place finish so that it can compete in the second round. Here, however, we will simplify our example by not considering the possibility that voters vote strategically and will assume instead that they simply vote for the candidate whose spatial position they most prefer. In this case it is evident that the strategy pair (1/6, 5/6) won't keep a third competitor out of the race. All such a competitor need do is locate a bit to the right of 1/6 or to the left of 5/6 and it is guaranteed a second-place finish with one third of the vote while holding the top vote-getter to under 50%. Between the first and second rounds, it might then attempt a coalition with the eliminated candidate in order to secure a majority in the runoff.

To preclude this possibility we will continue to assume as before that A and B are symmetric with respect to 1/2. If the third candidate enters, say, to the immediate left of A, its vote share will be approximately A whereas if it enters between A and B, its vote share will be (B – A)/2 as before. But now we want to see if the two initial competitors can position themselves so that a third entrant cannot come in second in the final vote count. First, to see what value A should be set at to keep the potential entrant from entering to the immediate left of A, notice that if it does so, it wins slightly less than the share A of the vote (A – 0 = A), whereas the candidate at A wins slightly more than 1/2 – A of the vote. Thus, if we set A equal to 1/2 – A, or simply A = 1/4, the third-party entrant cannot outpoll the candidate at A and surely doesn't outpoll the "unmolested" candidate at B. So instead suppose the entrant tries to locate between A and B so as to win the vote share (B – A)/2. Since this share is constant for any position between A and B, its best chance of knocking off, say, A in the first round is to position its platform as close to the right of A as possible without identically matching A's position. But this won't work as long as A > (B – A)/2 or A > B/3. But B = 1 – A, which means that if A = 1/4, B = 3/4 then a third candidate cannot knock off either competitor and be a distinct choice. Thus, (0.25, 0.75) is a Nash equilibrium since any deviation from it results in the entry of a third-party candidate and the elimination of one of the two original candidates from a runoff.

Clearly the preceding analyses of three-candidate elections treat some quite special cases: Preference distributions are uniform, the election concerns a single issue and voters are precluded from voting strategically. Nor have we considered elections with more than three candidates, and there is no reason to assume that conclusions reached when the number of candidates equals three carry over to when that number increases to four. If, for instance, there are four candidates in our plurality rule case, the reader should confirm that, although there is no Nash equilibrium with three candidates, there is an equilibrium with four (with two candidates at ¼ and two at ¾). We have, nevertheless, learned something despite the limitations of our examples. It is often argued, for example, that runoff elections encourage the formation of third parties if only so that those parties can forestall a winner on the first ballot and then subsequently trade its support for policy concessions. This may be true, but our analysis reveals that it is not the whole story: If the two preexisting parties show some strategic sophistication, and if third parties form in the hopes of actually winning the election, then they can at least minimize the incentives for third-party entry even if they cannot altogether eliminate those incentives. Our analysis also suggests that while the threat of third-party entry need not keep candidates in a plurality system from converging toward the electorate's median preference, a majoritarian system with a runoff might be more attuned with those who prefer that parties and their candidates offer the electorate distinct choices. Our examples, of course, only hint at such conclusions, but they once again illustrate the utility of game theory for uncovering the forces that operate in alternative electoral systems.

4.6 Candidate Objectives and Game-Theoretic Reasoning

In the 1960s and 1970s when various American industries were under assault by Japanese firms, most notably in the automotive and electronics industries, the charge was leveled by those experiencing an erosion of their market positions that the Japanese were "playing unfairly." By ignoring their shareholders and employing a longer-term strategy than the heads of American firms could adopt, given their dependence on Wall Street's quarterly evaluation of their corporations (which, so the argument went, depended on immediate profits), Japanese firms were afforded the opportunity to maximize short-term market share with an eye to long-term profits. The fear was that if one's competitors could "unfairly" focus on market share, they could forgo short-term considerations of profit in anticipation of ultimately monopolizing the market. This, in turn, brought into question those textbook economic market models that presumed profit maximization.

A similar issue arises in the modeling of elections with respect to assigning objectives to candidates and parties. Earlier we note that absent any uncertainty, and given the assumption of two candidates and an electorate in which everyone votes, a variety of alternative assumptions about objectives are necessarily equivalent. Under the stated conditions of the Median Voter Theorem, maximizing votes, probability of winning, plurality and expected plurality are all equivalent. That things might change, however, if we were to incorporate uncertainty or non-voting into our models are suggested by Figure 4.4. This figure offers two probability density functions, f_1 and f_2, defined over the range of pluralities enjoyed by, say, candidate 1, where, holding constant the strategy of the opponent, we suppose f_1 prevails if candidate 1 chooses the strategy x_1 and f_2 prevails if he instead chooses x_2. Since f_1 and f_2 are both drawn as symmetric densities, x_1 yields a higher expected plurality, but x_2, owing to the smaller variance of f_2, yields the greater probability of winning (the greater probability of a positive plurality). It would seem, then, that the strategy a candidate adopts will depend on which assumption best characterizes his objective: to maximize the probability of winning or expected plurality.

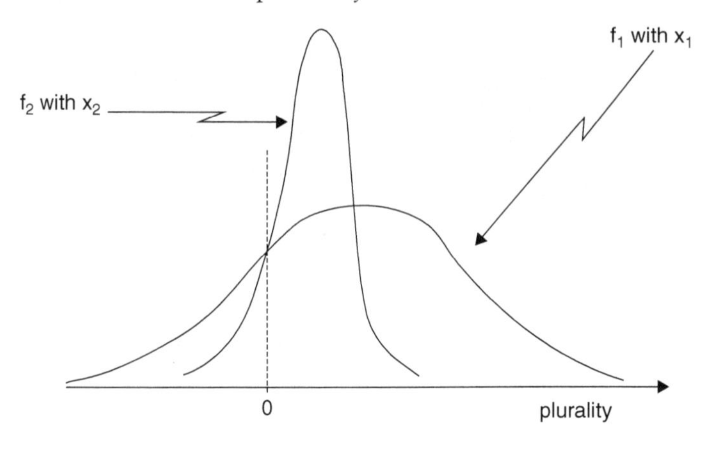

Figure 4.4 Questioning the equivalence of objective functions

In a world of uncertainty, maximizing one's probability of winning might appear to be the more natural assumption. But suppose a candidate, like the Japanese firms noted above, is concerned not simply with winning, but with trying not to lose, if that is to be the outcome, by as small a margin as possible because doing so might allow him to compete in a future election. However, any conclusions we draw about objective functions from Figure 4.4 are flawed since they take no account of the candidate's opponent and how that opponent might respond to the objectives of the candidate we are discussing. In other words, drawing conclusions from Figure 4.4 commits the error of confusing decision theoretic and game theoretic thinking. Asking whether alternative objective functions are equivalent or different, then, should be restated to ask whether the equilibria that prevail under one set of assumptions differ in any way from the equilibria that prevail under an alternative set of assumptions.

To explore this possibility, suppose again for purpose of illustrating things, that $Pl_i(c1,c2)$, candidate *i*'s plurality as a function of the two candidate's strategies, *c1* and *c2*, is a random variable. Thus, a candidate that maximizes probability of winning maximizes $Pr[Pl_i(c1,c2) > 0]$ whereas a candidate that maximizes expected plurality maximizes $E[Pl_i(c1,c2)]$. The strategy pair $(c1^*, c2^*)$, then, is an equilibrium for candidates maximizing probability of winning if and only if

$$Pr[Pl_i(c1^*,c2) > 0] > Pr[Pl_i(c1^*,c2^*) > 0] > Pr[Pl_i(c1,c2^*) > 0] \qquad (4.1)$$

for all *c1* and *c2*. Similarly, the strategy pair $(c1^{**}, c2^{**})$ is an equilibrium under expected plurality maximization if and only if

$$E[Pl_i(c1^{**},c2)] > E[Pl_i(c1^{**},c2^{**})] > E[Pl_i(c1,c2^{**})] \qquad (4.2).$$

Now suppose that the distribution of Pl_i is symmetric about its mean for all $(c1, c2)$ and suppose in addition that the election itself is symmetric—if the candidates switch strategies, they switch payoffs. These two assumptions imply that in equilibrium, the probability that plurality, $Pl_i(c1^*,c2^*)$, exceeds zero equals 1/2 and that expected plurality, $E[Pl_i(c1^*,c2^*)]$, equals 0. That is,

$$Pr[Pl_i(c1^*,c2^*) > 0] = ½ \text{ and } E[Pl_i(c1^*,c2^*)] = 0$$

From the assumption that $(c1^*,c2^*)$ is a Nash equilibrium, expression (4.1) requires that

$$Pr[Pl_i(c1^*,c2) > 0] > Pr[Pl_i(c1^*,c2^*) > 0] = ½,$$

But if the distribution of Pl_i is symmetric about its mean, then if $Pr[Pl_i(c1^*,c2^*) > 0] = ½$, it must be the case that $E[Pl_i(c1^*,c2^*)] = 0$ and whenever $Pr[Pl_i(c1^*,c2) > 0] > ½$, it must be that $E[Pl_i(c1^*,c2)] > 0$. Similarly, if $Pr[Pl_i(c1,c2^*) > 0] < ½$, it must be the case that $E[Pl_i(c1,c2^*)] < 0$. Putting these inequalities together reveals that if $(c1^*, c2^*)$ satisfies expression (4.1), it necessarily satisfies expression (4.2). A parallel argument establishes that if $(c1^{**}, c2^{**})$ satisfies expression (4.2), it necessarily satisfies (4.1). Thus, under the assumption that uncertainty occasions a symmetric distribution for the random variable Pl_i, if the election itself is symmetric and thus provides for no inherent bias favoring one candidate over the other, maximizing probability of winning and maximizing expected plurality are equivalent objective functions—if a strategy pair is an equilibrium under one objective function, it is an equilibrium under the other.

As with our other examples, we appreciate that the assumptions imposed here—notably the symmetry of probability densities and of the election—are limiting. But once again we also see how a decision theoretic perspective is misleading and that any answers to questions pertaining to the equivalence of candidate objectives require a game theoretic analysis.

4.7 The Strategy of Introducing New Issues

Candidates in competitive elections do not merely jockey for position in some issue space. They also attempt to increase the salience of issues on which they hold an advantage and to downplay those issues on which their opponents might be advantaged. Thus, in the 2012 American presidential contest, Obama emphasized Romney's past ties to his venture capital company, Bain, in the hopes that the public would view Romney as but another greedy out-of-touch-with-the-average-person capitalist. Romney, in contrast, emphasized the economy, with its 8% unemployment and declining middle class incomes. Interestingly, it would seem that both candidates danced carefully around events in Libya and the killing of the American ambassador there. For Obama, what happened was at best an embarrassment and display of administrative incompetence, if not outright deception, whereas for Romney, an emphasis on foreign policy typically benefits an incumbent in American politics.

The most stark example of the tactic of issue salience manipulation comes when a challenger attempts to unseat an incumbent when the incumbent otherwise holds an advantage on all issues of the day. To see what we mean here, suppose an electorate with three voters is initially concerned with but a single issue and suppose the incumbent has positioned himself during his tenure in office at precisely the median preference. A challenger, then, would seem to confront an insurmountable obstacle. If he or she adopts any other position, they will lose the election. But if they simply match the incumbent, then voters might reasonably conclude that there is no reason to vote against the incumbent because the challenger is offering nothing new. What, then, can the challenger do? The answer, historically at least, has been (for sufficiently skilled challengers) to attempt to introduce a wholly new issue that somehow divides the electorate in a way that eliminates the existence of a Condorcet winner and thereby renders the incumbent vulnerable.

If things were this simple, of course, we should never observe incumbents being reelected. So, for purposes of exploring some possibilities as a way to illustrate the value of a spatial perspective, suppose the two candidate's positions on the first (pre-existing) issue are fixed at x_C for the challenger and at x_I for the incumbent, where x_I is the median preference on that first issue and x_C is not equal to x_I. Thus, neither candidate is afforded the luxury of being able to adjust their positions on the first issue. Now consider Figure 4.5a whereby we assume that a second (vertical) issue has been added that leaves the ideal preference of the median voter unchanged but which shifts the ideals of voters 1 and 3 up. The question we can ask now is if the challenger is limited to positions on the vertical line passing through x_C, can the incumbent find a position on the vertical line passing through x_I that renders him invulnerable to the challenger. And for the circular indifference curves in Figure 4.5a, the answer to our query is yes. In that figure we have drawn two indifference contours, one for voter 1 and one for 2 such that both curves slightly miss the vertical line coming up from x_C. To win, the challenger must both retain the loyalty of voter 3 and win that of voter 1 or 2. But notice that the indifference curves for 1 and 2 intersect so that if the incumbent moves to any part of the darkened portion of the line coming up from x_I, the challenger cannot attract either voter. But now consider Figure 4.5b. Here we leave voter 3 as he was in Figure 4.5a, but assume that voter 2's preferences become ellipses owing to the salience (to him) of issue 2, and, more critically, render voter 2's preferences non-separable—2's indifference contours now correspond to "tilted ellipses." Under this construction, the challenger can adopt the position A and, because the indifference contours of voters 1 and 2 through A do not intersect on the dashed line coming up from x_I, the incumbent has no winning response (for elaboration, see Dean Lacy and Niou, "Nonseparable Preferences and Issue Packaging in Elections," in N. Schofield, G. Caballero, and D. Kselman, eds., *Advances in Political Economy: Institutions, Modeling, and Empirical Analysis*, London: Springer, 2013: 203–15).

What we've learned here, then, is that not any additional issue can be used to unseat an incumbent. To do so requires, first, an issue that moves some subset of voters off the old line characterizing the original issue. And second, the issue needs to "interact" in some way with the pre-existing issue so as to yield non-separable preferences for some voters.

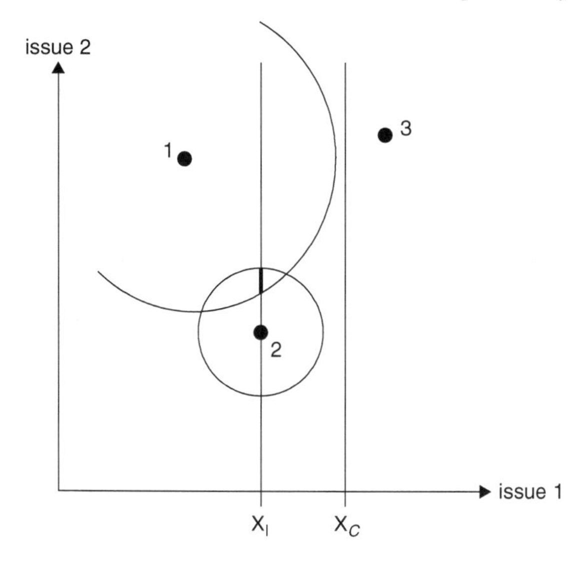

Figure 4.5a Issue-by-issue median invariance

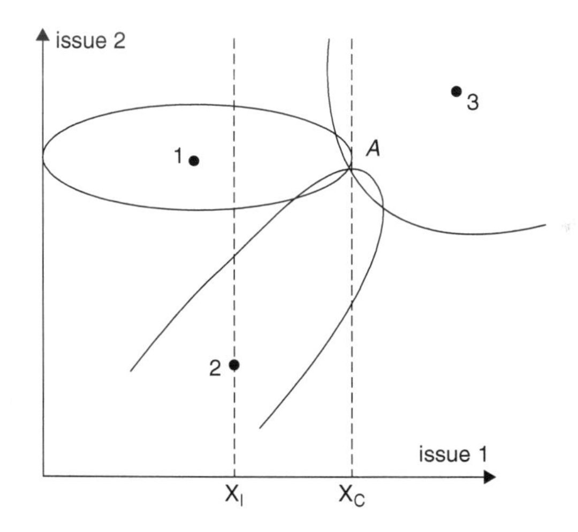

Figure 4.5b Issue-by-issue median variance

There is, of course, little that is game theoretic in this discussion since all we have done is to play with the geometry of spatial preferences. But what we do show here is that a more complete game theoretic model of elections would include strategies that, in addition to positions on the issues, also include whatever instruments are available to the candidates for the manipulation of issue salience. Unfortunately, we have little guidance as to what those instruments might be and thus how they function in a way that would allow for their formal mathematical representation. If we did, then the authors of this volume might abandon academic writing in favor of a career as political consultants.

4.8 Elections with Uninformed Voters

The election models that we've considered thus far in this chapter distort reality in at least one other supremely significant way—they assume that voters are perfectly informed about the candidates' issue positions, and they suppose that candidates know the criteria voters use to evaluate them. This is at best a curious assumption in light of public opinion polls telling us that citizens often vote without even knowing the candidates' names, never mind their issue positions. In an age when elections in particular and democratic institutions in general appear to be in ascendancy worldwide as the preferred mechanism for choosing governments and at least for rendering a government legitimate, it is indeed reasonable to ask whether democracies can function effectively in less than perfect information environments. What was the impact of Obama's strategy of denigrating Romney's character and candidacy in the 2012 presidential contest in lieu of revealing the policies he proposed to pursue if reelected? Why, as Republicans claimed prior to the actual balloting, might "left-leaning pollsters" attempt to make Obama appear more competitive than he might actually have been and why did Republicans care about any potential liberal bias in those polls? More generally, does the incomplete information of voters and candidates merely open the door to an insidious use of money as the primary currency of democracy? Do our informational shortcomings allow a hidden "power elite" to control the state so that it can generate policies that benefit it at the expense of everyone else?

Of course, given the small chance that a voter will be decisive in any mass election, and given the cost of information, we should not be surprised by the fact that voters are rationally uninformed. In choosing between learning about a candidate's likely performance if elected versus learning about which brand of television is the better buy, it is not unreasonable to suppose that learning about televisions is the more efficacious investment. Indeed, those voters who are informed often appear to approach politics as a sport as they gather information for the same reason that others memorize baseball batting averages, with persons of both types approaching their subject with the understanding that they have a nearly equivalent influence on outcomes. But consider the hypothesis that voters can use relatively costless cues in deciding for whom to vote. The party identification of a candidate has long been interpreted as such a cue. However, partisan associations are based primarily on historical events, and the particular hypothesis we want to explore here is whether citizens can learn to vote by using relatively costless information that is generated during the course of an election campaign—information such as interest-group endorsements and public opinion polls that report answers to such questions as, "If the election were held today, for whom would you vote?" The particular question we address is whether we should be disturbed by the fact that citizens, in lieu of reading every news item analyzing current events and in lieu of a thorough analysis of each candidate's record, often use seemingly irrelevant contemporaneous information to make their decisions about how to vote (e.g., "I voted for ___ because my brother-in-law preferred ___," or "I preferred ___ because ___ slanted the news in favor of the opponent").

For a hint as to how we might proceed here, consider America's 1964 Presidential election and its 1980 contest. In the 1964 election, which pitted the Democrat incumbent Lyndon Johnson against Senator Barry Goldwater, if we were to summarize Johnson's campaign by a single sentence it might be something like, "My opponent is a right-wing extremist and if you vote for him, he'll get us into a war." Similarly, the election of 1980 pitted the incumbent Democrat Jimmy Carter against Ronald Reagan, wherein the dominant theme of Carter's campaign could be summarized by the statement, "My opponent is a right-wing extremist and if you vote for him, he'll get us into a war." It would seem, then, that a voter who was uncertain as to whether the Republican challenger was or was not an extremist confronted an identical dilemma in both contests. There was, however, a difference between these two elections. Specifically, if an uninformed voter consulted the nightly network news on television, in 1964 he would consistently hear reports that gave Johnson a projected margin of victory of approximately 60% versus 40% for his opponent. Such numbers were consistent with the hypothesis (and Johnson's claim) that if Goldwater was not a right wing extremist, he was surely quite conservative in his proposed policies. In 1980, in contrast, those nightly news broadcasts reported a far more competitive contest—one in which, as the election approached, projected a dead heat or even a slight advantage for Reagan. This information, then, seemed wholly inconsistent with what Carter was saying about Reagan and might reasonably have led uninformed voters to conclude that Carter's characterization of his opponent was little more than electoral campaign bombast.

Taking our cue from the comparison of these two elections, and to approach the topic of elections with uninformed voters in a general way, suppose again then that an election concerns a single issue and that voter preferences over this issue look like the ones that set the stage for the Median Voter Theorem. That is, suppose each registered voter has a well-defined ideal point on the issue and his or her utility declines symmetrically as we move away from that ideal in either direction. Rather than assume, however, that all citizens know the candidates' positions on this issue, and following the line of research introduced by Richard McKelvey and Ordeshook (in the 1985 article, "Elections with Limited Information: A Fulfilled Expectations Model Using Contemporaneous Poll and Endorsement Data," *Journal of Economic Theory* June), suppose the electorate consists of two types of voters: (1) those who are informed about each candidate's issue position and (2) those who are uninformed about such things. Next, suppose that a poll is published announcing which candidate in a two-candidate election is ahead and by how much.

The preference for one candidate or the other of an informed voter should not be influenced by the poll. At most, such a poll is entertainment. But consider the poll's potential impact on uninformed voters in a way suggested by our comparison of America's 1964 and 1980 presidential contests. That the information contained in such a poll need not be irrelevant to their preferences over the candidates, consider an uninformed voter who regards himself as preferring a position to the right of "center" on the election's issue and who is concerned that the "conservative" candidate fits the extremist portrait painted by his or her liberal detractors. If asked to vote without any information, this citizen might simply choose randomly or abstain. However, if responses to the question "If the election were held today, for whom would you vote?" indicate that public preferences are approximately evenly divided between the candidates, then only two hypotheses are consistent with the poll's estimate. The first possibility, of course, is that few respondents know the candidates' positions and that the polls are meaningless. Barring this possibility, however, and supposing that a "reasonable" proportion of those polled know where the candidates are, then either (1) both candidates are extremists of the opposite sort or (2) both candidates are near the median respondent.

If our uninformed voter assumes that the poll contains true information about preferences and perceptions, then regardless of whether subcondition (1) or (2) holds, the most likely possibility is that he or she is closer to the conservative than to the liberal candidate: It cannot be simultaneously true that the poll's respondents are informed and that the liberal is a moderate and the conservative an extremist, because this circumstance would produce a poll that shows the moderate holding a significant lead. Nor can it be true for the same reason that the liberal is an extremist and the conservative is a moderate. Hence, if this citizen reasonably infers a preference for the conservative and if this or other similar people are polled in a subsequent survey—including the election itself—they should no longer choose randomly but should hold a preference for the conservative.

Now consider an electorate in which each citizen's preferences on the issue are described by simple distance from the citizen's ideal point, so voters vote for the candidate who they believe (perhaps incorrectly in the case of uninformed voters) is closest to their ideal. For example, suppose the electorate is divided evenly into uninformed and informed subparts and that the ideal point density for each subpart is as shown in Figure 4.6 (f_U(x) and f_I,(x) respectively, with f(x) being the overall preference density). With respect to the information available to voters, informed voters are defined as those who know the positions of the two candidates, say x_A and x_B. Uninformed voters do not know x_A and x_B but suppose, on the basis of such things as interest-group endorsements or the candidates' party labels, they know that $x_A < x_B$ (they know that candidate A is the "liberal," and B the "conservative").

If we look now at what might happen in a sequence of public opinion polls that ask, "If the election were held today . . . ?" informed voters have a clear preference. In particular, we have drawn f_I(x) in Figure 4.6 so that 30% most prefer A and 70% most prefer B. Suppose now that among the uninformed voters included in the poll (and keep in mind that pollsters do not generally sort respondents by whether they appear informed or uninformed), respondents are loath to admit that they are uninformed and, therefore, that they

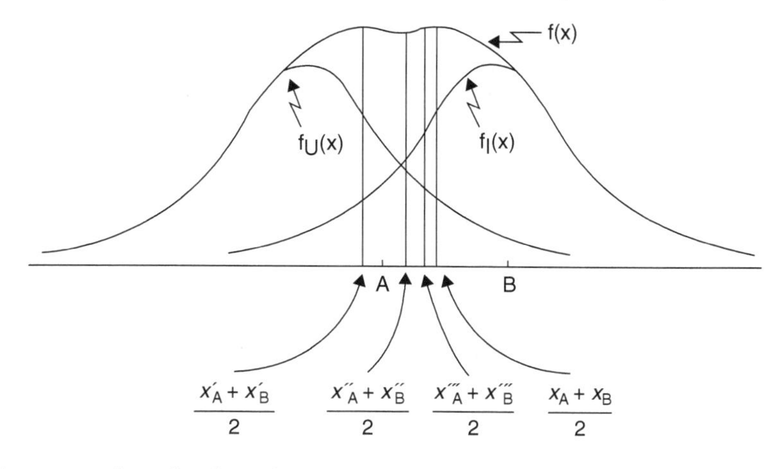

Figure 4.6 Informed and uninformed electorates

Table 4.1 Poll Sequence with Informed and Uninformed Voters

	Poll 1		*Poll 2*		*Poll n*	
	A	*B*	*A*	*B*	*A*	*B*
informed	30%	70%	30%	70%	30%	70%
uninformed	50	50	72	28	92	8
reported	40	60	51	49	61	39

respond randomly. If, for purposes of an example, we now suppose that there are approximately the same number of informed and uninformed voters, then an initial poll produces the result shown in Table 4.1 under poll 1. Of course, pollsters do not know who is informed or uninformed—they merely report the aggregate (average) result, which in this instance is a forty-sixty split between A and B.

To this point, the poll serves the interest of those who look at politics as a spectator sport. But now, to show how a poll might be used by an uninformed voter to make an informed decision, suppose that uninformed voters are armed with one additional piece of information—namely, their preferences on the issue relative to the population. That is, suppose each uninformed voter knows what proportion of the overall electorate prefers policies to the right and to the left of his or her ideal. This assumption, of course, credits uninformed voters with knowledge they are unlikely to possess in precise detail, but it is not unreasonable to suppose that they have an approximate idea of where they are relative to the overall population—more liberal, more conservative, much more liberal, etc. Our assumption merely formalizes this idea so as to allow a formal treatment. Specifically, if an uninformed voter assumes that all others are informed, then any such voter can (inaccurately) infer from the poll whether the midpoint between the candidates is to the left or right of his or her ideal point. And this inference, in combination with the knowledge of which candidate is to the left of the other, allows that person to infer which candidate is closer to their ideal. Thus, in our example, all uninformed voters with ideal points to the right of the fortieth percentile (at $(x'_A + x'_B)/2$) will infer that candidate B is closer to their ideal, and those to the left infer a preference for A.

Now consider what happens in a second poll. Informed voters, of course, are unaffected by the poll results because they know the candidates' positions and so they continue to divide 30–70 between A and B. But uninformed voters no longer choose randomly (although those between the 40th and 61st percentile are choosing incorrectly), and if all such voters to the left of the 40th percentile indicate a preference for A while all uninformed voters to the right of this percentile indicate a preference for B, then, as Figure 4.6 shows, uninformed voters divide 72–28. The overall poll as reported in Table 4.1 under poll 2, then, is 51% for A, 49% for B (corresponding to the midpoint $(x''_A + x''_B)/2$). At this point, uninformed voters with ideal points between the 40th and 51st percentiles should change their preference from B to A on the basis of a re-estimate of the position of the candidates' midpoint relative to their ideal (in which case the inferred midpoint now becomes $(x'''_A + x'''_B)/2$). That is, the link between polls and inferred preference introduces a dynamic to the process, and the question we should ask is whether this process ends at any particular point—is there a poll that reproduces itself and toward which this dynamic inevitably moves? Our answer to this question (ignoring some mathematical niceties) is that this process converges (poll *n as n goes to* ∞) to a situation in which all uninformed voters to the left of the *actual* midpoint vote for A and those to the right choose B. That is, in equilibrium, all uninformed voters act like informed ones.

To see the logic of this result, notice first that if everyone somehow infers the true midpoint, then any poll will necessarily reproduce itself and therefore it cannot cause any uninformed voter to revise his or her preference. Thus, everyone somehow inferring the correct midpoint between the candidates is an equilibrium to the adjustment process. That such a circumstance is the unique equilibrium follows from the fact that if the midpoint assumed by uninformed voters differs from what informed voters know, then a subsequent poll will average these beliefs (in proportion to the percentage of uninformed versus informed) and result in a new inferred midpoint that lies somewhere between reality and the old inference. A succession of such averages, then, converges to the true midpoint that was known originally only by informed voters. Thus, we have described a situation in which voters act like simple decision makers who condition their decisions on readily available information—in this instance, on a guess, derived from the poll, as to which candidate is closer to their ideal preference. Unlike a standard decision problem, though, information is endogenous so that actions in one period (poll) affect information in the next, which in turn affects actions in the next period, and so on.

Turning now to the strategic problem that confronts the candidates, in deciding what policy position to adopt at the beginning of their campaign, these candidates might contemplate appealing only to informed voters. However, if the candidates appreciate the role that polls can play, they should also appreciate that uninformed voters can eventually act like informed ones. Perhaps we should not be surprised, then, when we see candidates either attempting to manipulate the polls to their own advantage—a possibility that is especially troublesome in those countries in which the major sources of news have party affiliations—or denying the validity of polls that show them doing poorly. Overall, though, the candidates should plan on choosing the electorate's overall median preference, and in this way the relevance of the Median Voter Theorem can be reasserted.

Substantively, the preceding analysis suggests that although the attention paid to the mass media's predictions in an election campaign may be evidence that the electorate regards politics merely as a spectator sport, we cannot also preclude the possibility that parts of the electorate find useful information in those polls. We have reached this conclusion, of course, with a number of heroic assumptions, the most serious being that the election concerns a single issue and that uninformed voters know their relative placement on the issue. Thus, if this model is to be genuinely useful, we must generalize it to accommodate more issues as well as more realistic indirect information sources for voters. In fact, this model can be generalized in the following way: Rather than supposing that uninformed voters know which candidate is to the left and which is to the right, assume, in the case of an election that concerns a single issue, that there are two identifiable subpopulations in the electorate whose preferences are polled and reported separately, and assume that uninformed voters know how preferences are distributed within each subpopulation. Uninformed voters can then use the poll results from each subpart to estimate the midpoint between the candidates as well as their left-right orientation. This approach can then be generalized to n issues by merely supposing that there are $n + 1$ such subpopulations.

Any complete model, of course, should also incorporate the fact that many voters use retrospective information in deciding how to vote, as they answer such questions as, "Am I better off today than I was four years ago?" Nevertheless, our discussion does expand the relevance of the Median Voter Theorem, and it shows one circumstance under which uninformed voters can learn to vote correctly and thus, one circumstance under which the candidates will maintain their incentive to converge to the median preference under less than "ideal" conditions. In addition, the above model brings into question how we should evaluate voter knowledge. The usual measure is how accurately voters identify the policy positions of candidates or parties. The preceding model, however, tells us that such information is inessential to casting a correct vote. What voters need to know is which candidate is closest to them in terms of proposed policies, and learning that need not require knowing precise policy positions. In the above model, a voter can infer what he or she needs in order to cast a correct vote by being aware of his or her preferences relative to the electorate as a whole and paying due attention to learning which candidate is ahead in the polls and by how much. Not only is information of this sort less costly to secure, but models of the sort we have just discussed promise the possibility that there are other relatively costless sources of information voters might use to make judgments about candidates and parties. It is not difficult, for instance, to extend the above model to multiple dimensions and to suppose that it isn't the entire electorate being sampled by pollsters but rather only specific identifiable subgroups (e.g., farmers, urban residents, unemployed, Hispanics, small business owners, members of a specific political party, and so on) whose general policy orientations are known and who an uninformed voter can then use to sequentially estimate bisecting lines or (in the case of more than three dimensions) hyper-lines in order to estimate which candidate is closest to their ideal. Such models remain highly mathematically abstract and it is certainly a stretch to suppose that voters who are otherwise uninformed about party and candidate policy positions are nevertheless capable of undertaking the econometric and statistical tasks assigned to them in the preceding analysis. Nevertheless, that analysis suggests that democratic elections may be far more robust than studies of voter information might otherwise lead us to believe.

4.9 Other Applications

Because this chapter is intended to demonstrate the relevance of zero-sum games to politics, our applications here of spatial preferences focus on electoral competition between political parties or their candidates. There is, though, no reason to suppose that such preferences apply only to voters or that other alternative electoral schemes cannot be usefully studied with such preferences. With respect to this latter point, Romer and Rosenthal (in their 1978 article, "Political Resource Allocation, Controlled Agendas, and The Status Quo," *Public Choice* 33 and in their 1979 article, "Bureaucrats Versus Voters: On the Political Economy of Resource Allocation by Direct Democracy," *Quarterly Journal of Economics* 93) employ a spatial conceptualization of preferences to show how those in control of the alternatives voters might confront in public referenda can induce outcomes other than median preferences. Specifically, suppose a referendum on, say, school budgets, is conducted wherein if a budget proposed by a school board is not approved by the electorate, the school district's budget reverts to some predetermined value. Instead of choosing between candidates, voters then are choosing between the proposed budget and the reversion. And if school bureaucrats are intent on securing the largest budget possible, it is in their interest to make the reversion as draconian and unacceptable as possible—perhaps even $0. Romer and Rosenthal, in fact, go beyond this to study referenda in which school bureaucrats are allowed to call for a succession of budget referenda wherein the reversion takes hold only if proposed budgets are defeated at the polls a fixed number of times. This, then, establishes an interesting game between bureaucrats and voters wherein bureaucrats, unsure, perhaps, of the electorate's preferences, use the succession of referenda to see how large a budget is feasible, but where voters, aware of the self-interest of bureaucrats and interested in minimizing their taxes without undercutting a viable public education system, understand the game that is being played.

In a wholly different context, Klevorick and Kramer (in their 1972 Cowles Foundation Discussion Paper, "Social Choice on Pollution Management: The Genossenschaften," Cowles Foundation, Yale University) examine a particularly imaginative mechanism for controlling pollution along the Rhine River in Germany wherein regional pollution control boards, consisting of representatives from industry and from the towns and villages affected, vote on an effluent tax rate. The particular twist here is that preferences for alternative tax rates are derived mathematically from more fundamental preferences relating to, in the case of industry, profits and the costs of pollution control, and for towns and villages, from the impact of effluent charges on household budgets, and shown to have a spatial representation. However, the innovative feature of Germany's Genossenschaften is that the relative voting weights of industry versus residents of the affected towns and villages are endogenously determined as a function of taxes paid—the greater the taxes paid, the greater is one's voting weight. Assuming that majority rule operates over the sole dimension of tax rate, Klevorick and Kramer then establish that such a system has an equilibrium—that tax rates and voting weights do not necessarily cycle endlessly.

Romer, Rosenthal, Klevorick and Kramer thereby illustrate the application of spatial preferences to situations that go beyond simple plurality or majority rule elections, and indeed, there is no reason to even limit spatial preferences to voters. Later in this volume, for instance, we assume such preferences hold for members of a committee such as a legislature. We think, moreover, that by now the reader can see how a spatial structure might be applied to, say, understanding the decisions of a supreme or constitutional court wherein a court's members can be described as liberal or conservative, left-leaning versus right-leaning, sympathetic to government expansion versus more concerned with strict interpretations of individual rights. The discussion of such things, though, will be postponed until after we consider a more fundamental problem of collective action.

4.10 Key Ideas and Concepts

maxmin
minmax
Plott theorem
McKelvey's "Chaos Theorem"
McKelvey's $4r$ ball
redistributive politics
abstention from indifference
abstention from alienation
equivalent objectives
unimodal preferences
limited information elections

Exercises for Chapter 4

1. A (row chooser) and B (column chooser) must play the following 2 × 2 game:

8, −8	4, −4
2, −2	6, −6

Beforehand, B can pay four dollars to a third person to learn A's decision.

   a. Draw the full extensive form, assuming that A does not know whether or not B purchased the information.
   b. Portray the corresponding strategic form.
   c. Solve the game for equilibrium strategies.

2. Games A – C are all zero sum. (Payoffs are to row chooser.) What is the expected value of game A for row chooser?

	b_1	b_2
a_1	40	Play game B
a_2	Play game B	Play game C

Game A

	t_1	t_2
s_1	10	40
s_2	60	30

Game B

	t_1	t_2
s_1	40	30
s_2	30	60
s_3	40	20

Game C

3. Does the following zero-sum strategic form game possess a pure strategy equilibrium (games within cells are played if that cell is realized as a result of players' initial strategies)?

<table>
<tr><th></th><th colspan="2">b_1</th><th colspan="2">b_2</th><th>b_3</th></tr>
<tr><td rowspan="2">a_1</td><td colspan="2" rowspan="2">10</td><td>1</td><td>3</td><td rowspan="2">6</td></tr>
<tr><td>4</td><td>2</td></tr>
<tr><td rowspan="2">a_2</td><td>7</td><td>10</td><td>10</td><td>20</td><td rowspan="2">6</td></tr>
<tr><td>6</td><td>5</td><td>5</td><td>10</td></tr>
<tr><td>a_3</td><td colspan="2">8</td><td colspan="2">5</td><td>5</td></tr>
</table>

4. The Median Voter Theorem states that if an election concerns one issue, the median preference of the electorate is an equilibrium in the sense that it cannot be beaten in a majority vote and that in two-candidate contests, both candidates should adopt the median preference as their platform. We

also know that there is no such equilibrium with three candidates if those candidates choose their platforms simultaneously. However, is there a pure strategy equilibrium with four candidates? To simplify your argument, assume that the voter ideal points are spread out uniformly across the election issue, and assume either an even or odd number of voters, whichever facilitates your analysis.

5. Suppose two presidential election candidates, who maximize their probability of winning, must decide how to allocate three days among six states. Whoever allocates the most time to a state wins that state and whoever wins the most electoral votes wins the election. The states' electoral votes are as follows: 27, 27, 24, 18, 2, and 2. Assume that all ties are broken by a coin toss and that transportation technology renders days non-divisible.

    a. Does the corresponding two-candidate election game have a Nash equilibrium in pure strategies?
    b. Does your answer change if the three largest states have the same electoral weight?
    c. Does your answer to part (a) change if days are divisible?

6. Suppose there are six voters, 1, 2, 3, 4, 5, and 6, who are all concerned with the same issue, who have most-preferred positions that are ordered the same as their labels (i.e., 1 most prefers the leftmost position and 6 the rightmost position), and who have single-peaked preferences on this issue. Also, suppose that weighted voting is used and that voter i has i votes. If two plurality-maximizing candidates are free to adopt any position on the issue as their election platform, and if all of the conditions for the Median Voter Theorem are otherwise satisfied, which voter's ideal point will they adopt as their platform?
7. In addition to choosing a policy on the real line, suppose the candidates can also try to mask their positions by presenting themselves as lotteries—as probability distributions over the feasible policy space. Suppose that two parameters—the mean and variance—characterize a candidate's strategy (such as when that strategy corresponds to a normal density function), and that voter i's utility for position x is $u_i(x) = -(x_i - x)^2$. Describe the two-candidate election equilibrium if all other provisions of the Median Voter Theorem apply.
8. Consider a twenty-five voter electorate that is distributed across five districts (A, B, C, D, E). Five voters live in each district, and each voter is identified by his or her district (e.g., each "a" lives in A) and his or her ideal point on the policy $X = [1, 9]$. Suppose the ideal points of the voters are as follows:

$x = 1$	$x = 2$	$x = 3$	$x = 4$	$x = 5$	$x = 6$	$x = 7$	$x = 8$	$x = 9$
b	b	b	a	a	a	a	a	d
e	c	c	b	b	c	c	d	d
e	c	e	d	d	e			
				e				

All voters have single-peaked preferences. The electorate chooses a policy on X by using the following procedure: First, using majority rule, voters within each district choose a legislator from two possible candidates, where these candidates compete by choosing positions on X. Second, the five legislators, who are constrained to most prefer the policy they advocated in getting elected and to offer that policy as a proposal before the full legislature, meet and must decide, by majority rule, which point on X will be the new policy.

a. Tell which policy will be chosen, assuming that all legislators are sophisticated and choose an agenda at random from the set of possible binary agendas, and explain why.
b. How does your answer to (a) change if states A and D and the voters in them are eliminated?
c. How does your answer to (a) change if states C and E and the voters in them are eliminated?

9. Consider a nine-voter electorate in which voter i's ideal point on the issue equals i. Suppose a two-candidate election is held and that all of the assumptions needed for the median voter result hold except that candidate I must choose voter 3's ideal point as his stated policy. Candidate II has no such restriction.

   a. Who wins the election?
   b. Below are four possible descriptions of the outcome. Decide which of the following statements are true.

      i. The median voter's ideal point must be the outcome.
      ii. The median voter's ideal point cannot be the outcome, but another voter's ideal point can be the outcome.
      iii. The median voter's ideal point cannot be the outcome and neither can any other voter's ideal point.
      iv. Either the median voter's ideal point or another voter's ideal point can be the outcome.

10. Suppose an electorate consists of five votes with the following preferences over the three platforms that each of two candidates might choose:

voters 1 and 2:	*A*	*B*	*C*
voter 3:	*A*	*C*	*B*
voters 4 and 5:	*C*	*B*	*A*

Suppose each voter, after each candidate (simultaneously) chooses his or her platform, votes for his or her preferred candidate with probability p and for the opponent with probability $1 - p$, $p > 1/2$. If a voter is indifferent, $p = 1/2$. What is the final outcome (A, B, or C) if both candidates maximize their expected plurality?

11. Consider the following twenty-nine player game, where two players (I and II) are candidates and the other players, $(1, 2, \ldots, 27)$, are voters. Candidates

will compete for office by using one of four possible campaign strategies (A, B, C, D) to recruit voters who will vote for them. The outcome of the election will be determined by which of the two candidates has recruited the most voters. In the first stage of the game, the two candidates simultaneously choose one of the four campaign strategies. For some combinations of campaign strategy choices (BB, CC, DB), the candidates must debate. Candidates choose debate strategies simultaneously, but afterward they both learn the other's campaign strategy. Below is a table that shows the results of different combinations of campaign strategies. The cell values represent the number of voters recruited by candidate I (row chooser); 27 minus the cell value equals the number of voters recruited by candidate II (column chooser).

<table>
<tr><th></th><th>A</th><th colspan="2">B</th><th colspan="2">C</th><th>D</th></tr>
<tr><td>A</td><td>15</td><td colspan="2">3</td><td colspan="2">21</td><td>7</td></tr>
<tr><td rowspan="2">B</td><td rowspan="2">9</td><td>12</td><td>27</td><td colspan="2" rowspan="2">5</td><td rowspan="2">13</td></tr>
<tr><td>0</td><td>15</td></tr>
<tr><td rowspan="2">C</td><td rowspan="2">17</td><td colspan="2" rowspan="2">19</td><td>24</td><td>17</td><td rowspan="2">23</td></tr>
<tr><td>14</td><td>18</td></tr>
<tr><td rowspan="2">D</td><td rowspan="2">11</td><td>12</td><td>13</td><td colspan="2" rowspan="2">27</td><td rowspan="2">11</td></tr>
<tr><td>11</td><td>14</td></tr>
</table>

After the campaign, the winner of the election is determined by majority rule. Thus, the candidate that recruits a majority of voters, wins the election.

a. Which candidate wins the election?
b. If campaign laws are changed so that strategy D cannot be chosen, which candidate wins the election?

12. In a seven-person electorate, all election outcomes depend on the candidate's positions on one policy dimension, and the ideal point of voter i on this dimension, x_i, equals i. All voter utility functions are of the form $u_i = -|W - x_i|^2$ where W is the policy position of the winning candidate and x_i is voter i's ideal point. In this electorate, there are two political parties. Voters 1, 2, and 7 are in party X. Voters 4, 5, and 6 are in party Y. Voter 3 is in neither party. There are four possible candidates: X_1, X_2, Y_1, and Y_2. Party X uses majority rule to nominate X_1 or X_2. Party Y uses majority rule to nominate Y_1 or Y_2. All voters then choose from among the two nominated candidates to determine W, the policy position of the winning candidate. Assume that all assumptions necessary for the Median Voter Theorem hold except in the case where they conflict with the statement of the question. Use the concept of Nash equilibrium to find the possible locations of W in the following two situations.

a. Voters are nonstrategic when nominating a candidate.
b. Voters are strategic when nominating a candidate.

5 The Prisoners' Dilemma and Collective Action

5.1 The Prisoners' Dilemma

The primary Continental European powers in 1914—the "great powers" of France, Germany, Austria and Russia—each faced a dilemma. With tensions among them running high for a variety of reasons, each knew that whoever mobilized first for war would possess a powerful strategic advantage should war actually break out. Given the technology of warfare and transportation at the time, the mobilization of armies took weeks. Troops had to be mustered, armed and collected at disembarkation points, food and ammunition supplies gathered, and transportation networks coordinated to move men and material to the places conflict was most likely. It was also understood that being the first to mobilize meant that any war would be fought on someone else's territory, and it was also understood that mobilization signaled a belief that war was imminent, since no country could afford to maintain an indefinite state of mobilization without seriously damaging morale, its economy and perhaps even the survivability of its regime. Thus, while all states might have preferred to avoid war (a questionable assumption in the case of Austria), the current system of alliances was all that was needed to ensure that the dilemma was played out fully: If Austria attacked Serbia, Russia was bound to defend Serbia. Germany was not only allied to Austria in its defense, but was concerned that if it waited too long for the inevitable conflict to arise, Russia would complete the building-out and upgrading of its rail system, making mobilization of that giant even easier. France, in turn, was allied with Russia and in any conflict between Germany and Russia that led to German mobilization, France could not lag because Germany remained France's arch enemy and primary strategic threat. Britain was not tied firmly to any of these alliances, but it was the primary guarantor of Belgian neutrality, and it was through Belgium that Germany was most likely to advance on France. The Kaiser's decision, moreover, to upgrade Germany's fleet was, from the British perspective, a form of mobilization to which it was especially sensitive.

The dilemma to which we refer here has a name—the Prisoners' Dilemma (PD)—though it gets that name from a different, more fictionalized scenario. In fact, of all the games that have drawn the attention of social scientists and, in turn, given game theory visibility in the social sciences among those who otherwise have minimal interest in that theory, none can compete with the one we call the Prisoners' Dilemma. Its name derives from a scenario in which two criminals have in fact jointly committed a crime, but to successfully prosecute them, the prosecutor needs a confession. To that end he has them locked in separate cells so that they can no longer directly communicate and offers each the deal of a reduced sentence if they confess. Payoffs, stated in terms of years incarcerated, might then look as follows:

	Confess	Don't confess
Confess	−7, −7	0, −10
Don't confess	−10, 0	−2, −2

Figure 5.1 The classic Prisoners' Dilemma

Thus, by turning state's evidence and confessing, a prisoner can greatly reduce the sentence he must serve. If one is the sole prisoner to turn state's evidence, suppose the deal struck with the prosecutor is to have all charges dropped so that one's accomplice must serve a full prison term of 10 years. However, if both prisoners confess, each confession is worth less, in which case all that's secured is a reduced sentence of 7 years. If neither confesses, the prosecutor cannot convict for the ostensible crime, but the prosecutors, being who they are, are sufficiently enraged that they are able to secure a conviction for a minor offense with a prison term of 2 years. The incentive of both prisoners now is clear—regardless of what the other does, it's in one's interest to confess. Regardless of what blood oaths might have been sworn to before being locked in their separate cells, each prisoner can reason "I agreed never to rat on my partner, but if he finks out and I stick to the agreement, off I go for 10 years; and if he doesn't rat, I can escape prison entirely. It would be pretty nasty of me to rat, but what if he does and I don't? And he just might do that because he confronts the same incentive to do so as I do." The end result, of course, is a Nash equilibrium outcome that both prisoners prefer to avoid, (–7,–7), in favor of some other possibility.

Of course, the Prisoners' Dilemma would be of little interest if its application were confined to prisoners. But, as we've already suggested in referring to the motives of political elites leading up to WWI, this game has had broad application and allows us to understand such things as the reason why people might agree to the establishment of a state that has the power to coerce them, why collectivities sometimes seem to act in ways that are irrational from the perspective of the individuals involved, why various industries might even lobby to be regulated by government, why governments regulate and license barbers and taxi cab drivers but do little in the way of regulating automobile safety aside from requiring the installation of seat belts, why our actions can depend critically on whether we anticipate interacting with the same person again in the future, why the simple basic act of voting in a democracy can be a difficult thing to explain, why labor unions and various associations often offer low cost life insurance to their members, why attempts at reforming the economy of one society prove to be a success whereas similar reforms applied elsewhere lead to massive inefficiency and corruption, and why nations sometimes war even when they might prefer other means of resolving disputes.

The universality of Prisoners' Dilemma-type situations is no better illustrated than the example from Chapter 1 of the school of herring that forms a swirling ball when approached by predators, only to offer those predators a ready and efficiently consumed target. The optimal strategy for the school, of course, is to scatter itself in every direction, thereby thinning out so a predator can, at best, consume a fraction of the species and perhaps even exhaust itself in the process. Each individual fish in the school has two choices: Swim away from the others or attempt to move to the center of the slowly shrinking ball so as to not be the first to be eaten and perhaps hope that predators will sate themselves before the ball is fully consumed. It seems reasonable to suppose now that staying put and/or trying to move into whatever remains of the swirling ball yields a higher chance of survival than attempting to flee does, regardless of what all others of your species do. If some or all others attempt to flee, predators will be preoccupied picking off those swimming in their direction. If few or none attempt to flee, then by being one of the few defectors, one becomes an inviting target for predators. So in either case, it seems best to stick to the swirling ball. In this context we are reminded once again of Pastor Martin Niemöller's famous quote in reference to the crimes committed by the Nazis with which we introduce Chapter 2. Implicit in the moral lesson Niemöller offers is the idea that if people fail to act when not personally and immediately threatened, it is because the seemingly safe course is to keep one's head down—to try, as with the threatened school of fish, to swim to the center of the swirling ball. Of course, if everyone acts in accordance with this reasoning so that no one has an incentive to be the first to oppose the rise of authoritarianism, the end result is that everyone suffers.

As we suggest in the introduction to this chapter, an earlier dilemma gripped Europe's imperial powers at the outbreak of World War I. Summarizing in her seminal history of the outbreak of that war, *The Guns of August*, Barbara Tuchman wrote, "War pressed against every frontier. Suddenly, dismayed governments struggled and twisted to fend it off. It was no use. Agents at frontiers were reporting every cavalry patrol as a deployment to beat the mobilization gun. General staffs, goaded by their relentless timetables, were pounding the table for the signal to move lest their opponents gain an hour's head start. Appalled upon the brink, the chiefs of state who would be ultimately responsible for their country's fate attempted to back away but the pull of military schedules dragged them forward" (New York: Macmillan Co., 1962: 72). In its absolutely simplest form, then, as represented by a 2-person game, the circumstances confronting Europe's Great Powers did not look much different than the game in Figure 5.2, which is virtually identical to Figure 5.1.

Other examples of the Prisoners' Dilemma abound in the literature, both academic and classical. For readers of a more classical bent, consider Puccini's opera *Tosca* in which the chief of police (Scarpia) condemns Tosca's lover

	Mobilize	Don't mobilize
Mobilize	−10, −10	10, −20
Don't mobilize	−20, 10	0, 0

Figure 5.2 World War I dilemma

	Keep bargain	Double-cross
Keep bargain	−10, −10	−20, 20
Double-cross	20, −20	−15, −15

Figure 5.3 Puccini's *Tosca*

(Cavaradossi) to death but offers to save him in exchange for Tosca's favors. Tosca consents. The problem as posed offers Tosca two alternatives: keep the bargain, or double cross Scarpia by stabbing him when he comes to her. Similarly, Scarpia must decide whether to issue blank or live ammunition to the firing squad that Cavaradossi will face. If the bargain is kept, Tosca's satisfaction in getting her lover back will be marred by her surrender to Scarpia, and Scarpia's satisfaction is diminished by having to reprieve his rival. If Tosca double-crosses Scarpia and gets away with it, she wins most and he loses most; and vice versa. If both double-cross, both lose, but not so much as each would have lost had he or she alone kept the bargain. To see now how this situation corresponds to a dilemma, let us choose some arbitrary numbers that nevertheless seem reasonable and consistent with ordinal preferences, and let Tosca be the row chooser and Scarpia be the column chooser, so that we arrive at the game illustrated in Figure 5.3. Notice now that we can assign payoffs in such a way that both players have a dominant strategy—to double-cross the other. However, if both double-cross, then an outcome prevails, (−15, −15), that is inferior to both keeping the bargain.

5.2 Some Simple Dilemmas in Politics

The Logic of Term Limits: The political processes that mimic the Prisoners' Dilemma in form are vast and we can survey only a few. But first, the thing to keep in mind is that there is no reason to suppose that dilemmas are limited to the interaction of two persons. Consider, for example, the issue of legislative term limits as it arises in countries with single-member district representation. Here the thing to be understood is why voters would want to tie their hands and preclude the opportunity to reelect their representative to parliament or the legislature. After all, by imposing a limit on the number of terms a politician can serve, voters are foregoing the potential benefit of being represented by an experienced and ostensibly skilled individual. The answer to such queries, though, is offered by the following scenario. Suppose a legislature consists of n (odd)

	$X < (n-1)/2$ other legislators propose a program for their constituencies	Precisely $(n-1)/2$ other legislators propose a program for their constituencies	$X > (n-1)/2$ other legislators propose a program for their constituencies
Choose a legislator who will propose a benefit for the district	0	$B-(n+1)C/2n$	$B-(X+1)C/n$
Choose a legislator who will not propose a benefit for the district	0	0	$-XC/n$

Figure 5.4 Term limits and inefficient public programs

representatives, and suppose that in any given session of the legislature each representative can place on the agenda a proposal to fund a project for his or her district at national expense—a bridge, a highway, a publicly funded hospital, and so on. Suppose each such project costs C and yields the benefit B, where that cost is borne by all constituencies but the benefit accrues only to the constituency in question. Also, suppose that such projects are funded if and only if a majority of the legislature approves of them as a package. Now let the voters in each constituency choose between two candidates: The first is certain to propose such a project for his or her district whereas the second is opposed to such projects with the argument that they are an inefficient use of taxpayer dollars. The situation confronting voters, then, can be represented as shown in Figure 5.4 (our example here is taken from the 1985 article by Niou and Ordeshook, "Universalism In Congress," *American Journal of Political Science* 29(2) = 246–258).

To see how the entries in this table are calculated, if X, the number of legislators proposing a program for their constituencies, is less than $(n-1)/2$, then even if the constituency in question elects a representative who will propose a benefit, the number of legislators doing so remains less than a majority—in which case we suppose no such projects are funded. If precisely $(n-1)/2$ other representatives propose a benefit for their districts, then the voters of the district in question (row chooser) are pivotal for determining whether or not there is a majority favoring such programs. If those voters choose a legislator who will propose a program, they receive the benefit B, but at a cost of the $(n+1)/2$ funded programs, $(n+1)C/2$, which is shared across all n constituencies. Finally, if a majority within the legislature proposes and approves of such projects regardless of who the voters from the constituency in question elect, those voters will or will not get the benefit B depending on who they elect at a cost of their share of all programs approved. Notice now that electing a legislator who will propose a program (weakly) dominates the alternative if, when the constituency in question is pivotal, $B-(n+1)C/2n > 0$, or equivalently for large legislators (where $(n+1)/n$ is approximately equal to 1.0), if $B > C/2$. Thus, if $C > B > C/2$, voters in each constituency are caught in a Prisoners' Dilemma: All have an incentive to choose a representative who will propose an economically inefficient program for his or her constituents, but with everyone doing so, the final outcome has every constituency suffering a net loss of $B-C < 0$. And because it is a dilemma, no constituency has an incentive to unilaterally vote otherwise. Term limits, so the argument goes, short-circuit this tendency to inefficiency by removing the incentive of legislators to sponsor the narrow interests of their constituents by removing their opportunity to pursue reelection.

Vote Trading in Legislatures: The preceding discussion of the ostensible logic of term limits recognizes the fact that, in representing their constituents, individual legislators are often in the business of seeking votes for things that, while beneficial to their constituents, are, *ceteris paribus*, opposed by a majority of legislators. Thus, legislation often proceeds by trading votes—a legislator gives up control over one or more pieces of legislation in exchange for support on something else. To see, now, the sorts of dilemmas that vote trading can yield, suppose three legislators (1, 2, and 3) are contemplating the disposition of bills A, B, C, D, E, and F, where their payoffs from each bill's passage are as shown in Table 5.1 (a bill's defeat pays 0, and let payoffs be additive across bills).

Table 5.1 A 6-Bill Opportunity for Vote Trading

Legislator	*Payoff from Bill's Passage*					
	A	*B*	*C*	*D*	*E*	*F*
1	3	3	2	−4	−4	2
2	2	−4	−4	2	3	3
3	−4	2	3	3	2	−4

Notice that if all three legislators vote their preferences, all bills pass in a majority vote and each legislator (or, hopefully, their constituents) earns +2. Suppose, however, that legislator 1 sees that his loss from having D pass exceeds his gain from having C pass and that exactly the opposite is true for legislator 2. Thus, 1 might propose that he and 2 trade votes on these bills, with 1 voting against C and 2 reciprocating by voting against D in order to defeat both measures. Having consummated this trade, 1 sees a similar opportunity with legislator 3 with respect to bills E and F, whereas 2 sees just such an opportunity with 3 on bills A and B. If all trades are made, though, no bills pass, and each legislator's payoff, 0, is less than what he or she gets if no trading had occurred. Given the inefficiency of unrestrained vote trading, suppose each legislator considers refraining from such activity, hoping that others will do the same. This possibility yields the game in Figure 5.5, where "trade" means exhibiting a willingness to trade with anyone who signals the same willingness, "refrain" means being unwilling to trade with anyone, and where profitable trades require the agreement of two legislators. As this figure shows, trading is a dominant strategy, so the outcome (0, 0, 0) seems unavoidable (barring, of course, coalitional deals to deliberately exclude one legislator from making any deals).

	Legislator 3			
	Trade		Refrain	
	Legislator 2		Legislator 2	
	Trade	Refrain	Trade	Refrain
Trade	0, 0, 0	4, –4, 4	4, 4, –4	2, 2, 2
Refrain	–4, 4, 4	2, 2, 2	2, 2, 2	2, 2, 2

Figure 5.5 A vote trading dilemma

America's Electoral College: Although, as we noted earlier, the U.S. Constitution affords each state the authority to determine how its Electoral College vote can be apportioned among the candidates competing for the presidency and vice presidency, the states themselves, in an effort to increase their importance in early presidential contests, converged incrementally to winner-take-all systems whereby the party slate with a plurality of the popular vote in a state wins all of that state's electoral vote. Over time, however, this has left any number of relatively uncompetitive states feeling as if they are being ignored during a presidential campaign. Why, for example, campaign in California, despite its 55 votes in the Electoral College, when the outcome there between Democrats and Republicans seems pre-ordained? Why not instead campaign in New Hampshire, with its 4 electoral votes, since the outcome there can hang on who visited that state last? As a consequence, a number of states, including California, have joined a movement to change the system so as to either elect electors by district or allocate a state's Electoral College vote proportionally in accord with the popular vote. The proverbial fly in the ointment here, however, is that uncompetitive states such as California are largely controlled by the dominant party and a shift to proportional allocation can only benefit the opposition. Thus, heavily Democratic California has little desire to unilaterally defect from current arrangements unless it is joined by a populous state such as Texas that favors the opposition. The individual states, then, are trapped in a dilemma, with no one or two of them willing to defect from the status quo. As a consequence, a number of them (totaling 132 Electoral College votes as of 2012) have passed legislation mandating a shift to proportional allocation ONLY IF the Electoral College weight of the states signing onto this compact totals a majority, 270 Electoral votes. It remains to be seen whether the states here are caught on the horns of a constitutional dilemma wherein the U.S. Constitution prohibits them from "enter[ing] into any Agreement or Compact with another State . . ." (Article I, section 10), but it's clear they recognize that the incentives of the status quo have them playing a Prisoners' Dilemma.

Grading on a Curve: For an example of an admittedly contrived dilemma that perhaps strikes closer to home for the readers of this volume, consider again the example, briefly discussed in Chapter 1 to illustrate the difference between decision and game theoretic decision making, of the professor who grades on a curve and, aiming for an average grade of B, grades on the basis of a mid-term and final exam weighted equally. Now consider the unlikely event that every student in the class receives precisely the same grade of 85 out of 100 (a B) on the midterm exam. The dilemma confronting students now is whether to study for the final. If none study and instead engage in a variety of more pleasurable social activities, each will receive the same grade of B because the overall curve for the class will not change. So suppose the students, meeting before or after class, agree among themselves not to study. Now comes the night before the exam when each student must decide, in the privacy of their dormitory room, whether or not to abide by the agreement. Here, though, each can reason thus: "If I study and few others do, I'll get an A whereas if most others study I'll keep my B. And although I'll keep my B if I don't study and all others abide by the agreement, even if only a few of my classmates defect and study, my grade will drop to a C." The students, then, are in a dilemma wherein studying, despite whatever agreements they've reached beforehand, is the dominant strategy.

Gulags, Famines and Vote Fraud: Other instances of the Prisoners' Dilemma can be both more real and far more sinister. We have, for example, the following observation about Stalin's purges during his reign of terror in the 1930s:

> Stalin and [NKVD head] Yezhov wanted "the direct physical liquidation of the entire counter-revolution" . . . [and] the revised quotas were sent back down from Moscow to the regions . . . Here Stalin and Yezhov anticipated the execution of 79,950 Soviet citizens by shooting and the sentencing of 193,000 more to 8 to 10 years in the gulag . . . The killing and imprisonment quotas were officially called "limits" though everyone involved knew they were meant to be exceeded. . . . No NKVD officer wished to be seen as lacking élan when confronting "counter-revolution" . . . [and] by the end of 1938 the NKVD had executed some 386,798 Soviet citizens.
>
> (Timothy Snyder, *Bloodlands*, NY: Basic Books, 2010: 81)

Despite the evident dominant strategy confronting regional NKVD officials, it is not altogether clear whether this situation corresponds identically to a Prisoners' Dilemma because we can only speculate about the preferences and perceptions of those officials, though it is worth remembering that many were themselves subsequently liquidated after they washed the blood from their hands. An equivalent circumstance arose with Stalin's induced genocidal famine in Ukraine in the early 1930s whereby, depending on who does the counting, between 3 and 10 million Ukrainians died of starvation. There regional Communist party officials were initially induced to over-report grain harvests so as to make it appear that their efforts at collectivization had borne fruit. But those reports subsequently became quotas that they either could not meet or could meet only by starving the local population after confiscating whatever grain they held in reserve, including the seed grain used for future planting. Ultimately, of course, those regional officials, unable to meet the artificial quotas they themselves had inadvertently set, were liquidated or were to be found on a slow train to the gulag.

It would seem, however, that Russia has a propensity to repeatedly implement dilemmas of this sort. Most objective analysts agree that the 2004, 2008 and 2012 presidential elections there (as well as the intervening parliamentary contests) were marred by massive vote fraud wherein upwards of 20% of the "ballots" awarded to Putin or his 2008 puppet, Medvedev, were either fraudulently cast or simply did not exist. The evidence is thin, however, that the Kremlin itself engineered this fraud, and one might even argue that although Putin sought a mandate for himself or his party, the actual level of fraud exceeded what was deemed desirable by the Kremlin's apparatchiks. Shortly after his initial election to the presidency in 2000, though, Putin put in place a system of incentives that paralleled those confronting NKVD officials under Stalin. In seeking firmer control over the agencies of the state, Putin eliminated regionally elected governors and replaced them with Kremlin-appointed officials. Thus, no less than for the NKVD under Stalin, the newly appointed governors' fates rested entirely in Putin's hands. But how were those appointed officials to provide a measure of their adherence to "the party line"? The answer was: What better way to indicate one's subservience and loyalty to the powers that be than by reporting overwhelming electoral support for those powers? What regional official wants to be viewed as a weak supporter of Putin when others act differently? Thus, Putin's cadre of governors had an incentive to pursue any and all methods available to inflate Putin's vote so as to yield any number of election districts that, as in the old Soviet Union, reported 100% turnout rates with 100% of the vote going to Putin.

Catch 22: Prisoners' Dilemmas can also arise in humorous contexts, and it was surely this that Joseph Heller had in mind when he penned this conversation between Major Major and the World War II bomber pilot Yossarian in his cynical portrayal of the idiocy of war in his novel *Catch 22* (NY: Simon & Schuster, 1961: 125):

> [MAJOR MAJOR] Colonel Cathcart is our commanding officer and we must obey him. Why don't you fly four more missions and see what happens?
>
> [YOSSARIAN] I don't want to.
>
> [MAJOR MAJOR] Suppose we let you pick your missions and fly milk runs . . . That way you can fly the four missions and not run any risks.
>
> [YOSSARIAN] I don't want to fly milk runs. I don't want to be in the war anymore.
>
> [MAJOR MAJOR] Would you like to see our country lose?
>
> [YOSSARIAN] We won't lose. We've got more men, more money, and more material. There are ten million men in uniform who could replace me. Some people are getting killed and a lot more are making money and having fun. Let somebody else get killed.
>
> [MAJOR MAJOR] But suppose everybody on our side felt that way?
>
> [YOSSARIAN] Then I'd certainly be a damned fool to feel any other way.

There is, in fact, a more serious side to this Prisoners' Dilemma, which Kings, Queens, Prime Ministers, Presidents and Generals have long understood. While appeals to "God and country" might motivate ordinary individuals to enlist and wear a uniform, those appeals fray badly when bullets fly overhead and have as much impact on the motive to put oneself in harm's way as Major Major's words had on Yossarian. The solution, then, is to establish an *esprit de corps* whereby the ordinary soldier's loyalty is not necessarily tied to a particular regime or geography but rather to those immediately around him—to his fellow combatants. Heroic deeds have been committed not because a soldier seeks to defend some abstract ideology or to sustain a regime that he may or may not approve of but rather because "he doesn't want to let his buddies down." With one's motives personalized in this way, the Dilemma is transformed into something else.

Arms Races: More serious dilemmas include arms races, wherein both sides of an implicit conflict find it impossible to reduce their spending on arms for fear that the other side will not do so. Consider, then, the standoff between President Reagan and Premier Gorbachev at their 1986 Reykjavik summit where a new agreement on arms control was thwarted, ostensibly, by Reagan's insistence on pursuing the development of an anti-missile defense shield known at the time as Star Wars. Gorbachev's objection was evident, despite the opinion of his military that an effective shield would almost certainly be impossible to build and if built could be overcome by a new generation of ICBM warheads. Star Wars, Gorbachev argued and doubtlessly believed, would only exacerbate the arms race between the US and USSR, which was a race the Soviet Union could ill afford and could pursue only by undermining Gorbachev's plans for reform of a dismally performing Soviet economy. Thus, while both sides to the dispute saw the benefit of curtailing the proliferation of atomic weaponry, that proliferation seemed not only unavoidable but a problem magnified by a system, however imperfect, designed to defend against such weapons.

For an admittedly contrived numeric example of this dilemma, suppose each of two national leaders can either increase spending on armaments or decrease spending. If both choose the same strategy, the security of both is unchanged and worth, say, 100. Suppose further that the cost of any increase in armaments is 75. However, if one country disarms unilaterally, suppose it is overcome by its adversary, which gains 275, which is an outcome that is worth −100 to the vanquished leader. The strategic form of this situation is portrayed in Figure 5.6, which reveals that both persons have increasing arms spending as their dominant choice.

The imperatives of the Prisoners' Dilemma in this case suggest that if asked, "Are two countries more likely to reach an arms agreement (1) if they more earnestly and honestly approach arms negotiations, (2) if negotiators learn to trust what the other side says, or (3) if domestic political-economic considerations render the costs of pursuing arms buildups unacceptable?" we must answer that it is only the last possibility that can lead to a resolution because it is the one

	Increase arms	Decrease arms
Increase arms	25, 25	200, −100
Decrease arms	−100, 200	100, 100

Figure 5.6 A simple arms race

that changes the situation's payoffs. What needs to be appreciated here is that we cannot resolve such dilemmas via mere communication and agreements to coordinate. Regardless of what agreements are reached beforehand, once the participants in the dilemma are left alone, the choice of a dominant strategy seems inevitable.

Low Price Guarantees and Market Collusion: For an admittedly fanciful example, although perhaps one that suggests that not all government regulation of competitive markets is in our interest, imagine that a country's buggy-whip industry is in trouble: Given a declining interest in buggy whips owing to the new-fangled invention of the automobile, there are far too many producers of buggy whips (exactly two), so the price for such items has declined to $6 each, while inflation has driven costs of manufacturing to $4/whip. At a $6 price tag, 400 whips/month/firm can be sold. Both producers deem this profit margin unacceptable, so they've met to form a cartel to fix prices at $8/whip, in which case 250 whips/firm will be sold. However, if either firm cheats on this agreement and charges $6, it will capture the entire market of 800 whips. Abiding by the time-honored political maxim, "I seen me chances and I tooks them," suppose you are a legislator with sufficient power to force the government to regulate the price of whips "in the public interest"—no one, of course, wants "inferior" quality whips flooding the market, especially those made in some distant Asian country. However, rather than simply act to "protect jobs," you ask yourself what the upper limit might be on the campaign contributions you can demand of the two firms in the whip industry. To answer this question, Figure 5.7 portrays the strategic form confronting the firms. If both firms charge a market price of $6, each earns $800 (400 whips sold at a profit of $2/whip) whereas if both charge $8, each earns $1,000 (250 whips sold at a profit of $4/whip). A defecting firm earns $1,600 and the other earns nothing. For an enterprising legislator, then, there is an opportunity to "earn" up to $200 from each firm as gratitude for government enforcement of a whip cartel—the difference between a competitive $800 profit per firm versus a government-sanctioned cartel profit of $1,000.

This simple example, which by now may appear to the reader to be a mere repetition of the same game with different payoffs, reveals something about the marketing strategies we sometimes observe among competing retailers of consumer goods. Suppose the two firms would prefer not to be a party to our legislator's attempt at extortion and that both instead offer the following deal to their customers: "If you find a comparable whip cheaper elsewhere, we will match the price and include in it a 10% discount." Actually, this scheme works best for retailers selling products not of their manufacture but where, instead,

	Charge $6.00	Charge $8.00
Charge $6.00	$800, $800	$1,600, $0
Charge $8.00	$0, $1,600	$1,000, $1,000

Figure 5.7 Cartelizing the buggy whip industry

all retailers in the market sell similar or identical products. Since explicit collusion on price (price fixing) is generally illegal (since if it were legal, how could legislators extort?), the "low price guarantee" serves the same purpose. Specifically, it removes any incentive retailers might have to defect from any implicit agreement. If anyone defects, competitors offering such a guarantee will, in effect via that guarantee, automatically lower their price to match that of the defector (and with the 10% premium, even undercut the price charged by the defecting firm). Thus, low price guarantees are a way for retailers to police the pricing agreements they might make among themselves without the need for government sanction or intervention. They have the advantage, moreover, of seemingly offering the customer a guarantee that they are not being ripped off while at the same time avoiding the need to bribe a legislature.

Foot Binding: Although its precise origins are unknown and thought to have begun during the Soong Dynasty (960–1279), the practice of foot binding was widespread throughout China by the 18th and 19th centuries. This was a practice whereby a young woman's feet were deliberately deformed by daily (and quite painful) binding so as to leave her unable to walk normally but which was deemed essential to render her eligible to marry into a rich family. Indeed, regardless of a family's wishes, if they hoped to marry a daughter to anyone of position, it was essential that her feet be deformed to correspond with convention. However, in the latter quarter of the 19th century, a number of Natural Foot societies were established, often at the instigation of foreign missionaries, in the attempt to rid China of this cruel practice. It would appear, though, that society was trapped in a Prisoners' Dilemma wherein however much individual parents might have wished to spare a daughter years of pain and a lifetime of virtual imprisonment owing to her inability to walk normally, it was difficult or impossible to defect from a centuries-old standard of physical beauty and consign a daughter to a marriage-less life and even servitude. Serious inroads into the practice in urban centers weren't made until the government of the Republic of China in 1912 officially banned the practice; and even still, as late as the 1940s, Mao and the Communist Party reported finding the practice still in effect in rural China.

Canoe Races, Overthrowing the King and Gymnastics: Imagine the following scenario: Two 2-man canoes are in a race. However, this race only determines which two rowers will compete later against each other in a footrace for the ultimate prize. It might seem that each pair of rowers has an incentive to submerge whatever competitive issues they might have between them in order to win the first race so that they can compete ultimately for the prize. Notice, however, that the rowers in both canoes each have some incentive to slack off during the canoe race in order to conserve their energy for the footrace. Indeed, we can imagine this incentive growing stronger if, instead of 2-man canoes, we substitute 8-man rowing shells so that slacking off becomes more difficult to detect. Failing to exert maximum effort runs the risk of losing the rowing contest and thus foregoing the opportunity to engage in the footrace, but if everyone else in your boat slacks off, you'd be a fool to not do the same since otherwise you are sure to lose the footrace even if your boat wins the rowing contest.

The rowers, then, are in a Prisoners' Dilemma, albeit one that is somewhat muted by the desire to win the rowing contest and not be detected as a slacker. Our example, though, has some real world parallels. In 208 BC imperial China, during the reign of the son of the first emperor of China, the descendants of the former rulers of the kingdoms of Yan, Zhao, Qi and Wei sought to overthrow the Qin Dynasty and restore their states. However, much like our rowers, the heads of the various armies had an incentive to slack off in their attack on the capital of Qin in order to conserve their strength for the ultimate contest among themselves as to who would assume the throne to become the new emperor. In this case, a solution to the Dilemma was found by agreeing that whoever reached the Qin capital of Xianyang first would take Qin's place.

One needn't revert to ancient history for examples of similar dilemmas. Consider, for instance, the competition among members of the same party in a primary election for that party's nomination to compete for the ultimate prize of, say, president. The problem here is that during the primary campaign, each contestant has an incentive to undercut their competitors by, perhaps, emphasizing those opponents' character flaws. But if all primary election competitors do this, they damage the party's ultimate nominee in the general election. Indeed, how many times do we see opponents in a general election using the words of a competitor's fellow party members—words that arose during the fight for each party's nomination? Thinking back to our rowers, now, we note that three Olympic sports conduct both individual and team contests: gymnastics, swimming and track and field. In gymnastics medals are awarded for a team of gymnasts as well as for individual performances, while in swimming and track and field, medals are awarded for both individual competitions and relays. What we find interesting here, though, is that the gymnastics team competition precedes the individual ones whereas in track and field and swimming, the relays come later. One might suspect, then, that the same issues that arose for our rowers might arise here: If a swimmer or runner were to race in a relay first, he or she might slack off a bit so as to conserve energy for the individual events. This possibility is forestalled, however, by holding those individual events first. Why then is gymnastics different? The answer, we suspect, is that the incentives would be fundamentally the same were it not for one modification of the rules. Specifically, only the top two scorers from each country's team are allowed to compete for the individual all-around gymnast medals. Thus, slacking off opens the door to being precluded from the next round of competition (for elaboration, see Emerson M.S. Niou and Guofu Tan, "External Threat and Collective Action," *Economic Inquiry*, 43(3): 519–530, 2005).

Markets, Public Goods and Why We Have Government: Suppose two people are contemplating installing pollution control devices on their respective cars, where each device costs C and provides a benefit of B in terms of a device's contribution to clean air. However, suppose the benefit accrues equally to both persons since they both breathe the same air. If benefits are additive (i.e., if the benefit of two installed devices is $2B$ to both persons) then the game between them looks as follows:

	Install	Don't Install
Install	$2B - C, 2B - C$	$B - C, B$
Don't install	$B, B - C$	0, 0

Figure 5.8 A pollution control dilemma

Suppose finally that $2B > C > B$. That is, the cost of each device exceeds the benefit accruing to a single individual so that neither person has an incentive to install such a device on his or her own. Such devices are beneficial only if both persons install them. Thus, with $2B > C > B$, our two motorists are caught in a Prisoners' Dilemma so that even if they agree beforehand to install such devices, each has an incentive to cheat to realize the benefit B without incurring any cost. The solution to this problem is, of course, well known. It is in the interest of both players to find or establish a third player who will oversee the situation, most likely by implementing a system of automobile inspections and fining those who fail to have the appropriate equipment on their cars. The idea here is that although this third person or entity is given the authority to coerce through fines, no fines will be collected if the desired outcome is achieved, provided, of course, that the newly created entity does not exceed its authority.

As simple as this example is, it reveals one of the fundamental problems associated with unregulated markets and with collective action generally. The characteristic of the commodity here of a pollution control device is that it exhibits an *externality*, in this case a positive one. If one person "consumes" cleaner air by purchasing and installing such a device, he or she automatically or incidentally provides the benefit of clean air to everyone else. However, in deciding whether or not to make that purchase, we can suppose that a person will base their decision on the benefits and costs they alone experience. The cleaner air in this case is what economists call a *public good*—a good that, when consumed by one person can be consumed by others at zero additional cost. Public goods stand in contrast to *private goods*, which are ostensibly those that, when consumed by one person, have no impact whatsoever on the welfare of others.

In practice, the debate as to whether a good is public or private can sometimes be a contentious one since often that dispute concerns the proper role of government. Consider for example a law that requires motorcyclists to wear helmets. When such laws are imposed, it is not uncommon for many if not most motorcyclists to object with the argument that their safety is their own private concern and if they wish to take the chance of severe bodily injury they should be free to do so. The counter argument is that if a subset of motorcyclists fail to wear helmets, they necessarily raise the insurance rates for everyone, including those who do not ride motorcycles. Here, then, some see a Prisoners' Dilemma warranting government intrusion while others see no dilemma at all but instead another instance of government trying to over-regulate our lives.

Regardless of which side of the debate one takes here, the existence of such debates reveals that public goods and their associated externalities come in a variety of forms. For the converse of a pollution control device, suppose you decide that an inexpensive way to heat your home is by burning rubber tires in a fireplace. Tires, being a petroleum product, most likely burn quite nicely and we would guess, do an effective job of heating one's home. However, the smoke and soot produced are not likely to be appreciated by one's neighbors. Tires, in this case, or at least those in one's fireplace, are a public bad because they engender a negative externality—dirty air that your neighbors must then consume. Thus, just as we might choose to regulate pollution control devices on cars by authorizing the state to inspect and fine, we are equally likely to authorize that same agent to regulate against the burning of tires for heating homes.

The variety and form of public goods and externalities is even greater than our examples suggest. Consider a public park. Here we have a potential public good that is nevertheless subject to exclusion: We can build a fence around it and admit only those who pay a fee. However, once we build a park of even modest size for one person, we can allow others to use it at zero marginal cost. The park, though, is not a perfect public good since it is subject to exhaustion—in this case, to overcrowding. Hence, the need perhaps for that fence and entrance fee. Public parks, though, do illustrate a pervasive problem that warrants collective action of some sort—specifically, they remind us of a class of goods that yield a common resource dilemma. Areas that in one era are productive fishing grounds can become depleted from over-fishing, water resources that benefit thousands if not millions of acres can become depleted by over-irrigation, forests that prevent soil erosion can become depleted to the extent that people view wood as a free good, and fertile grasslands can become dust bowls from overgrazing. Each of these situations corresponds to what social scientists refer to as the *Tragedy of the Commons* whereby some shared resource—a public good—is depleted by the actions of individuals acting rationally to produce an otherwise "irrational" (collectively undesirable) outcome. And although we do not wish to get into the debate as to its causes, a significant portion of the scientific community argues that today the most threatening instance of this "tragedy" is global warming.

Some public goods can be transformed on the basis of technology. Consider lighthouses, where the warning a traditionally designed lighthouse provides is a public good—the light it shines can be "consumed" by any passing ship. Leaving provision of lighthouses to the private sector, then, is likely to yield an inefficient outcome since whoever builds it cannot charge for its warning—once the warning is made available to one, it is available to all (and is not subject to exhaustion from consumption). We should not be surprised, then, to find the provision of lighthouses a task commonly assigned to governments. But warnings need not take the form of a bright light—they can also be radio signals, the meaning of which can be learned only if one has subscribed to a service and given a password. It may not be warnings that are consumed on various websites, but subscribing to a website's service (which, presumably, can be consumed at zero marginal cost for any number of additional subscribers) for a fee is hardly uncommon. Website services that are free, in contrast, illustrate some of the private-sector inventiveness that often occurs when dealing with public goods. Consider the production of TV programs. Cable and satellite services may be required now to watch the vast majority of available programs, but major networks continue to function in most countries, often without government ownership or subsidy despite the fact that their transmissions are virtually perfect examples of public goods—once the signal is sent into the air, that signal can be consumed for free by anyone in range of its transmission. The innovation here (if one wants to call it that) is a simple one—advertising; charging companies a fee for advertising their product along with whatever is intended to entertain or inform people. This, of course, is the same solution reached by countless websites as well as newspapers sold on the street at below the cost of production.

Sometimes externalities take a less than physical form. Consider the commercial airline industry, which is perhaps one of the most heavily regulated industries today, at least with respect to the matter of safety. There is, of course, little doubt that the flying public favors that regulation since who, after all, wants to fly in a plane piloted by an exhausted, untrained or drunk pilot sitting in front of an instrument panel that "doesn't quite work"? But it is also the case that the industry itself prefers to be regulated in this way even though it greatly increases operating costs. Airlines know that absent regulation, they would find themselves in a Prisoners' Dilemma whereby each, in the competition for customers, would have an incentive to reduce operating costs by slacking off on aircraft maintenance. But if just one airline cheats and its plane crashes with hundreds of passengers killed, there is a negative externality that impacts all air carriers in the form of an increased fear of flying on the part of the general public. Airlines, apparently, have a hard enough time being profitable, and if planes began dropping out of the skies with some regularity, no doubt passengers and profits would vanish. Air safety, then, is a public good for nearly everyone involved and every reputable airline has an interest in having the government regulate it and all other airlines in order to ensure that that good is in fact provided.

5.3 Cooperation and the Problem of Collective Action

In one respect the actions of the passengers of United Airlines flight 93 on September 11, 2001, were remarkable while from another perspective they were not. In thwarting the attempt of the terrorists onboard to divert the flight and fly the plane into either the U.S. Capital or the White House, the actions of the passengers were clearly heroic because they did so while dooming themselves. On the other hand, a cynic might argue that their actions were unremarkable because they knew by then that they were doomed regardless of what they did—information about the terrorist attacks on New York City's World Trade Center and the Pentagon became known when the terrorists allowed the passengers to call home. Nevertheless, attacking the terrorists as they did remains an uncommon example of coordinated collective action. How many times do we read of people in a bank, when confronted by an armed robber, meekly lying down as ordered while the robber scoops up whatever cash he can carry despite the fact that if all patrons rushed the robber they could almost certainly thwart the crime? A similar dilemma becomes evident when we ask the question, "How do dictators survive when the vast majority of their populations prefer to do away with them?" What, for instance, is the mechanism wherein a loathed dictator such as Nicolae Ceausescu rules Romania with an iron fist for his own private benefit, building absurd palaces his country can ill afford, and where in a seeming instant after years of mis-rule, the population suddenly turns to overthrow his regime? An equivalent situation arose, perhaps, in Nazi Germany. As British and American troops crossed the Rhine in 1945, the nearly uniform plea of German citizens in one village after another was to disclaim membership in the Nazi party and proclaim opposition to Hitler's rule, asserting even that they viewed those troops as liberators. The troops themselves were hardly prepared to treat Germans the same as Belgians or the Dutch, and for the most part discounted such assertions, especially after they'd seen the death and labor camps that permeated the countryside. However, we ought to keep in mind that Hitler never, in any election, garnered majority support of the electorate. Thus, while it's possible to assume that a majority of the population had been emotionally caught up in Germany's early military triumphs, as casualties mounted and the tide of war turned that enthusiasm surely began to dissipate so that preferences after 1942 reverted to what they had been earlier. Nevertheless, the common retort of Allied troops to individual Germans was, "Then why didn't you stop him?" However, with SS units still active and fanatical Hitler Youth mercilessly killing anyone who exhibited less than unswerving loyalty to Hitler, it is difficult to see how any individual acting unilaterally could have much of an incentive to oppose the regime. One has to wonder, then, whether a considerably larger percentage of Germans professing to think of themselves as having been liberated by Allied armies were in fact sincere in their pleas.

Rulers, in sustaining their regimes, commonly take advantage of the difficulties of those they rule to coordinate. For example, in discussing British imperial rule, Stevenson (*The First World War as Political Tragedy*, NY: Basic Books, 2004: 101) notes that

> In a population of 300 million [in 1914] British India contained in peacetime some 1,200 white officials in the Indian Civil Service, 700 white police officers, and 77,000 British soldiers along side 173,000 Indian troops. Similarly, a few hundred British administrators and 4–5,000 white soldiers, with 13,000 indigenous ones, governed 12.5 million Egyptians . . . British leaders knew that their eastern empire . . . "depends on prestige and bluff."

Prestige and bluff work, of course, only if the masses cannot coordinate to challenge the status quo.

There is, perhaps, no more dramatic example of the failure of the members of a culture to coordinate with its very survival threatened than that of the Jews of Europe in the 1930s and early '40s. As Hitler and his henchmen incrementally moved on Jews within Germany and thereafter in subjugated countries, there may have been occasional points of resistance (not the least bit aided by the powers that be in London or Washington), but for the most part disbelief was followed by resignation and acquiescence. And although a general simultaneous uprising by all Jews in Europe might still not have been much of a match for the Wehrmacht or the Einsatzgruppen's death brigades, such an uprising would surely have made Hitler's policies more apparent to the rest of the world. However, to expect such an uprising is to ask people to act contrary to individual incentives even when those incentives or motives allow for altruism. Individual resistance holds no promise of positive rewards if one has little confidence that one's actions will encourage most if not all others to follow or will even be seen by anyone other than those who are positioned to administer the ultimate punishment. And when communication and coordination are rendered difficult or impossible, the dominant choice is to merely attempt to survive in the hopes that someone or something will change the "game."

Each of the preceding examples and scenarios illustrates an *n*-person Prisoners' Dilemma wherein it's in everyone's interest to engage in collective or coordinated action but where no one wants to be the first to move for fear that others will not follow and that they will be the uniquely punished individual (or at least the first to be punished). But coordinated collective action poses other problems, illustrated in part by our contrived canoeing example. In the case of the 2-man canoe, slacking off might be difficult to do because one's actions are likely to be noticeable. In an 8-man shell, on the other hand, the subtle slacking of a single individual might not be so easily detected and might not even greatly threaten the boat's likelihood of winning the race. An equivalent large-*n* problem arises in several other examples of the Prisoners' Dilemma considered thus far. Consider automobile pollution control devices. If the situation concerned only two people, then we can imagine ways in which they might police themselves to maintain any cooperative agreements. Perhaps they could simply go with each other to observe the installation of the devices on their cars. But if *n* equals millions, such a solution is impractical, with the problems of enforcement exacerbated by the following reasoning: "My lone defection is likely to go unnoticed and the damage to the environment caused by my car will, in the greater scheme of things, do little to degrade the overall quality of the air." Similarly, if we consider again our example of two buggy whip manufacturers, we can again imagine a small handful of companies finding ways to cooperate and to safeguard against defections. But suppose there are hundreds if not thousands of manufacturers. In that case, sustaining a collusive agreement might become exceedingly difficult. Indeed, consider the licensing of taxicabs. From one perspective, such licensing makes sense—at least to the extent that it provides a guarantee that the driver has some minimal capabilities when it comes to driving and is using a vehicle that meets a few standards of safety. But what then justifies governments acting to regulate prices as well? Why preclude the opportunity of taxi drivers to offer their customers a bargain—a posted or even ad hoc discounted price?

The answer here lies in the fact that defection is too easy and thus taxi drivers or their companies find it difficult to collude on price without the intervention of an outside agent—the state. Licensing drivers, moreover, has the additional advantage of precluding entry so that "defectors" cannot come from outside the industry. We should not be surprised, then, to see governments regulating the prices of things like taxi fares and acting as an agent for the firms themselves, but not regulating, say, the price automobile manufacturers charge, where the entry of "exogenous defectors" is more difficult (although we do see such implicit government sanctioned and enforced collusion when those defectors are foreign firms. Here the government can disguise the assistance it gives to domestic manufacturers in the form of tariffs and import quotas with the argument that they are protecting domestic jobs).

Mancur Olson, in his seminal volume *The Logic of Collective Action* (Cambridge, MA: Harvard University Press, 1965), identified the problems associated with public goods, externalities and Prisoners' Dilemmas when the number of participants grows large. But what he also saw was the many ways in which collectivities invent mechanisms for circumventing those problems. For example, consider Lenin's organization of the Communist Party in Bolshevik Russia. The specific problem with which he had to deal was that individual interests had not yet been supplanted by the sought-after collective interest. Thus, individuals still had an incentive to shirk from their class responsibilities with respect to such things as reporting counter revolutionaries and acting in the workplace in society's interests rather than their own. Lenin's solution was to construct a party with a well defined hierarchical structure wherein small cadres of people would act as a unit and report to some higher authority, which in turn consisted of a cadre of individuals that would itself report to some yet higher authority, and so on and so forth up the chain of command. In this way, rather than require that collective action be organized on a mass basis wherein individuals have an incentive to shirk under the assumption that their shirking would go unnoticed, he recast things as a set of smaller games where monitoring would work more effectively because individual defections from more limited definitions of group goals would be more noticeable. Lenin's solution, then, was not much different in principle from that employed by armies wherein soldiers become attached to those immediately around them as opposed to fighting for grand abstract principles.

In contrast, Olson observed that entities such as labor unions and voluntary associations such as the AARP (the American Association of Retired Persons) did not have Lenin's or the military's solution available to them. The dilemma for a union, for instance, is that unless employees of some firm are required by law to belong to the union, individual workers have an incentive to view whatever pay raises and benefits the union negotiates with management as a public good and to refrain from joining the union and paying dues. A similar problem confronts a lobbying entity such as the AARP. Why pay for a membership if the benefits the Association provides through national law accrue to all retired persons regardless of whether they are members or not? In this case, both entities—unions and voluntary associations—find a solution in terms of privatizing some of the benefits of membership. Thus, it is quite common to find such organizations offering their members *and only their members* such things as life or health insurance at a reduced group cost. Of course, it is not unknown historically to see entities such as labor unions finding another way to privatize the benefits of membership via the proverbial brick through the window of those who refuse to join the union or a strike or to ostracize the defector at the workplace.

Before we leave the subject of the problems associated with collective action, let us consider an example that illustrates formally the problems of collective action as it relates to large groups. Consider an extreme form of a public good which, when provided by one person, is fully provided and benefits everyone equally. This situation might correspond to the case of the crowd standing before the loathed Ceausescu of Romania with everyone aware that if any one person rises up to yell a denunciation of him, the rest would almost certainly follow. In its 2-person variant, if we ignore the preferences of elites like Ceausescu and his entourage as well as any complications arising from the presence or non-presence of a Soviet army, such a situation might look as presented in Figure 5.9.

Notice that this game is much like our pollution control example in Figure 5.8 except that here one person's action is sufficient to provide the good and benefits are not additive. Thus, as long as $B > C$, this game is not a Prisoners' Dilemma but instead has two pure strategy Nash equilibria in which one person provides the good and the other free rides. It also has the mixed strategy equilibrium $((1 - C/B, C/B), (1 - C/B, C/B))$. Now, however, suppose there are n players and that each, seeing the coordination problem that confronts them, abides by a mixed strategy $(p, 1 - p)$. If that strategy is an equilibrium, each person's pure strategies must yield the same expected value, since otherwise, as we already know, the person would have an incentive to abandon the mixed strategy and adopt the pure strategy yielding the greater expected return. For the game at hand, it is clear that E(provide) = $B - C$ since that payoff prevails in every cell regardless of the decisions of the $n - 1$ other players. As for E(don't provide), the person in question receives B unless all other $n - 1$ persons fail to provide. And the probability that all others fail to provide when abiding by their mixed strategies equals $(1 - p)^{n-1}$. Thus, the probability that at least one person provides equals $1 - (1 - p)^{n-1}$, in which case

$$\text{E(don't provide)} = B[1 - (1 - p)^{n-1}] + 0(1 - p)^{n-1}$$

or simply $B[1 - (1 - p)^{n-1}]$. Setting E(provide) = E(don't provide) yields, after some algebraic manipulation

$$p = 1 - (C/B)^{1/(n-1)}$$

	Provide	Don't provide
Provide	$B - C, B - C$	$B - C, B$
Don't provide	$B, B - C$	0, 0

Figure 5.9 A public goods game

Thus, the larger is n, the less likely will any one person unilaterally act to provide the benefit, whatever it might be. What interests us, however, is the likelihood that no one chooses to provide as n increases. Since the likelihood that any one person doesn't provide is $1 - p = (C/B)^{1/(n-1)}$, the likelihood that all n persons fail to do so is this probability raised to the nth power, or

$$(C/B)^{n/(n-1)}$$

Since $n/(n - 1)$ decreases from 2 to 1 as n increases from 2 to ∞ and $C/B < 1$, this probability necessarily increases as n increases (for example, if $C/B = ½$, then for $n = 2$, this probability is ¼ whereas as n increases, the probability that everyone chooses "don't provide" converges to ½).

There are other ways to model the decreased likelihood of some public good being provided as the number of relevant persons increases. One possibility is to assume, as suggested earlier, that the likelihood of detecting a defection decreases as n increases. Alternatively, we can suppose that the public good shows diminishing marginal returns as more people contribute so that if a great many contribute, no individual's contribution will yield a private benefit that exceeds its private cost. For a numerical example, suppose the public benefit of the public good under consideration is $(x - 1)/x$ when x people contribute. Thus, at $x = 2$, the benefit is ½ and increases at a decreasing rate to 1 as x increases. Suppose the individual cost to contributing is C. Thus, with x people contributing, any person making a contribution enjoys a net benefit of $(x - 1)/x - C$, whereas if that person shirks, his benefit becomes $(x - 2)/(x - 1)$ but at no cost. Some simple algebra, then, shows that a person should shirk if $C > 1/x(x - 1)$. On the other hand, a person would nevertheless enjoy a positive benefit from contributing if $(x - 1)/x - C > 0$. A person is thereby caught in a Prisoners' Dilemma if $(x - 1)/x > C > 1/x(x - 1)$ - if he prefers to shirk but still experiences a net benefit when contributing. Taking C out of this dual inequality reveals that the dilemma exists as long as $x > 2$. Thus, regardless of the number of people benefitting from provision of this good, no more than two people would contribute and the rest would shirk. There is, of course, nothing magical about the number 2 here as it is merely the product of the functional form assumed for the benefit of the good in question as a function of the number of people contributing to that good. Nevertheless, what we see here is how a public good with diminishing marginal value as its supply increases can yield a situation in which a significant share of the relevant population shirks and attempts to pass off responsibility for some minimal provision of the good to some small subset of people.

5.4 Escaping the Dilemma: Repetition and Reputation

Bricks through windows, fines, government intervention, low cost group life insurance policies and monitoring systems that parallel Lenin's Bolshevik revolution are but a few of the ways people contend with Prisoner-type Dilemmas.

Correspondingly, understanding how people overcome dilemmas is a major component of political theory and leads immediately to speculation about the role of leadership, about the incentives for politicians themselves to entrepreneur collective action, and about the potential role of preexisting institutions. However, in contemplating such things we should consider the fact that no Prisoners' Dilemma is isolated from all other things in the world. For example, abiding by agreements in legislative vote trading scenarios may confront legislators with a dilemma, but it certainly is not a dilemma in which legislators are likely to participate only once nor are they likely to avoid interacting with their fellow legislators subsequently in other contexts. Thus, among other things, we should consider the possibility that the expectation of repetition of the dilemma alters people's incentives to choose one strategy over another. Similarly, tariff negotiations also yield a dilemma wherein each country party to any agreement most likely would prefer, *ceteris paribus*, to defect from any agreement, but as with vote trading, countries must deal with each other on a continual basis, so a defection in one instance has implications for subsequent action and negotiations. Arms races are analogous to tariff negotiations, where here the incentives to defect depend on the ability of the participants to detect a defection, to respond to a defection by the other side, and the likelihood that both countries might prefer to negotiate further arms limitations in the future.

> **The Legal Case of John P. Calvo**: John Calvo was the publisher of the *Spartanberg Express* of Spartenberg Georgia, but in July of 1861 he set sail out of Charleston harbor to act as a privateer, intercepting Union ships so as to secure their cargo for the Confederacy in America's Civil War. Unfortunately for Calvo, on August of that year he was caught by the Union frigate *Wabash* while sailing the captured *Mary Alice* to North Carolina. Calvo was immediately sent to New York, imprisoned and charged with the capital crime of piracy, the penalty for which was execution. Seeking to avoid this rather draconian punishment, Calvo wrote to President Lincoln, arguing that he was not a pirate but a privateer, which was a distinction with a difference since privateers were treated as prisoners of war and thus their treatment regulated, at least in part, by international rules of war. Calvo's argument, though, was not based on mere legal principle—something which nations can too easily twist to serve immediate self-interest. Instead, he pointed out to Lincoln that if the Union treated privateers as pirates rather than as soldiers, escalating recriminations would follow wherein the Confederacy would do the same with members of the Union's Navy that it held or would in the future hold as prisoners.

The dilemma confronting Lincoln here would not arise if Calvo were an isolated case. Clearly Lincoln preferred to dissuade the South from engaging in privateering and implementation of a harsh penalty in the Calvo case would contribute to that. On the other hand, as America's Civil War began to unfold,

Calvo quite correctly noted that the Union and the Confederacy were playing a repeated game and while it might be in the interest of each "player" to treat a prisoner as an ordinary criminal if the game were played but once, they were about to play an indeterminate number of times. And repetition allows for threats, promises, rewards and sanctions. A nearly identical situation prevailed in World War II wherein Germany's Luftwaffe established its own prisons for captured Allied airmen shot down over Western Europe and where those inmates in general received better treatment than other Allied prisoners (not to mention those from the Soviet Union). The explanation was that the Allies held any number of captured Luftwaffe airmen, and the expectation was that reciprocity would prevail. The reader should not be surprised, then, to learn that Calvo in May of 1862 was discharged as part of a prisoner exchange between North and South.

The Ogallala Aquifer: Much of the water that midwest America uses for irrigation, including the states of Texas, Kansas, Nebraska, Wyoming, Oklahoma, Colorado and New Mexico, comes from the Ogallala Aquifer—one of the world's great sources of fresh water. However, beginning in the 1950s when the aquifer was first tapped, the demands upon it were such that the level of the water began dropping at the rate of a foot per year, which threatened its ability to meet irrigation needs. Thus, beginning in the 1980s both state and federal governments began exploring ways to preserve the water table. One method was to directly regulate the amount of water individual farmers could use, but American farmers are a rather cantankerous group who do not take kindly to government inspectors tromping through their fields monitoring things or insisting that regulatory gauges be installed on their irrigation equipment. Another, potentially more effective, means was to institute new farming methods—for farmers to abandon traditional methods of deep plowing in favor of more water conserving technologies. The problem here was that moving to those technologies required new and not inexpensive equipment. Thus, while farmers could agree that the aquifer needed to be preserved and that they would all be better off if everyone shifted to the new technologies, they found themselves in a classic Prisoners' Dilemma—to purchase the requisite equipment and thereby increase individual production costs, they would only leave themselves vulnerable to the lower costs of farmers who defected from an agreement to abandon the old technologies. A variety of governmental approaches were taken to encourage a shift in technology, but one in particular seemed to bear fruit even before it was supposed to. Among the things tried, doubtlessly with a measure of idealism that often accompanies a bureaucracy's actions but without an appreciation of the fact that farmers were no less conscious of the need to conserve than were bureaucrats at their desks, was teaching elementary schoolchildren about conservation. The idea here was that when those children became adults and farmers they would be more environmentally conscious than their parents and grandparents. The impact of these programs, however, seemed

> immediate. Indeed, as a 1993 *National Geographic* (183, March: 80–109) article on the subject noted, nothing was as effective as a grandchild sitting on his grandfather's lap asking if the water would be there for him to farm when he grew up: "a worried grandchild is worth five visits from a regulatory official" (p. 103).

John Calvo and midwest American farmers might seem to have little in common, and we do not assume that the reader yet sees the relevance of elementary school education programs to the subject of Prisoners' Dilemmas. Nevertheless, among other things, both examples suggest that it is an imprudent research strategy to treat many of our examples of Prisoners' Dilemmas as we have—as strategic situations isolated from the temporal flow of events—and that any complete analysis should examine the context of that game and the larger extensive form in which it is imbedded. And both examples, in fact, illustrate a critically important potential solution to the Dilemma.

Of course, a Prisoners' Dilemma, like any game, can be part of nearly any larger scenario. But our earlier discussion of vote trading is a useful place to begin a deeper study of such things. Specifically, we note that unless all the bills upon which legislators might trade votes are placed in a single bill that is voted on once, up or down, legislation is generally considered sequentially. Thus, individual legislators confront the problem of whether or not to abide by an agreement or to renege if the legislation with which they are most concerned has already been disposed of. Alternatively, if that legislation is to be considered late in a legislative session, one must then be concerned that others will keep their side of the bargain after the bills that most concern them have been disposed of. The problem this discussion points to is but another one of the arguments against term limits. If legislators are likely to be reelected—or at least if electoral rules do not prohibit them from seeking reelection—a defection in one legislative session can be punished in subsequent sessions. That is, barring term limits, legislators are not playing a one shot game as portrayed in Figure 5.5, but rather are playing that game a potentially indeterminate number of times. Thus, one scenario in which we might embed a Prisoners' Dilemma has the Dilemma being replayed by the same participants. However, before we charge ahead and try to write down an extensive or a strategic form for a repeated game, we must decide whether it is repeated a finite and known number of times or whether it is repeated an unknown or infinite number of times. This choice is critical, because if a game is repeated a finite number of times so that the terminal nodes of the corresponding extensive form would be apparent and well defined, we can analyze that form by backward reduction. On the other hand, if the Dilemma is repeated an indefinite or even infinite number of times, there is no point at which we can begin the backward reduction process. Indeed, if only one branch leads to a repetition, then backward reduction cannot be attempted and we must use other analytic tools to analyze the situation.

Interest in the repeated Prisoners' Dilemma does not arise merely because it reveals interesting analytic problems. Rather, we can attribute that interest to the fact that people, whether observed in the experimental laboratory or in everyday activity, seem able to use the fact of repetition to "solve" this Dilemma for themselves and avoid mutually disadvantageous outcomes. Legislators do not wholly succumb to granting every constituency a benefit so as to wholly bankrupt the state (although the evidence is mounting that this is not always the case in places like California); the trading relations of states do not necessarily evolve into an anarchy in which each state imposes the highest tariff barriers sought by the interests within them; people do contribute to charities even though they might each be better off by defecting to a pattern of non-contribution; and subjects in the laboratory when playing the Prisoners' Dilemma frequently choose to cooperate.

This last observation—from the experimental laboratory—suggests that repetition of the Dilemma is an important variable in understanding the patterns of cooperation that do emerge. This observation gains theoretical sustenance, in turn, from the idea that if a person defects to a dominant strategy in an early play of the Dilemma, others can sanction that person in successive plays by refusing to cooperate. If, for example, a legislator defects from a vote-trading agreement, then legislators can refuse to trade with that person subsequently, and if a firm defects from the cartel price, then others can threaten to defect as well, in which case the gains from unilateral defection are lost.

A rather graphic example of the effects of repeated play is provided by a particular event during World War I called "The Christmas Truce of 1914." Despite the mayhem and horrors of trench warfare in that conflict, where British and German troops faced each other across the desolate fields of Flanders, a sign was raised the day before Christmas from a German trench that read "You no fight, we no fight." On the British side, soldiers from the Queens Westminster Rifles popped out of their trenches and then quickly retreated, but no shots were fired. Similar events occurred along the line during the afternoon, whereupon thousands of troops from both sides joined in the desolate no-mans-land between them to spend Christmas Eve and Day playing soccer, sharing food, cigarettes and song. One British soldier even sat so a German barber could give him a haircut. Both the British and German high commands were furious when learning of the truce—soldiers on the front lines were supposed to kill each other, not sing and dance. The truce, though, represented a larger problem for each side's high commands. Their great fear was that when given the order to attack, the troops would respond with "Why?" And the Christmas Truce threatened precisely that, so that instead of meeting the goals set by Kings, Tsars, and Kaisers—namely, for their troops in the field to slaughter each other with maximum efficiency—the horror of horrors of peace would spontaneously prevail. That fear is understandable. If one looks at the motives of individual troops in their muddy, rat-infested trenches, their primary objective is not to kill each other for King, Kaiser or country, but rather to survive. And what better way to survive than to somehow convince the men facing you that if they don't try hard to kill you, you won't try hard to kill them. Viewed from the myopic perspective of a single day, those troops on both sides confronted a Prisoners'

Dilemma wherein the dominant strategy was to kill the enemy and then go home. But they weren't playing this game only once—they played it repeatedly, day after day, and therein found a solution to their Dilemma. Specifically, in its repeated form, cooperation could emerge in which mortars would be aimed imperfectly and rifles aimed high. If the British or Germans aimed well, the other side would respond in kind, but if either aimed poorly or didn't fire at all, this action was reciprocated. As a consequence, the commands of both sides confronted the threat of two armies who might refuse to kill each other with the efficiency demanded of them. In short, repetition allowed for more complicated strategies than simply "cooperate" or "defect"—it gave players the opportunity to signal a willingness to cooperate as well as the opportunity to punish each other subsequently if either defected from cooperation.

Mere repetition alone does not "solve" the Dilemma for its participants. Specifically, suppose a 2-person Dilemma is to be played some *finite* number of times and that the number of repetitions (as well as all other aspects of the situation) is common knowledge. Both persons know, then, that defection is dominant (subgame perfect) in the last play of the game—by assumption, nothing follows this last play and so the last subgame corresponds to a single play of the dilemma. But this means that it is futile to cooperate on the next to the last trial in the hopes of inducing cooperation on the last. Thus, both persons should defect on the next to the last play as well. The repetition of this argument establishes that *the only subgame perfect equilibrium to the finitely repeated Prisoners' Dilemma is defection at every opportunity.* This situation changes importantly, however, if we suppose that the dilemma can be replayed an indefinite number of times. The first thing that changes is that we can no longer use backward reduction—either there is no "last game" or the identity of that game is unknown. But the second thing that changes is that the number of strategies available to people becomes infinite (e.g., cooperate on every round; cooperate on every ith round; cooperate until the other person defects, in which case defect for j rounds; and so on). This multiplicity of strategies, in turn, opens the door to the existence of a plethora of equilibria and a reemergence of the problem of coordination to achieve any particular equilibrium.

However, before we can adequately address the issue of coordination, we must consider a third thing that changes when we allow indefinite repetition. In particular, although we might assume that people merely accumulate payoffs in the finitely repeated dilemma, we cannot make this same assumption in the infinitely repeated case since, if all payoffs are positive, payoffs accumulated ad infinitum imply that every strategy yields the same total—infinity (or minus infinity as the case might be). Moreover, such an assumption makes little behavioral sense, because it is only reasonable to suppose that people weigh payoffs received early more than they weigh payoffs that will be realized only in some distant future. Surely those mud-soaked troops sitting in their rat-infested WWI trenches gave greater weight to surviving today and tomorrow than they did to the promise of survival a week hence, and greater weight to the promise of survival a week hence to survival a month later. Similarly, when making a financial investment, a profit today of $X is almost certainly associated with greater utility than $X earned next month, and $X earned next month is afforded greater utility than $X earned next year. It is reasonable to suppose, then, that people discount future payoffs and evaluate strategies by the present value of the payoffs they realize (notice that we can suppose that even if people do not discount, there is always some probability, p, that the next play will be the last. Some simple algebra shows that this assumption leads to the same analysis we are about to undertake).

To represent this assumption formally in the context of a 2-person strategic interaction, suppose $u_{it}(s_1, s_2)$ is person i's utility in period t when 1 uses the strategy s_1 and 2 uses the strategy s_2. Then using the same formula we use to calculate present values in accounting, i's discounted payoff from s_1 across periods $t = 1, 2, \ldots$ and so on, can be expressed as

$$u_i(s_1, s_2) = u_{i1}(s_1, s_2) + ru_{i2}(s_1, s_2) + \ldots + r^{t-1}u_{it}(s_1, s_2) + \ldots$$

where r, $0 < r < 1$, is the rate with which person i discounts future payoffs. Admittedly, such an expression is formidable if the utility number in each term is different from the rest, but if all terms are the same, then we can use the following mathematical identity (which applies as long as $0 < r < 1$) to simplify the analysis:

$$1 + r + r^2 + r^3 + \ldots + r^n + \ldots = 1/(1 - r).$$

The reader might legitimately wonder at this point how we arrived at this expression. In fact it is quite simple, and worth reviewing since it allows us to introduce an idea that we make considerable use of later. First, notice that if we have a value X that reoccurs an infinite number of times but is discounted by r each time, we can write the discounted value V as follows:

$$V = X + rX + r^2X + \ldots = X + r(X + rX + r^2X + \ldots) = X + rV$$

Where we give V the label *continuation value*. That is, after realizing X in the first period, we anticipate realizing the discounted value of the endless stream of payoffs. Solving for V, then, gives us

$$V = X/(1 - r).$$

To see now how we use this identity, suppose, for a numerical example, that we take the simple Prisoners' Dilemma portrayed in Figure 5.10a and assume that it is played repeatedly an infinite or indeterminate number of times. If both players choose to cooperate all the time, both persons receive an infinite but discounted stream of payoffs of 10. Thus, the discounted present value of $10/(1 - r)$. On the other hand, if both persons defect to their dominant choice, then their payoff is an infinite repetition of zeros, or merely zero. Which strategies are in equilibrium in this repeated game depends on what other strategies (out of the infinity of possibilities) we allow the players to consider. However, since we are merely interested in illustrating the fact that infinite (or indeterminate) repetition alone can induce cooperation as an equilibrium, consider these four strategies: (c_1) cooperate always; (c_2) defect always; (c_3) cooperate as long as the other person cooperates, but defect forever when the other person defects for the first time; and (c_4) play tit-for-tat. The strategy of tit-for-tat, c_4, requires that a person cooperate on the first trial, and then on trial t it has that person matching what the other person chose on trial $t - 1$. Of course, two tit-for-tatters or a tit-for-tatter and someone choosing c_1 or c_3 never defect and thereby they both realize the payoff of $10/(1 - r)$ from their strategies. On the other hand, a tit-for-tatter playing against a person who chooses c_2 gets a payoff of -5 in the first round, and zero thereafter, whereas the opponent receives $+20$ in the first round and 0 thereafter. These calculations, then, allow us to describe the strategic form of the game by using these repeated strategies, and Figure 5.10b gives this form for the four identified strategies. The argument that repetition can induce cooperation is made, now, by observing that any combination of c_3 and c_4 such as (c_4, c_4) corresponds to a Nash equilibrium provided that $10/(1 - r) \geq 20$, or equivalently, that $r \geq ½$ – so cooperation is an equilibrium if neither person discounts the future too greatly (if r is not "too small").

	Cooperate	Defect
Cooperate	10, 10	–5, 20
Defect	20, –5	0, 0

Figure 5.10a A simple Dilemma

	c_1	c_2	c_3	c_4
c_1	$10/(1-r)$, $10/(1-r)$	$-5/(1-r)$, $20/(1-r)$	$10/(1-r)$, $10/(1-r)$	$10/(1-r)$, $10/(1-r)$
c_2	$20/(1-r)$, $-5/(1-r)$	0, 0	20, –5	20, –5
c_3	$10/(1-r)$, $10/(1-r)$	–5, 20	$10/(1-r)$, $10/(1-r)$	$10/(1-r)$, $10/(1-r)$
c_4	$10/(1-r)$, $10/(1-r)$	–5, 20	$10/(1-r)$, $10/(1-r)$	$10/(1-r)$, $10/(1-r)$

Figure 5.10b An infinitely repeated version of Figure 5.10a

It is important to emphasize that we have merely established that repetition **can** induce a cooperative equilibrium; we have not shown that it necessarily leads to cooperation. First, for such an equilibrium to exist, the players must give the future sufficient weight—the discount parameter r must be sufficiently high. We can speculate, then, why this observation explains the impact of elementary school conservation programs on the willingness of America's Midwestern farmers to incur the cost of shifting to new farming technologies in order to preserve the Ogallala aquifer. Those costs are largely short term whereas the benefits were long term. It is reasonable to hypothesize, however, that the "worried grandchild" on the grandfather's knee expressing a concern about the viability of farming in 20 or 30 years lowered the discount farmers placed on the future (i.e., raised the value of r), and with this change in values, cooperative outcomes became sustainable as equilibria.

There is, though, a second reason why we cannot use our example to argue that cooperative outcomes will necessarily arise from repeated play even if discount parameters meet the requirements for cooperation to be in equilibrium. Specifically, notice that the strategy pair (c_2, c_2)—(defect always, defect always)—is also an equilibrium regardless of the value of r. Thus, repetition adds equilibria, but it does not rid us of the "undesirable" one. Infinite repetition, in fact, leads to new problems. Because infinite repetition allows an infinity of pure strategies, equilibria multiply to keep pace. And we now know through various *folk theorems* (named as such because it is uncertain who first proposed or established them) that

> *In an infinitely repeated game, any outcome that gives each player what that player can guarantee for himself if he plays the game without coordinating with anyone else—any outcome that satisfies the security value of each player—can correspond to an equilibrium.*

where by *security value* we mean the worst outcome a player can realize if all other players act against his or her interest. As a result of such theorems, game theorists have sought to limit the strategies they think are most likely to be considered by players. *Stationary strategies*, for example, require that a player make the same choice every time he encounters strategically equivalent information sets—information sets that encompass the same number of decision nodes such that the nodes in such sets can be matched with identical subsequent subgames. Defecting always and cooperating always are stationary, but tit-for-tat is not since it requires a player to condition his action on the opponent's previous decision. *Trigger strategies* have players playing cooperatively until someone defects, at which point the strategy "triggers" a permanent punishment. The strategy c_3 is a trigger strategy. Trigger strategies are a special case of *punishment strategies*, of which tit-for-tat is the most familiar example. Finally, game theorists have examined strategies that allow only certain degrees of complexity. All of the strategies in Figure 5.10b are simple because they do not require that a player store much information about past moves; however, a strategy that says, "In the event of a defection punish for five rounds," requires a player of more complexity because the player must now count periods. We will not review this research because most of it requires complex analytic structures. We merely point to its existence in order to demonstrate that game theory is not yet a closed area of inquiry.

Even if we set aside the analytic complexities associated with an infinitely repeated game, one objection to the above "solution" to the Prisoners' Dilemma is that it is difficult to imagine a situation in which people play an infinite sequence of games with anyone, never mind with the same person and the same game. But this fact allows us to introduce and understand another idea, that of *reputation*. Imagine a situation in which a person in fact does play a PD game an infinite or indeterminate number of times, but not necessarily with the same person any two times in a row. Thus, it is not possible for any one person to punish a defection with a strategy of tit-for-tat by defecting themselves on the next play of the game since if they did so they'd be punishing the wrong person. On the other hand, suppose defectors can be identified and in some way "labeled" whereby it is known to everyone whether a specific individual cooperated on their last play of the game. In this instance they can be punished by whoever plays with them next. Moreover, we can imagine even more complex arrangements in which a failure to punish someone who warrants punishment is itself deemed a defection. In this way society might attempt to construct a system of self-enforcing social norms.

> **Ancestors and Genealogy**: If we allow ourselves the opportunity to engage in a bit of wild speculation, we note that the authors of this volume have long been fascinated by the fact that it is not uncommon in Asian society, China especially, for a person to be able to trace back their ancestry 30, 40 or even 50 generations. This contrasts sharply with European societies in which it is not uncommon for a person to barely know the identities of their great grandparents. But consider how two abstract societies might evolve; in one, there is a well-developed legal system available for the enforcement of contracts and, in the other, little or none at all. In the first, then, punishments can be formally applied, whereas in the second something must act as a substitute. And that substitute might just be the promise (threat) that any defection from an agreement today will place a heavy burden on one's progeny generations into the future. Thus, even if discount rates are "high," the potential for a long series of inter-generational punishments nevertheless renders the threat of punishment as compelling as that provided by any cadre of lawyers.

The door is now open, of course, to a variety of complex modeling exercises in which we explore things such as the emergence and evolution of social norms, the role of computational complexity and the different ways people learn to coordinate in repeated games. More on that later. Instead, to further illustrate the role of reputation in repeated PD situations, consider legislators who from time to time strike deals with each other to support each other's pet legislative projects. If those projects are much like the private constituency-specific proposals we considered earlier when arguing for why voters might prefer term limits, then clearly legislators have a short-term incentive to defect from any agreement (especially if such legislation is voted on sequentially). That incentive increases, moreover, if a legislator thinks there is a good chance that those with whom he has struck bargains might not be in the legislature next session either because of electoral defeat or retirement. But if a legislator himself intends to make a career of being a representative, he surely has an interest in maintaining a reputation as one who keeps a bargain because otherwise he may find it difficult if not impossible to come to agreements with others at some future date. Those others may simply refuse to "do business" with him because he is deemed unreliable or to enforce a legislative norm of keeping bargains once agreed to.

When the Mob Ruled Vegas: To conclude this discussion with yet one more example, consider casino gambling in Las Vegas. We know that the casinos are, like airlines, heavily regulated although not with the issue of anyone's safety in mind. Rather, those regulations are intended to keep the casinos honest—no loaded dice, no unbalanced roulette wheels, and no slot machines preset to never pay off. But why would casinos agree to be stringently regulated? Indeed, during the period when "the mob ruled Vegas" why didn't the mob subvert whatever regulatory authority the state of Nevada established so as to render its authority impotent? Surely the likes of Moe Dalitz, Frank Costello, Lucky Luciano, Meyer Lansky, and Bugsy Siegl, if they preferred to do so, weren't above using whatever means were available to them, some not very nice, to have themselves regulated in ways that allowed for crooked casino gambling. Instead of cheating the casino's patrons, the mob chose to cheat the government by skimming its profits and under-reporting income. The mob, however, as well as today's suit and tie corporate casino CEOs, understood one thing: If the casinos earned a reputation for crooked gambling, then it would be impossible to attract the type of customer they preferred—middle class America. Despite images to the contrary, the great mass of disposable income in America lies in its middle class, which is unlikely to make Vegas a vacation destination if the roulette wheel is fixed, the dice loaded, the cards marked and the slot machines too heavily biased in favor of the house. A reputation for honesty is a public good for the casinos and when given a choice between cheating their customers versus cheating the government, cheating the government was the more profitable thing to do. Still, why be regulated by an independent state agency? Surely the casinos, acting on their own, could in the era of mob rule, regulate themselves. It may be true that each casino or mob family had an incentive to defect and rig a game here or there, but self-regulation in the era of Bugsy Siegl and the like required little more than a bomb in a car or a ride to the desert to give a defector a permanent home under the tumbleweed and sand. However, not only would middle income America stay away from rigged casinos, they might be even less inclined to visit Vegas if it earned a reputation for having bodies continually discovered here and there with an occasional car exploding in casino parking lots. The solution to all problems—detecting and punishing defectors—was to be visibly and straightforwardly regulated by a state agency that was wholly independent from any casino influence and that worked at sustaining its own reputation for being cleaner than the Pope.

5.5 Constitutional Design and A Recursive Game

The preceding discussion and analyses of the Prisoners' Dilemma point to two general "solutions" for avoiding the consequences of being caught in an unsavory equilibrium: Either change the game or somehow embed it in a repeated context. Insofar as the first solution is concerned, it should be evident by now that Prisoners' Dilemmas in their various forms are commonly employed to justify government intervention into unregulated markets and, in a broader context, to justify the very existence of the state itself. Recall our discussion of public goods and the example of installing anti-pollution devices on cars. As we note, a solution to the Dilemma here has the players establishing or hiring an agent (government) who can inspect cars and levy fines in the event a car is found without the appropriate device so that by thus altering the game's payoffs we render the situation something other than a Dilemma. This solution, though, immediately takes us to the central problem of democratic politics: Empowering an agent qua government to resolve society's Dilemmas requires that we somehow ensure that the agent acts as intended. As Madison eloquently stated the problem: "*In framing a government which is to be administered by men over men, the great difficulty lies in this: you must first enable the government to control the governed; and in the next place oblige it to control itself*" (*Federalist* #51).

The usual sought-after method for controlling government—at least democratic governments—is via a well-crafted constitution that defines the structure of the state and specifies the limits of its powers. But this merely pushed the problem back a step to understanding how constitutional provisions are themselves enforced, especially those that constrain what the state cannot do. One commonly asserted answer is that those provisions are enforced in a democracy by the courts in combination with voters sanctioning incumbent candidates in elections. But this also only pushes the problem back because we now must specify how the authority of the courts is maintained and why incumbent politicians do not simply subvert competitive elections with fraud or the outright elimination of potential competitors (which we know they in fact sometimes prefer to do).

It is not uncommon for people to think of constitutions as a contract—a social contract. However, this conceptualization runs afoul of the search for an ultimate enforcer of constitutional provisions. In the world of business, both legal and illegal, there are readily identifiable sources of enforcement, ranging from the courts to the gun, whereby defectors from a contract can be sanctioned. But the things that enforce such contracts are either unavailable or are themselves the product of a constitution's design when it ostensibly establishes the institutional structure of a *democratic* state and specifies that state's legitimate functions. Interestingly, however, game theory and the various folk theorems just alluded to provide an alternative hypothesis about constitutional enforcement that, at the same time, give some guidance as to proper democratic constitutional design. The specific hypothesis we are about to offer and elaborate on, in fact, derives from the second "solution" to Prisoners' Dilemmas offered in the literature: namely, repeated play. Earlier we show how infinite repetition of the Dilemma introduces strategies such as tit-for-tat that admit of equilibria which are not present in the single or finite play of this game and which include wholly cooperative or efficient outcomes. But now we have a new problem: The infinitely repeated dilemma has a great many equilibria, including the old one in which the players maintain an inefficient outcome. What is required to ensure that this solution "works properly," then, is a mechanism that coordinates people to a specific cooperative (efficient) equilibrium.

Returning then to the issue of constitutional enforcement, suppose that instead of viewing democratic constitutions as social contracts, we instead conceptualize them as equilibrium selection devices—as mechanisms for political-social coordination. With this perspective, a democratic constitution's enforcement derives from its ability to coordinate society to equilibria of a specific sort—namely, to equilibria that pertain to the structure of the state and, with Madison in mind, to the state's intended functions.

To begin to see what we mean, consider the provision in the U.S. Constitution that requires that two thirds of the Senate approve of treaties negotiated and proposed by the President (Art. 2, Sect. 2) or the provision that allows the Senate and House to over-ride a Presidential veto of legislation with a two thirds vote in each chamber (Art. 1, Sect. 7). One question we can ask is: How important is it that the fraction in both cases be two thirds? Might the republic have foundered if it were three fifths, or if it were three fifths for any special vote in the House but two thirds in the Senate, or vice versa? Or instead of stating the requirement as a fraction, what if the U.S. Constitution had a number in decimal form, such as 0.69, which, after all, is pretty close to two thirds? The fact is that while the specific fraction or number chosen might make a difference with respect to a particular item of legislation or treaty, it probably is of little consequence in terms of the republic's long-term survival. Instead, what's important is that a specific number be agreed to beforehand and that it not be subject to negotiation as we move from issue to issue. Absent such an agreement, the choice of a number would merely inherit the instability that would characterize the specifics of the policy being addressed. Recall that if the majority rule cycle A > B > C > A pertains, then any one of these outcomes can be made the final majority choice by the judicious selection of an agenda—by the choice of which two alternatives are paired against each other first, with the survivor pitted against the third outcome. In this case, if that same committee were to vote on which agenda to use, the majority preference over agendas would inherit the intransitivity that exists over the alternative outcomes themselves. This problem, however, is general, and absent a prior agreement over rules and voting procedures, we should anticipate interminable and dysfunctional debate over those procedures.

What is also interesting about the U.S. Constitution as originally drafted is that aside from identifying a quorum in the legislature so as to preclude it from conducting any business if too few legislators are present, it never speaks of majority rule and specifies a rule only when something other than a simple majority is deemed necessary. What, then, we might ask precludes Congress from adopting a policy of approving legislation only if some super-majority votes for it (as it does when closing debate by implementing cloture) or what precludes the U.S. Supreme court from requiring that its decisions be binding only if the opinions of the justices are, as we typically require in jury trials, unanimous? The answer, of course, is *nothing*, so we can then also ask whether the drafters of the U.S. Constitution left the door open to some dangerous procedural manipulations by not requiring majority rule in those cases. Did they forget to dot the i's and cross the t's?

The answer, if history is any judge, is no because the delegates in Philadelphia assumed a society-wide consensus existed as to when simple majority rule was legitimate. Stated in game theoretic terms, although there may have been a multitude of potential equilibria when it came to voting rules, society was already coordinated to the equilibrium of using majority rule in most instances and thus their document needed to address—and thereby coordinate to—a different rule only in those instances where something other than a simple majority vote was deemed advisable. This same argument can be applied to other parts of the constitution. The convention in Philadelphia in 1787, for instance, discussed a number of ways of selecting a chief executive, for appointing federal judges, and schemes of legislative representation. In each case, there were passionate preferences over the alternatives, but once again it is reasonable to suppose that in terms of the republic's survival, it was less important as to which alternative was chosen in each case and more important that they simply chose *some* alternative.

From this perspective, constitutions are best conceptualized not as contracts but as *equilibrium coordinating devices*. And it is this conceptualization that explains the source of their enforcement as well as some of the generally accepted rules for writing national constitutions. Specifically, if they do in fact coordinate to an equilibrium, then once they have attained legitimacy, they become self-enforcing. Their provisions are adhered to because unilateral defections decrease the payoffs of defectors either directly or in the long term via constitutionally sanctioned punishments. The view of constitutions as equilibrium selection and coordination devices, we might add, also helps us grapple with the ill-defined word "legitimacy." Specifically, we can say that a constitution is legitimate if each member of society expects that all other members will abide by its provisions. A constitution can be said to be legitimate, then, if it successfully coordinates expectations about the beliefs of people to an equilibrium.

This discussion, though, leads to one last question; namely, what have we gained by viewing constitutions as equilibrium selection devices as opposed to contracts? We cannot answer this question fully here since this volume is not specifically about constitutional design. But one answer is provided by noting that one rule of good constitutional design that appears to have widespread acceptance is "the shorter the better." Overly long national constitutions that read like the complex contractual documents drawn up by lawyers are usually deemed suspect at best and doomed to failure at worst. If a constitution is to be an effective coordination device, then it needs to be relatively understandable to the population generally. Having it be understandable only to specialists and some small cadre of lawyers, or to require convoluted interpretations, makes it difficult, if not impossible, to coordinate people's expectations as to what they can expect from the state and what are the legitimate versus the illegitimate actions that the state may take when regulating society. In this respect, we can compare constitutions to social norms, which also coordinate society. So by way of example, consider the norm "give an old lady a seat on the bus." When this norm is taught to us, it is taught using no more complex a specification than those nine words. But imagine such a norm being written in the language of contracts. First we must define "old" and then in all likelihood make provision for exceptions. Perhaps the norm should be applied also to middle-aged women carrying heavy packages. But now, how heavy must those packages be? Is 10 kilograms heavy enough? Who determines a package's weight? Shall we place scales on buses to determine when to apply the norm? What if the package didn't appear to be heavy but was nevertheless overly cumbersome? How cumbersome does it have to be? At this point of course, the norm begins to degenerate into silliness and is no longer likely to work. And by *work* we mean if we were to find ourselves seated on a bus and an elderly lady stood before us, at what point would we feel guilty by remaining seated? Clearly, the norm here that is likely to work best is the one our parents taught us. The same is generally true of constitutions.

A great many other things coordinate people, many of which fall under the general rubric of social norms. And here we find yet another reason for keeping a constitution short. Specifically, the longer, more expansive and more detailed a constitution, the more likely it is to "step over the line" and try to regulate things that are regulated by preexisting norms. A society's norms, however, are likely to consist of things that have evolved over millennia so that they have a greater hold on people's beliefs and actions than any newly crafted document. Thus, in any conflict between a pre-existing norm and a written constitution, we can expect the norm to prevail and, in the process, to erode the perceived legitimacy of the document. Consider again the Chinese practice of foot binding, which we discussed earlier as an example of a Prisoners' Dilemma. There, despite the formation of innumerable Natural Foot societies, the 1902 edict of the Dowager Empress Cixi and the formal prohibition of the practice in 1912 by the Nationalist government, the practice was reported to have continued in rural regions as late as the 1940s. It took nearly three-quarters of a century to rid the country of this standard of erotic beauty, thereby illustrating the difficulty one encounters with changing social norms. Thus, to imbed such a prohibition in a national constitution, however well intended, would most likely only undermine the perceived legitimacy of the entire document. If one part of a constitution is deemed ineffectual, it can only damage the ability of its other parts to coordinate in other domains.

This view of constitutions also tells us that James Madison was wrong when he argued against the inclusion of a Bill of Rights in the U.S. Constitution. Once again, there is nothing in the document per se that enforces those rights. If we were to again argue that the court does so, we must again ask why anyone would abide by the court's decisions. The real answer here about enforcement is that the listing of rights coordinates people's expectations not only about what they believe are legitimate rights, but also about what they think others will think and how others will act if those rights are violated by the state. Bills of rights, then, coordinate to equilibria wherein no one has an incentive to unilaterally act to oppose them—most importantly political elites who find rights an "inconvenience" when applied to the rest of us. The role of the court then becomes that of interpreting those rights in circumstances that may or may not have been anticipated by the original authors with the assumption that their edicts will be binding because people are already coordinated to yet another equilibrium—namely, one in which the courts are the ultimate referee in their interpretation.

To this point, our argument about the coordinating function of constitutions has been verbal, and it is perhaps useful then to offer an analytic illustration of things. So consider the issue of secession and whether, in a federal state, federal subjects should be constitutionally allowed or barred from seceding. The usual argument is that such words in a constitutional document are simply that—mere words on paper—and that federations survive or dissolve depending on how "profitable" they are to their individual components. However, let us consider an abstract scenario in which a federation consists of three members, 1, 2 and 3, where member 1 is contemplating secession. Thus, let 1's alternative strategies be

S: secede from the federation
S': don't secede

Members 2 and 3, in turn, have two strategies:

P: If member 1 secedes, punish it so as to maintain the federation
P': If member 1 secedes, don't punish it and allow the federation to dissolve

As things stand right now with the federation intact, suppose its overall "value" (measured in some way) is π, where member i is afforded the share a_i. Thus,

		Member 3: P		Member 3: P'	
		Member 2		Member 2	
		P	P'	P	P'
Member 1	S	$b_1\pi$, $b_2\pi$-C, $b_3\pi$-C	π_1, π_2, π_3-c	π_1, π_2-c, π_3	π_1, π_2, π_3
	S'	$a_1\pi$, $a_2\pi$, $a_3\pi$	$a_1\pi$, $a_2\pi$, $a_3\pi$	$a_1\pi$, $a_2\pi$, $a_3\pi$	$a_1\pi$, $a_2\pi$, $a_3\pi$

Figure 5.11 Game of secession

the status quo payoff vector for the federation is $(a_1\pi, a_2\pi, a_3\pi)$, where $a_1 + a_2 + a_3 = 1$. If the federation collapses, which we assume occurs if member 1 secedes and is not punished, the payoff vector (π_1, π_2, π_3) prevails. However, to forestall this collapse in the event that 1 attempts secession, players 2 and 3 must both act to punish, in which case we can suppose that the payoff vector $(b_1\pi, b_2\pi - C, b_3\pi - C)$ is realized, where $C > 0$ is the cost incurred by 2 and 3 in applying the punishment and the *b*'s correspond to a new allocation of the federation's "value." Finally, if only one member acts to punish 1, the punishment is ineffectual, the federation dissolves, and the unit acting unilaterally to punish incurs a cost of $c > 0$. Thus, modeled as a $2 \times 2 \times 2$ strategic form, we have the game in Figure 5.11.

Consider the last column of this table, where what is being compared for member 1 is the value of seceding without being punished, (S,P',P'), versus maintaining the status quo, (S',P',P'). Thus, (S,P',P') is an equilibrium if $\pi_1 > a_1\pi$ and $\pi_j > \pi_j - c$. Since this second inequality is necessarily satisfied, (S,P',P') is an equilibrium if it is more profitable for member 1 to see the federation dissolved than maintained without any attempted secession. At the same time (S',P',P') is an equilibrium if the opposite is true with respect to member 1. Thus, whether or not the federation survives would appear to depend on the usual economic arguments about its value to individual members. However, notice that (S',P,P), which corresponds to member 1 not seceding because it will be punished by members 2 and 3 if it tries to do so, is also an equilibrium whenever $a_1 > b_1$. Nothing we have said about this example, though, precludes the possibility that $b_1 < a_1 < \pi_1/\pi$, and if this condition is satisfied then we have two diagonally opposite equilibria, one in which member 1 secedes and is not punished and one in which it doesn't secede because it will be punished if it attempts to do so. Whether the federation survives and member 1 refrains from seceding, then, depends on whether members 2 and 3 can coordinate to punishing 1 if it attempts to secede. And it is this coordination that a constitution can provide.

Modeling secession as a single play game, however, is admittedly a bit of a stretch, since if a federal subject chooses not to secede in one period, there is no reason why it cannot reconsider its decision later. So by way now of introducing the idea of a *recursive game*, suppose we complicate things by assuming that if the federation is not dissolved either because player 1 does not attempt to secede or because it does but is punished for doing so, the game is repeated. Thus, two branches of the game's extensive form double back on each other. To see how such situations are treated, suppose, in any iteration of the game, that 1 secedes with probability p_1 and that 2 and 3 punish 1 with probabilities p_2 and p_3 respectively. Infinite repetition, of course, allows for an infinity of strategies, but here we consider only one type—*stationary strategies*. In fact, the primary reason for extending our example of secession is to introduce this type of strategy, which requires that a player make the same choices at every iteration of a game. Thus, with a stationary strategy, a player does not condition his or her choice on the past, so tit-for-tat in the repeated Prisoners' Dilemma is an example of a strategy that is not stationary. Figure 5.11 continues to describe the one period payoffs to the players but by way of some additional useful notation let V(S) be the payoff to 1 of the stationary strategy of attempting to secede and V(S') be the expected payoff of not seceding. Similarly, let $V_j(P)$ and $V_j(P')$ be player *j*'s expected payoffs of punishing and not punishing, $j = 2$ or 3, if and when player 1 chooses to secede. Finally, following the first play of the game, let v_1 be the *continuation value* of the game for member 1 -its payoff from all future plays of the game, which we discount in the first period by δ (with V_j defined in the same way for $j = 2$ and 3). Whether 1 chooses S or S' depends, of course, on which is greater, V(S) or V(S'). Thus, with stationary strategies,

$$v_1 = \max\{V(S), V(S')\}$$
$$v_2 = v_3 = \max\{V(P), V(P')\}$$

Our notation, in combination with the entries in Figure 5.11, is sufficient to calculate V(S), etc. Specifically

$$V(S) = p_2p_3(b_1\pi + \delta v_1) + (1 - p_2p_3)\pi/(1 - \delta)$$
$$V(S') = a_1\pi + \delta v_1$$
$$V_j(P) = p_1p_k(b_j\pi - C + \delta v_j) + p_1(1 - p_k)(\pi_j/(1 - \delta) - c) + (1 - p_1)(a_j\pi + \delta v_j)$$
$$V_j(P') = p_1\pi_j/(1 - \delta) + (1 - p_k)(a_j\pi + \delta v_j)$$

If we narrow our search for equilibria to pure strategies, we know that

$$p_1 = 1 \text{ if } V(S) \geq V(S') \text{ in which case } v_1 = V(S)$$
$$p_1 = 0 \text{ if } V(S) < V(S') \text{ in which case } v_1 = V(S')$$
$$p_j = 1 \text{ if } V(P) \geq V(P') \text{ in which case } v_j = V(S)$$
$$p_j = 0 \text{ if } V(P) < V(P') \text{ in which case } v_j = V(P')$$

There are eight possible pure strategy stationary equilibria, with the conditions for each to be an equilibrium presented in Table 5.2 (where X assumes various complex algebraic values, all of which are complex but of little consequence to

Table 5.2 Potential Equilibria to the Recursive Secession Game

Cases	*Player 1*	*Player 2*	*Player 3*
(0, 0, 0)	$a_1 \geq \pi_1/\pi$	$c \geq 0$	$c \geq 0$
(0, 1, 0)	$a_1 \geq \pi_1/\pi$	$c < 0$	$C \geq X$
(0, 0, 1)	$a_1 \geq \pi_1/\pi$	$C \geq X$	$c < 0$
(0, 1, 1)	$a_1 \geq b_1$	$C < X$	$C < X$
(1, 0, 0)	$a_1 < \pi_1/\pi$	$c \geq 0$	$c \geq 0$
(1, 1, 0)	$a_1 < \pi_1/\pi$	$c < 0$	$C \geq X$
(1, 0, 1)	$a_1 < \pi_1/\pi$	$C \geq X$	$c < 0$
(1, 1, 1)	$a_1 < b_1$	$C < X$	$C < X$

our conclusions). From this table, however, we can see that (0, 1, 0), (0, 0, 1), (1, 1, 0) (1, 0, 1) and (1, 1, 1) cannot be stationary equilibria since otherwise we violate the assumptions that $c > 0$ and $a_1 > b_1$. So as before, we have the two equilibria (0, 0, 0) and (1, 0, 0) in which player 1's decision to secede or not depends solely on whether remaining in the federation is more profitable or valuable than going it alone. And then, as before, there is the additional equilibrium, (0, 1, 1), in which 1 does not secede since, if it does, it knows it will be punished. It is the coordination of 2 and 3 to this equilibrium that, presumably, a constitution can accomplish.

It is, of course, unsurprising that we reach the same conclusion treating Figure 5.11 as a recursive game as we did when treating it as a one shot game, given the restrictions we place on potential equilibria—namely, stationary pure strategies. Our example here is intended primarily to introduce the notions of continuation value and stationary strategies—ideas to which we return in subsequent chapters. Our example is also quite limited in allowing only member 1 to contemplate secession. However, lest the reader think our example is too far removed from reality to convey any substantive lessons, imagine what might have happened had the U.S. Constitution explicitly prohibited secession or explicitly allowed it. If it had explicitly prohibited it, we can ask whether Robert E. Lee, with his sense of duty and legality, would have turned down Lincoln's offer to lead the Union Army against secessionist South Carolina? Absent Virginia and Lee's leadership, the American Civil War arguably would have been a brief and far less bloody affair. Conversely, if the Constitution had explicitly allowed secession, could Lincoln have mobilized the North to contest a South intent on separation, especially when perhaps a majority of the North had little sympathy for those who argued for the abolition of slavery? Similarly, imagine a Soviet Union in 1989 in which Premier Gorbachev, in lieu of rendering secession legal by encouraging the Politburo to amend the Soviet Constitution to lay out a specific, albeit tortuous process for secession, had left things as they were. Would the Baltic Republics have felt encouraged to declare their separation from the USSR as they obviously did? Similarly, we can view the dissolution of the Warsaw Pact as a sequence of secessions, with Poland leading the way. But keep in mind that all that changed prior to the Pact's dissolution was the Kremlin (and Gorbachev in particular) explicitly stating that it would hereafter abide by the "Sinatra doctrine" of letting each member state "do it their way." Soviet troops were still present and still capable of punishing defectors as they had done in 1956 (Hungary) and 1968 (Czechoslovakia). Only the words had changed, but that change was sufficient to change expectations and, thus, the equilibrium members of the Pact assumed they were in. We wouldn't, though, want to leave the reader with the impression that members of the Pact, immediately upon hearing of the ostensible change in Kremlin policy, assumed a new equilibrium had prevailed. The powers that be in the Kremlin, after all, might change their mind and one couldn't also preclude the possibility that the personnel of the Kremlin itself might change, peacefully or otherwise. What ensued, then, upon Gorbachev's announcement of a new policy with respect to its allies was a gradual and hesitant shift in the actions of the Pact's member states, with the Kremlin's reaction being carefully observed and with everyone on the alert for a signal that Gorbachev's words had been somehow misinterpreted or were but a part of Gorbachev's public relations agenda with respect to the West. In Hungary, for example, the events of 1956, originally officially called a "counter-revolution" were, in January 1989, renamed a "popular uprising," whereupon Hungarian officials waited to see the Kremlin's response. The subsequent absence of a response was taken to mean that the Sinatra Doctrine was, in fact, real. Later, in Chapter 7 we will examine such signaling more closely, but our point here is simply that words can coordinate to one equilibrium or another since they coordinate expectations and beliefs as to what strategies are legitimate and which ones others are abiding by.

5.6 Evolutionarily Stable Strategies and Corruption

One of the ostensible mysteries of economic reform following the collapse of the Soviet Union is why states that were a part of the USSR's Central European empire largely succeeded in their reform efforts whereas the component parts of the Union itself, Russia included, have been far less successful at establishing markets that avoid the inefficiencies of massive corruption.

Corruption, of course, takes a great many forms, from election fraud to the outright buying of favors from incumbent politicians, and it also has many sources, from a weak or non-existent rule of law to a mass media controlled by the state and thus unlikely to do the investigative work that uncovers official malfeasance. There are, though, intermediate causes, and here let us consider the situation that existed in ex-Soviet states following the USSR's dissolution. Clearly, a great many individuals hoped for a rapid transformation of the economies of, say, Russia and Ukraine into something that mimicked Western Europe or the Scandinavian states, but soon enough Western investors learned that the obligation of contracts meant something different in ex-Soviet states than what they had anticipated or hoped for. As many an investor learned, former Soviet apparatchiks and members of the old nomenklatura regarded contracts as something binding only until a better offer came along or they were able to extract another bribe to maintain an agreement. The particular problem was that Western investors, having previously operated in political systems with a strong legal code, invested with long-term planning horizons wherein they were willing to make investments even if profits did not materialize years down the road. In contrast, those raised under a Soviet regime in which the law favored whoever had access to the powers that be, operated with far shorter planning horizons if only because there was the inherent uncertainty as to who would be favored tomorrow by those powers. What rendered this situation especially frustrating was that there did exist a core of native ex-Soviets willing to operate by the West's rules and who preferred the establishment of a meaningful rule of law in their countries whereby long-term contracts could be enforced. Soon enough, though—as early as the late '90s—those same persons acted no differently than their apparatchik and nomenklatura counterparts.

The suggestion here, then, is that post-Soviet states consisted of two types of individuals—those who placed a high discount on the future and those with a low discount. And in this context it is easy to imagine a situation in which even those who might have preferred operating with a long-term planning horizon, found themselves in a game in which the only viable alternative for them was to operate with a short-term horizon. To illustrate, consider the 2-person game in Figure 5.12:

	Short term	Long term
Short term	20, 20	40, 0
Long term	0, 40	35, 35

Figure 5.12 Prisoners' Dilemma of planning horizons

With numbers here meant only to convey relative preferences, the idea is that if both persons operate with a short-term horizon, they are precluded from engaging in those mutually beneficial long-term investments that require enforceable contracts and their profits are minimal. On the other hand, if one of them adopts a long-term perspective, there's a good chance he can be taken advantage of by someone who gives little weight to the future and the sanctity of contracts—hence the asymmetric payoffs in the corresponding cells. Finally, if both adopt a long-term perspective, then presumably both can realize the full potential of efficient markets.

Obviously, the game in Figure 5.12 is a Prisoners' Dilemma, but by itself it can at best offer a rather simple-minded and incomplete explanation for why a society attempting to develop a viable rule of law and efficient markets might fail to do so. There is obviously something missing here because this example does not explain why Central Europe, in contrast to the USSR (with the exception of the Baltic Republics), succeeded in its economic reforms. To offer an explanation for this difference requires that we introduce a new idea—that of evolutionary stability.

Evolutionary Stability: Setting aside the issue of corruption for the time being, if we return once again to the infinitely repeated Prisoners' Dilemma, we note that its repeated version illustrates a problem that plagues many of the abstract models people use in the attempt to understand politics from a game theoretic perspective. Specifically, equilibria need not be unique and, as the folk theorems establish, equilibrium outcomes can encompass a great many possibilities with distinctly different implications within any one game. Our discussion in the preceding section argues that one mechanism for choosing an equilibrium is the coordinating function of a sustainable national constitution. The mathematical game theorist's approach, though, has been somewhat different, devoting a considerable research effort at formulating refinements of Nash's original idea with the hope of narrowing predictions in some way. This research agenda has not been altogether successful because many games have multiple and equally reasonable equilibria that no mathematical trick can or in fact should refine. After all, the existence of different cultures, of different solutions to common problems across cultures, and even so fundamental a thing as the existence of different languages suggest that we cannot do away with the possibility of equally plausible multiple equilibria. We are left, then, with noting that aside from such explicitly engineered creations as political constitutions, we have not yet fully addressed the general question of how people coordinate to ensure that some equilibrium ultimately prevails or the processes whereby one equilibrium as opposed to some other ultimately dictates outcomes. It is one thing to say in the context of the repeated Prisoners' Dilemma that cooperative equilibria exist, but it is quite another thing to assert that people will somehow achieve such an outcome.

The problem at hand, then, can be restated to ask why people might be predisposed to one strategy (pattern of behavior) rather than another. We can ask, for example, why certain norms persist in legislatures and why freshmen legislators, who may not share these norms, are nevertheless socialized to accept them (as against the possibility that some norm carried to the legislature by one or more freshmen becomes dominant). We could also ask why some standards of behavior but not others differentiate cultures. To address such questions, let us take a biological view and think of strategies as gene types in which a process of natural selection reinforces one strategy as against another. The classic illustration of what we mean by this is offered by a situation called Hawk and Dove.

Hawk–Dove: Suppose whenever two individuals confront each other, each must choose between being a Hawk versus a Dove. Being hawkish means to fight until injured or until the opponent retires; being a Dove is to strut about, but to retire whenever confronting a Hawk. Thus, if a Hawk meets a Dove, the Hawk extracts all value, V, from the situation and the Dove gets nothing. Suppose further that the winner in a physical confrontation between two Hawks gets V and the loser suffers a loss of D. Finally, if two Doves meet, they merely split V equally. Hence, if similar opponents have equal probabilities of winning and losing, then Figure 5.13 portrays the expected payoffs of two competing animals:

	Hawk	Dove
Hawk	$(V-D)/2$, $(V-D)/2$	V, 0
Dove	0, V	$V/2$, $V/2$

Figure 5.13 Hawk–Dove payoffs

Viewing this example as a single-play game, notice that (Dove, Hawk) and (Hawk, Dove) are equilibrium pairs if $D > V$, and only (Hawk, Hawk) is an equilibrium if $V > D$. However, rather than interpret this figure as the strategic form of a two-person game let us merely take Figure 5.13 as a table that tells us what payoffs are realized whenever two types of creatures encounter each other.

Now suppose that all creatures in a society are of the same type and are genetically predisposed to choosing one particular strategy so that all interactions are between pairs who abide by the same strategy—pure or randomized. But suppose also that on occasion (with the small probability p) "normal" strategies confront mutants (the norm held by the unsocialized freshman legislator?)—creatures who choose a different strategy. If, for example, society consists only of Doves, then the expected payoff to a Dove is

$$0p + (1-p)V/2 = (1-p)V/2 \tag{5.1}$$

whereas the payoff to a mutant Hawk is

$$p(V-D)/2 + (1-p)V = V - p(V+D)/2 \tag{5.2}$$

Some simple algebra reveals that as long as V exceeds pD, expression (5.2) is greater than (5.1), so on average mutants will "perform" better than non-mutants. If we now suppose that these payoffs measure the ability of a gene to reproduce itself (i.e., genes with higher payoffs reproduce more successfully than genes with lower payoffs), mutants will proliferate at the expense of Doves. On the other hand, if society consists of Hawks and it is Doves who are rare mutants, Hawks experience an expected payoff of

$$(1-p)(V-D)/2 + pV = [V - D + p(V+D)]/2 \tag{5.3}$$

whereas the payoff to a mutant Dove is

$$(1-p)0 + pV/2 = pV/2 \tag{5.4}$$

Thus, as long as V exceeds D, mutant Doves will suffer relative to Hawks. That is, with $V > D$, being a Hawk is *evolutionarily stable.*

To generalize this example, consider the following definition:

s is an ***evolutionarily stable strategy*** *if, for any other strategy s',*

$$u(s, s) > u(s', s) \tag{5.5}$$

or

$$u(s, s) = u(s', s) \text{ and } u(s, s') > u(s', s'). \tag{5.6}$$

To see this definition's relation to our argument that a society of Hawks is evolutionarily stable whenever $V > D$, notice first that the expected return from using the strategy Hawk, $E(\text{Hawk})$, can be written

$$E(\text{Hawk}) = (1-p)u(\text{Hawk, Hawk}) + pu(\text{Hawk, Dove}) \tag{5.7}$$

whereas the payoff from Dove is

$$E(\text{Dove}) = (1-p)u(\text{Dove, Hawk}) + pu(\text{Dove, Dove}) \tag{5.8}$$

where the various values of u are given in Figure 5.13. Expressions (5.7) and (5.8) are, of course, the same as (5.3) and (5.4) written more generally, so if all Hawks is evolutionarily stable, we require that $E(\text{Hawk}) > E(\text{Dove})$, or

$$(1-p)u(\text{Hawk, Hawk}) + pu(\text{Hawk, Dove}) >$$
$$(1-p)u(\text{Dove, Hawk}) + pu(\text{Dove, Dove}).$$

Since p is very small (a Dove is the rare mutant), this inequality is satisfied if

$$u(\text{Hawk, Hawk}) > u(\text{Dove, Hawk}),$$

which is what expression (5.5) requires and which, the reader might note, parallels the definition of a Nash equilibrium. Occasionally, however, equality might hold here as when a Hawk has no advantage over a Dove when confronting another Hawk. But we surely wouldn't want to label the situation evolutionarily stable if at the same time a Hawk does have an advantage over a Dove when confronting Doves. Thus, if equality in the above expression holds, then evolutionary stability requires

$$u(\text{Hawk, Dove}) > u(\text{Dove, Dove}),$$

which is what expression (5.6) requires.

One case now remains to be considered—namely, when $D > V > pD$ (where pD is approximately 0 for very small p). Keeping in mind that we are not interpreting Figure 5.13 as a game matrix, but merely as a table revealing the payoffs players receive when confronting someone of the same or different type, since neither all Doves nor all Hawks is an evolutionarily stable situation if $D > V > 0$, consider the possibility that the non-mutant (normal) members of the species each abide by a mixed strategy $(q, 1 - q)$. Once again, if mutants (who use the strategy $(r, 1 - r)$, $r \neq q$) arise with some small probability, then (after some algebraic manipulation in conjunction with Figure 5.13),

$$u(q,q) = [V - q^2D]/2$$
$$u(r,q) = [rV + V - qrD - qV]/2$$

In accordance with expression (5.5), if we set $u(q, q) > u(r, q)$ we get

$$(q - r)(V - qD) > 0.$$

We do not know, however, whether q is greater than or less than r. Indeed, if the mixed strategy $(q, 1 - q)$ is evolutionarily stable, it must be that neither strategies with $r > q$ or $r < q$ can gain a greater payoff. So the only circumstance in which we are assured that the preceding expression holds universally is if we set $q = V/D$. But now equality holds in the first part of expression (5.6), so it is the later part of expression (5.6) that must be satisfied. So computing $u(q, r)$ and $u(r, r)$,

$$u(q,r) = [qV + V - qrD - rV]/2$$
$$u(r,r) = [V - q^2D]/2$$

Expression (5.6)'s requirement that $u(q, r)$ exceed $u(r, r)$ becomes, after some algebraic manipulation,

$$(q - r)[V - rD] > 0,$$

which can be rewritten as

$$D(q - r)[V/D - r] > 0$$

Substituting $q = V/D$ gives

$$D(q - r)^2 > 0,$$

which is necessarily satisfied whenever r is not equal to q, because D is assumed to be positive and since the squared term is necessarily positive.

A natural question to pose at this point is to ask about the relationship between evolutionarily stable strategies and Nash equilibria, and here our result is straightforward: If we interpret the payoff structures used to represent the interaction of alternative types or species, such as Figure 5.13, as a symmetric strategic form game, then *an evolutionarily stable strategy corresponds to a Nash equilibrium, but not every Nash equilibrium corresponds to an evolutionarily stable strategy*. To see the first part of this statement, expressions (5.5) and (5.6) combine to state that s is evolutionarily stable only if $u(s, s) > u(s', s)$, which is simply the requirement that a player has no incentive to defect unilaterally from s. A Nash equilibrium need not correspond to an evolutionarily stable strategy, on the other hand, because of the second part of expression (5.6)—evolutionary stability imposes a somewhat stricter requirement on strategy pairs than does the definition of a Nash equilibrium. Notice, however, that any Nash equilibrium in which a player's payoffs decrease (as opposed to merely remain unchanged) by a unilateral defection is evolutionarily stable.

Let us now return to our original problem of the repeated Prisoners' Dilemma in order to assess what kind of solution the notion of evolutionary stability provides there. Before we do so, however, let us generalize (5.1) and (5.2) in a way demanded by the context of the analysis. Specifically, what we must keep in mind is that we are not analyzing merely the interactions of two specific individuals; instead, we are looking at how a particular strategy performs in a "population of strategies." Thus, it is possible that there are multiple simultaneous mutations occurring, and the ability of one strategy to predominate against another will depend not only on how the strategy does against a particular mutation but also on how each performs against other common "adversaries." To develop an appropriate notation that handles this possibility, let s_j ($j = 0, 1, 2, \ldots, n$) denote a particular strategy and let p_j denote the probability that any particular creature employs strategy s_j. Further, let s_0 denote the "normal" (non-mutant) strategy (which means that p_0 is very much larger than p_1, p_2, and so on), and let s_n denote the particular mutant strategy that concerns us. Then the extension and general form of expression (5.7) becomes

$$P_0U(s_0, s_0) + p_nU(s_0, s_n) + \Sigma p_jU(s_0, s_j), \tag{5.9}$$

and the extension and general form of expression (5.8) is

$$p_0U(s_n, s_0) + p_nU(s_n, s_n) + \Sigma p_jU(s_n, s_j) \tag{5.10}$$

where both summations are over all values of j except $j = 0$ and n. The particular difficulty with the Prisoners' Dilemma, now, is that there are a great many strategies for which $U(s_0, s_0) = U(s_n, s_n)$ and $U(s_0, s_n) = U(s_n, s_n)$ —Tit-for-Tat and Always Cooperate are two such examples, since neither ever defects against the other. One may hold an advantage over the other, however, because it performs better against a third type of strategy that appears in (5.9) and (5.10) only after the summation signs. Suppose there are three strategies: Tit-for-Tat (TFT); Tit-for-Two-Tats (TF2T, which allows for being the victim of two consecutive defections before it retaliates with a defection); and Suspicious-Tit-For-Tat (STFT, which begins with a defection and plays TFT thereafter). Notice now that TF2T can invade a population in which everyone plays TFT if it performs better against STFT than does TFT—if $U(\text{TF2T}, \text{STFT}) > U(\text{TFT}, \text{STFT})$. To see that this inequality in fact holds, notice first that TF2T versus STFT yields the sequence of choices (Cooperate, Defect), (Cooperate, Cooperate), (Cooperate, Cooperate), . . .—or, using the same numbers used to generate Figure 5.10b, the sequence of payoffs −5, 10, 10, 10, and so on. On the other hand, TFT versus STFT yields (Cooperate, Defect), (Defect, Cooperate), (Cooperate, Defect), (Defect, Cooperate), . . .—or, the payoff sequence −5, 20, −5, 20, −5, and so on. Thus, as long as the discount rate does not weight the second-round payoff of 20 too greatly at the expense of subsequent payoffs, TF2T can invade TFT.

We can use the preceding argument, now, to establish that no pure strategy is evolutionarily stable for the Prisoners' Dilemma in every possible environment. No pure strategy, of course, is dominant and thus no such strategy is at least as good and sometimes better than any other. Suppose X and Y are two strategies that earn the same payoff against each other as they do against themselves. Then there exists a strategy Z that is better against X than it is against Y, and there exists another strategy W that is better against Y than it is against X. Then, if X is the normal strategy and if Y and Z are possible mutants, Y can invade because of its relative advantage with respect to Z. On the other hand, if Y is the normal strategy and X and W are the mutant types, X can invade Y. What such an exercise tells us, then, is that the types of strategies likely to prevail as evolutionarily stable will be mixed strategies (see the 1987 article by R. Boyd and J. Lorberbaum, "No Pure Strategy is Evolutionarily Stable in the Repeated Prisoners' Dilemma," *Nature* 327, May).

Despite this somewhat pessimistic conclusion with respect to the Prisoners' Dilemma, we can still see here in the notion of evolutionary stability a way to approach such issues as learning, the evolution of social norms and endogenous processes of social coordination. It is, admittedly, a bit farfetched to imagine norms and social coordination arising out of some pure biological process whereby those with the "wrong" norms or who fail to coordinate effectively in everyday mundane circumstances die off at a greater rate than others or fail to reproduce at the same rate as anyone else. But consider the following thought experiment: Imagine that the human brain contains countless heuristics that direct our day to day actions, including heuristics as to when to apply other, higher level heuristics when encountering a new circumstance. Through trial and error, a person learns, consciously and unconsciously, which heuristics or ideas work best and which work poorly, with the poorer performing ideas being discarded or at least applied with decreasing frequency. If we think of these ideas or heuristics, then, as competing genes, the notion of evolutionary stability can be seen to have broader application than to mere biological processes. It is the application of this perspective to the issue of corruption and economic reform that we consider in the next subsection.

Corruption: Returning now to the issue of corruption and the relative success of countries at making a transition to competitive markets, we can begin with the supposition that post-Soviet states consisted of two types of individuals—those who placed a high discount on the future and those with a low discount. Referring back to our discussion of the repeated Prisoners' Dilemma, then, those with a high discount acted as if they had a low value of r (denoted r_{low}) whereas those with a low discount on the future acted as if they held a high value for r (denoted r_{high}). It would seem wholly reasonable, now, to suppose we can take the payoffs of the Hawk/Dove scenario and simply transplant the labels r_{low} and r_{high} for Hawk and Dove respectively. That is, a person who discounts the future highly and who thereby exhibits a low value for r will be quite willing to forego honoring any contract in favor of an immediate payoff and will, when dealing with a person who discounts the future less, be able to take advantage of that person. Two persons with long-term perspectives will each realize the full potential of their investments, $V/2$, whereas two persons with heavy discounts of the future will realize only a diminished payoff, $V - D$. Recall now that if $V > D$, the evolutionary strategy was to be a Hawk. Thus, here if $V > D$, it is evolutionarily stable to discount the future greatly. Mutant r_{high}'s—those who place a high value of maintaining long-term contracts—cannot invade a society in which the predominant culture is to operate as it was in the Soviet past with short-term planning horizons.

Of course, not all components of the USSR faired as poorly in their transition to efficient markets as did Russia and Ukraine. The Baltic states of Estonia, Latvia and Lithuania seem to have performed quite well, as did, as we noted earlier, the Central European states of the former Soviet Empire, especially Poland, Hungary, and both parts of the former Czechoslovakia. Thus, we cannot say that it was simply the prior existence of a Soviet-style Communist regime that explains the differences we observe. At the time of this transition it was not uncommon for journalists and other commentators to say that the difference was that Russia and Ukraine had been under Communism for 70+ years whereas the Baltic states and those in Central Europe only suffered a post-World War II experience with Communism. This "explanation," however, is no explanation at all; it is merely an empirically observed fact. It can, though, be made a part of an explanation if we suppose that with 70+ years under Soviet rule, there were few if any persons with the experience of having functioned in a "normal" Western-style economy; but with only 40 or so years of being a Soviet satellite, and with a significant diaspora willing to return to their countries once those countries freed themselves from a Soviet hegemony, there were a great many more persons in those states who held Western values or perspectives when it came to investments and the obligation of contracts. Thus, in these states it was the r_{low}'s who filled the role of mutant.

There remains, though, a problem with this analysis insofar as explaining the experience of Central Europe and the Baltic states. Specifically, even if r_{low}'s are mutants, as long as $V > D$, r_{high} cannot be evolutionarily stable. Mutant r_{low}'s will be able to invade a population of r_{high}'s. However, suppose the payoffs in the preceding table are themselves discounted under the assumption that they merely represent the one-period payoff of a potentially infinite sequence of payoffs. In other words, suppose we express payoffs as in Figure 5.14.

Now notice that even if $V > D$, if r_{low} is sufficiently low relative to r_{high}, there are two equilibria if we interpret the preceding table as a game, which opens the door to all r_{high}'s being evolutionarily stable. Suppose, moreover, that the

0 payoffs are not strictly 0 but some small amount α, in which case our table becomes

Now even if $V > D$, if r_{low} is sufficiently low and r_{high} sufficiently high, only r_{high} is evolutionarily stable. For example, if $V = 5$, $D = 3$, $r_{low} = 0$ and $r_{high} = 0.9$ then Figure 5.15 becomes Figure 5.16, in which case all α need be 0.1 or greater for r_{high} to be the unique evolutionarily stable equilibrium.

	r_{low}	r_{high}
r_{low}	$(V - D)/2(1-r_{low})$, $(V - D)/2(1-r_{low})$	$V/(1-r_{low})$,0
r_{high}	$0, V(1-r_{low})$	$V/2(1-r_{high})$, $V/2(1-r_{high})$

Figure 5.14 Hawk-Dove with discounted payoffs

	r_{low}	r_{high}
r_{low}	$(V - D)/2(1-r_{low})$, $(V - D)/2(1-r_{low})$	$(V - \alpha)/(1-r_{low})$, $\alpha/(1-r_{high})$
r_{high}	$\alpha/(1-r_{high})$, $(V - \alpha)/(1-r_{low})$,	$V/2(1-r_{high})$, $V/2(1-r_{high})$

Figure 5.15 Hawk-Dove with non-zero payoffs and discounting

	r_{low}	r_{high}
r_{low}	1, 1	$5 - \alpha$, 10α
r_{high}	10α, $5 - \alpha$	25, 25

Figure 5.16 A numerical example

None of this, of course, proves that an evolutionary approach holds the key to understanding why one society moves in the direction of becoming corrupt and another does not. There remains, for instance, some dramatic variance in the economic performance of the countries that existed under Communism for an identical period of time, especially if we include the former components of Yugoslavia. In 2011, for instance, Croatia's GDP growth was 0% whereas Estonia's was 7.6%. Nevertheless, such an analysis gives us a more promising approach to a theoretically satisfying explanation than does doing little more than making reference to the length of time one society versus another lived under Soviet-style Communism.

5.7 Key Ideas and Concepts

Prisoners' Dilemma
cooperation
collective action
public goods
externalities
reputation
social norm
tit-for-tat
trigger strategy
discount rate / discount factor
continuation value
recursive game / stationary strategy
coordination
evolutionarily stable equilibrium
evolutionary game

Exercises for Chapter 5

1. For what values of x is the following game a Prisoners' Dilemma?

x, 1	3, −4
1, 3	2, 3

2. What is the minimum fine someone could levy such that the following game is not a Prisoners' Dilemma (payoffs are to the row chooser and column chooser faces an equivalent choice)? What minimum fine renders the game not only not a Prisoners' Dilemma but establishes the "efficient" strategy as dominant?

0	9
−1	3

3. If free competition reigns in an industry, 20 million units of that industry's products will be sold by each firm at a net profit of $1 per item. But if they collude to set a higher price, each firm will sell 15 million units at a net profit of $2 each. If one firm defects to a lower price, its sales will soar to 35 million units while every other firm will sell nothing, and the creditors will begin to circle overhead. Senator Billie Bob proposes a licensing agreement whereby each member of the industry must pay a tax of $.20/item to produce the product at the fixed price that the cartel prefers—ostensibly to insure that "destructive competition" does not "leave hard-working Americans unemployed." What is the upper limit on how much money he can extract from each firm in the form of campaign contributions to his party?
4. If the cartel price for some commodity earns each firm $10 million, if each firm can earn $3 million in the competitive market, and if each firm can, by

defecting from the cartel, capture the entire market, which of the following games best illustrates this situation (all numbers are in millions)?

a.

10, 3	12, 0
0, 12	3, 10

b.

10, 10	20, 0
0, 20	3, 3

c.

0, 0	10, 0
0, 10	3, 3

d.

10, 10	0, 20
20, 0	3, 3

e. none of the above

5. You direct Consolidated Smoke and you must decide whether or not to agree to meet the president of Acme Sludge so that the two of you can fix prices for your similar products, in which case your corporations each earn $220 million. Both of you recognize however that the situation is a Prisoners' Dilemma: at the market price, you each earn $90 million, while if only one defects from the agreement, his corporation earns $300 million and the other corporation earns "zip." Being competitive entrepreneurs with your MBAs, neither of you trusts the other to maintain any agreement reached. An additional danger is that federal antitrust investigators (with probability .4) will detect your agreement, negate the price fixing scheme, and impose a fine of $50 million each. Congressman I. M. Crass, however, proposes to offer legislation that will make the cartel legal and enforceable in a court of law, and that will provide the regulatory teeth to maintain it; he demands some assistance in the next election—say, $50 million from each firm. The problem is he wants his money up front, before the legislature votes, and he can promise only a fifty-fifty chance that the proposed legislation will pass. Assuming that you make the best decision possible in the circumstances, what are your firm's expected profits?
6. Three students, 1, 2, and 3, have enrolled in Political Theory and the instructor has announced beforehand that he will award one A, one C, and one F (no pass-fail option this quarter). Homework does not count towards the final grade and the students have earned scores of 55, 65, and 70, respectively, on their midterm exam. Each student associates a payoff of 2, 1, and −1, respectively, with receiving an A, C, or F, and each knows that if they study at maximum effort they will earn an 80 on the final exam, and a moderate-low effort will yield a grade of 40. Suppose that, *ceteris paribus*, each student prefers as little effort as possible, but is willing to study if

doing so changes that student's grade for the better. A student's final grade is based on his or her relative standing as determined by the sum of midterm plus final exam grades.

a. Assuming that all three students must decide whether or not to study without being informed about the action taken by any other student, portray the situation in strategic form, with letter grades denoting the outcomes in each cell.
b. Is there a determinate final outcome?
c. Suppose students 1 and 2 can observe 3's action beforehand. Does this change their strategies and the final outcome?

7. Two farmers must share an irrigation system, which they use by alternating their access to it day by day. The farmer whose turn it is to extract water on a particular day must choose between taking the allotted share (for a benefit of 0 to himself and a cost of 0 to the other farmer) versus taking more than the allotted share (for a benefit of B to the farmer in question and a cost of C to the other farmer). However, the farmer who must otherwise sit idly by for the day can choose to inspect his opposite number's activities at a cost to himself of K. If an excessive extraction is detected, the farmer is empowered to fine the offender an amount F, which can be kept as compensation for any economic injury.

   a. Assuming that all parameter values exceed zero, and taking a myopic one-day view, for what parameter values is there a pure strategy equilibrium in which the farmer inspects with certainty?
   b. Assuming that there is no pure strategy equilibria, what is the mixed strategy equilibrium?
   c. Suppose that one of the farmers is to be picked at random as the one to use the irrigation system, and suppose the parameters are set such that (take more than allotted share, inspect) is a pure strategy equilibrium. Can we raise the value of F so that the farmers prefer that there not be any pure strategy equilibrium over what they would expect to get from playing the game with the old parameter values?

8. Two lobbyists, 1 and 2, each seek legislation that is diametrically opposed to the legislation sought by the other. Since we cannot construe this legislation as being "in the public interest," they must decide when to contribute money to a legislator's campaign war chest. (Assume legislators "come cheap," so the size of the contribution can be ignored in the calculation of a lobbyist's payoff.) The legislator sees no reelection threat and intends to convert this war chest into a personal retirement fund. But in the science-fiction world of this problem, suppose the law frowns on all such activities and that as a consequence, neither lobbyist dares to make a second offer if his or her first one is refused. The exact nature of the legislator's behavior, however, is unknown, so both lobbyists employ the following model: If the lobbyists commit at times t_1 and t_2 $(0 < t_1, t_2 < 1)$, respectively, and if $t_i < t_j$, then the probability that the contribution leads the

legislator to support i is t_i. If the legislator does not commit to supporting i at this time, he either supports the legislation sought by j with probability t_j or he supports no one with probability $1 - t_j$. The possibility exists, then, that the legislator never commits and that neither lobbyist gets what he wants. Assume that the payoff of no legislation to a lobbyist is equivalent to an even chance that one or the other lobbyist gets his or her way.

a. If lobbyist i ($i = 1$ or 2) makes the first move (by choosing $t_i < t_j$), lobbyist j learns this fact at that time and also learns the legislator's response to i's offer. Letting a strategy be the time at which to make the offer, does this situation have a pure strategy equilibrium, and, if so, what is it?
b. How is the situation changed if an offer and its rejection (but not its acceptance) cannot be observed by the competing lobbyist?

9. Show that if the "normal" strategy in a society playing the Prisoners' Dilemma is TFT, and if the allowable mutants are STFT, ALLD (always defect), and TF2T, then TF2T can invade if the probability of ALLD is low compared to STFT, whereas the opposite is true if ALLD is more common.

6 Agendas and Voting Rules

6.1 Agendas and Voting

Earlier chapters offer several examples of legislative and parliamentary bodies giving coherence to their deliberations and voting by establishing formalized voting rules whereby a set of alternatives is considered and voted on in some sequential fashion until a final outcome is arrived at. Agendas, though, need not be formal well-defined parliamentary procedures dictated by such things as Robert's Rules of Order. For example, as Convention delegates saddled up or stepped into their horse-drawn carriages in 1787 for the trip to Philadelphia, a young James Madison had already squirreled himself away in that city to study and refine his proposal for a radically new national charter. Called the Virginia Plan by historians, the delegates ultimately rejected many of Madison's ideas, including things he deemed critically important, such as having states represented in both legislative chambers, the House and the Senate, in proportion to their populations; a provision for a federal legislative veto over state laws; and a president selected by the national legislature. Nevertheless, having formulated a comprehensive proposal, much of the subsequent debate that summer was dictated or influenced by Madison's draft. Indeed, we should keep in mind that the delegates were, in fact, directed by the then-U.S. Congress to consider only amendments to the current Articles of Confederation whereas Madison's draft tossed the Articles into the trash and started anew. Madison, then, largely set the agenda for the Convention, albeit not in a formally defined way.

Whether formal or informal, game theory, and especially the analysis of extensive forms, seems especially well suited to the study of agendas and addressing a number of issues that arise with respect to them. These issues include examining the extent to which outcomes can be manipulated by choosing one agenda over another, assessing the power of agenda setters and the opportunities for voters to thwart the intentions of setters, the role of a strategic misrepresentation of preferences, the opportunities for manipulating outcomes by either inserting or withdrawing alternatives from an agenda, and the various forms agendas take in different legislative bodies and how those forms might bias outcomes in one direction or another.

Before we get to such substantive matters, however, we should first return to a reconsideration of McKelvey's result, discussed in Chapter 4, concerning dominated outcomes and his "*4r ball*." In doing so, it is also useful to introduce yet another abstract idea that is commonly employed in academic treatises on agendas, *the uncovered set*. Briefly, if S is a set of feasible outcomes, and if x, y and z are three outcomes in S, then *x covers y* with respect to S (which we can write as $x\ C_S\ y$) if and only if x defeats y (generally, in our case, by a majority vote) and for each z in S that y defeats, x defeats z as well. The uncovered set is then the set of outcomes that are not covered (there are variations of this definition to treat such things as indifference and ties, but this definition, which essentially establishes an equivalence between being uncovered and being undominated, is good enough for our purposes as it conveys the essential idea). The close correspondence between uncovered and undominated sets, now, establishes the connection to McKelvey's analysis—namely, *with spatial preferences, the uncovered set necessarily lies within the 4r ball defined by his result.*

Focusing on one special class of agendas called amendment agendas most easily illustrates the relevance of these ideas to the study of voting with formally prescribed *agendas*. An amendment agenda can be represented as a finite vector $(a, b, c, d, \ldots.)$ to indicate that the rule of voting is "a against b, the winner against c, the winner of that vote against d, and so on." Academia being what it is, the fact that amendment agendas have such a simple representation accounts, perhaps, for the fact that it is the most widely studied form despite the fact that virtually no parliamentary or legislative body employs it when considering more than three alternatives. For example, suppose one of the chambers of the U.S. Congress is considering five alternatives: The status quo, q; a bill, b, reported to the floor out of committee; an amendment to the bill, b_a; a substitute bill, s; and an amendment to the substitute, s_a. As we note in Chapter 2, the prescribed order of voting is as follows: First, the substitute is "refined" (i.e., a vote is taken between s and s_a); then the bill reported out of committee is "refined" (i.e., a vote is taken between b and b_a); next, a vote is taken between the winner of the first vote (s or s_a) and the second vote (b or b_a), with the winner then put against the status quo q, where the status quo corresponds to no bill passing. Such an agenda, then, requires a somewhat more complicated representation, such as $((s, s_a), (b, b_a), q)$.

Despite the limited substantive significance of amendment agendas, their analysis nevertheless reveals much about agendas in general, how formalized voting procedures can be used to manipulate final outcomes, and the sorts of questions we might try seek to answer generally by formalizing an analysis of other agenda forms. It doesn't hurt, moreover, that the analysis of amendment agendas is especially simple once we have their vector representation. To see what we mean, consider the agenda (a, b, c, d, e) and suppose that all voters are naïve and fail to look ahead to the consequences of their actions. Then all we need do to determine the final outcome is to determine the majority preference relation between a and b and to choose a winner, then between that winner and c, and so on, with the survivor corresponding to the *naïve voting outcome*. On the other hand, suppose all voters act strategically by looking ahead to see where their votes might lead. In other words, suppose voters work backwards up the extensive form of the game tree or voting tree used to describe the situation. In this case, our vector representation provides a convenient means of identifying the final outcome, *the strategic voting outcome*. Looking again at (*a, b, c, d, e*), suppose *d* beats *e* in a majority vote. Then we leave *d* in the vector and move on to *c*. Suppose *c* defeats *d* but loses to *e*. Since *c* fails to defeat every "surviving" alternative that follows it in the agenda, we eliminate *c*, and move on to *b* with the assumption that our agenda is now (*a, b, d, e*). If *b* beats both *d* and *e* we leave it in the vector, whereupon unless *a* defeats *b, d* and *e*, the strategic outcome is *b*; otherwise, it is *a*.

This might all seem a strange construction, but to understand its logic, let us look at the agenda in Figure 6.1 and consider what must be the case for *a* to emerge as the final outcome under strategic voting. Assuming that *d* beats *e*, that *c* loses to *e* and that *b* beats *d* and *e*, we have underlined the strategic equivalents of each node except those pertaining to *a*. What we see here is that for *b* to survive and challenge *a* in the initial vote, it must defeat the alternatives in the agenda that follow it and that survive, as well as the last alternative on the agenda, *e*, since if *b* is to emerge as the eventual outcome then regardless of whether voting is sincere or strategic, it must defeat this last entered alternative. Now consider *a*. Like *b*, if it is to emerge as the final outcome, it must defeat *e*. If it does so, it must then defeat *d* provided that *d* defeats *e*. In the second vote, *a* will be paired against *c*, but *a* need not defeat *c* since, in this example, *c* is "bumped off" by *e*. Thus, for *a* to emerge as the final outcome under strategic voting, it must defeat *b, d* and *e*—all those alternatives that defeated the "surviving" alternatives that follow it in the agenda.

Some simple results follow from this analysis of an amendment agenda's vector representation. First, it should be obvious that

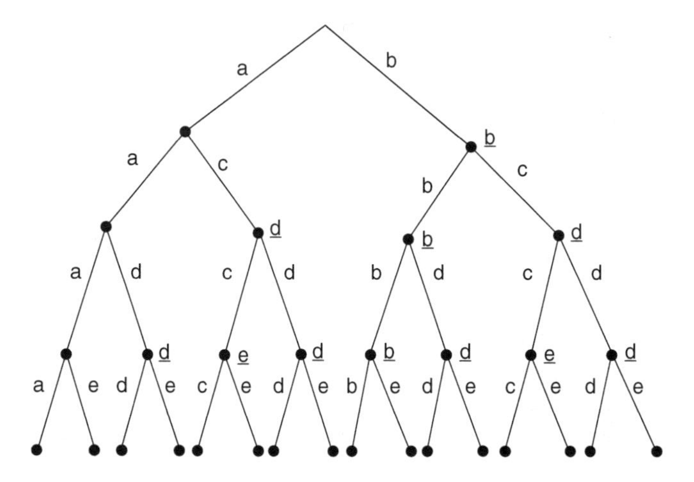

Figure 6.1 A 5-outcome amendment agenda

> *if any alternative on an amendment agenda is a Condorcet winner, it will emerge as the final outcome regardless of whether voters are naïve or strategic.*

Once the Condorcet winner is paired against something, it will defeat that something and anything considered subsequently. But what if there is no Condorcet winner? What if, in fact, the majority preference relations over some or all outcomes on the agenda are intransitive? Consider again an agenda such as (a, b, c, d, e) and for purposes of argument suppose, in applying our previous method of analyzing an agenda using its vector representation, that c has thus far survived by beating both d and e. But suppose c is *covered* by b (i.e., b beats c and everything that c beats; notably, d and e). Then b survives and c cannot emerge as the strategic voting outcome. Alternatively, if c covers b, which minimally requires that c beat b, then b cannot survive. This, then, illustrates the following fact:

> *if x and y are any two alternatives on an amendment agenda A, and if $x\ C_A\ y$ (if x covers y with respect to the other alternatives on the agenda A), then y cannot emerge as a sophisticated voting outcome.*

By itself, though, this result is of limited value since it tells us something about the outcome of an agenda only after the alternatives to be voted on have somehow been set exogenously. And as is often the case in committees and legislatures, political conflicts and disagreements often occur beforehand when it is being decided what alternatives are to be voted on. Madison's Virginia Plan, for instance, suggested no formal agenda; its impact was in offering specific proposals and defining the issues to be considered. Nevertheless, the preceding result about the uncovered set does tell us something about what kind of strategy legislators might pursue when attempting to defeat something on an agenda or the strategies they might pursue when making proposals for alternatives to be considered later in the voting. For example, it tells us that if it has already been established that y will be voted on and if a person finds y distasteful, then if that person can introduce into the agenda something that covers y, this will ensure under strategic voting that y will not emerge as the eventual outcome regardless of what else is proposed.

There is, of course, little game theory here aside from the simple idea of working backwards up voting trees or extensive forms to deduce eventual outcomes when all persons look ahead to the likely consequences of alternative actions. But suppose we ask a somewhat more complicated question. Specifically, suppose we ask what outcomes are likely to emerge if the things that can be put onto the agenda have a spatial representation so that the available alternatives are infinite in number and where people's preferences are like the one's we illustrate in earlier chapters when dealing with spatial voting games. To answer such a question obviously requires a model of the process whereby alternatives can be entered onto the agenda. One possibility is that those voting employ some strict form of parliamentary procedure that dictates who, when and what types of alternatives can be offered. Alternatively, we have the example described by Charles Plott and Michael Levine in 1978 of the California flying club that was considering an upgrading of the airplanes in its inventory ("A Model of Agenda Influence on Committee Decisions," *American Economic Review* 68,1). As they described the situation, any decision about upgrading the club's fleet was a complex one and entailed a number of interdependent choices, including which planes to refurbish, which to sell, how many new planes to buy, and what brand and types of planes to buy if any. Seeking to make their deliberations coherent, the president of the club assigned (somewhat unadvisedly, given his preferences) the design of the agenda to Plott and Levine, who then set out to devise an agenda that served their preferences as opposed to those of the club's president.

It is interesting to note at this point that our two scheming academics volunteered to establish an agenda not simply to make the club's decision process coherent, but also so they could manipulate the result to achieve an outcome close if not identical to their most preferred alternative. And it was only as the consequences of the agenda began to unfold that the club's president realized that Plott and Levine had been something other than purely altruistic. We leave it to the reader to do a bit of study on their own to learn how events ultimately unfolded. All we will add here is that the president would have been well advised to get a copy of this chapter (had it been available at the time) before relinquishing control of the design of an agenda to Plott and Levine.

One implication of Plott and Levine's experience is that the process whereby agendas are established—the selection of alternatives to be voted on and the manner in what that voting is to occur—can be an important part of politics, especially in those legislative assemblies such as the U.S. Congress that proceed by formalized parliamentary rules. And since such situations are likely to be conflictual, with different "players" preferring different outcomes and thus different agendas, it is important that we look at such strategic situations more closely. And here perhaps the most important intellectual contribution to date is R.D. McKelvey's 1976 analysis of endogenous agenda formation ("Intransitivities in Multidimensional Voting Models and Some Implications for Agenda Control," *Journal of Economic Theory* 12). Briefly, formulating endogenous agenda formation as a game, each player's strategy in McKelvey's analysis is an alternative, chosen from a spatial representation of the alternatives, to be entered onto the agenda. Thus, with n players, we let $\mathbf{s} = (s_1, s_2, \ldots, s_n)$, where s_i is player i's strategy and each strategy is a point in the policy space. McKelvey then assumes that the players choose their strategies without knowing each other's choices and the order in which the alternatives entered will be voted on. Letting $U_i(\mathbf{s})$ be the utility or payoff player i associates with a particular strategy vector, and letting $A(s_i)$ be the set of all possible agendas that can be formed if player i chooses s_i, then (taking some liberties with notation and mathematical details concerning continuous preferences and alternative spaces), McKelvey defines $U_i(\mathbf{s})$ thus:

$$U_i(\mathbf{s}) = \min_{\mathbf{s} \in A(s_i)} u_i(X^*)$$

where X^* is the strategic voting outcome of the corresponding agenda. This then defines a game in which each player chooses something to enter onto the agenda simultaneously with all other players and where payoffs are evaluated pessimistically as the worst outcome that might arise given one's choice. McKelvey then proves a rather nice result: namely, that there exists a Nash equilibrium in pure strategies to the described game.

We will bypass McKelvey's imaginative but complex proof, and turn instead to some subsidiary and substantively interesting results. Notice that if X^* is covered with respect to the set of all feasible alternatives, those players who prefer an alternative that covers X^* will have an incentive to shift to that alternative as their strategy. Thus, only uncovered outcomes can correspond to a Nash equilibrium here. And if that is the case, then the set of alternatives that can emerge from the above described agenda setting process must lie in the space's uncovered set, which we know from an earlier result of McKelvey, must lie within his $4r$ ball. So suppose we return to our peculiar example of the grocery store in which people's shopping carts are filled by some sort of collective majoritarian process. Earlier we established that if that process entailed choosing two voters to serve as candidates who are given an incentive to emerge victorious, the final platforms of those candidates would lie inside that $4r$ ball, because in eliminating dominated strategies they would eliminate from consideration as campaign strategies all covered alternatives. Now we see that if, instead, all consumers within the grocery store are required to act as a committee and if they voted using an agenda established as described above, the domain of possible outcomes is the same as if they chose an outcome via an electoral competition between candidates. This is not to say, of course, that the identical outcome would be chosen under the two different procedures. But it does tell us that there is a degree of institutional invariance with respect to final outcomes, where the extent of that invariance depends on the size of the $4r$ ball, which itself is a function of the extent to which the distribution of preferences corresponds to a symmetric function.

Once again, we emphasize that our discussion here considers only one possible way to model the endogenous construction of an agenda. Nevertheless, it does illustrate how game theory can contribute to a comparative assessment of political institutions. Here we see two specific institutional forms that are essentially equivalent if individual preferences render $4r$ "small." In this case, then, the question of whether or not those two forms are interchangeable becomes a question about the nature of individual preferences in specific substantive contexts and the various aspects of reality that might act to keep those forms from performing as described in their abstract representations. The situation is not much different, then, when we say that a falling ball and a feather will fall at the same rate in a vacuum even though we know that our real life experiences with balls and feathers do not match this theoretical prediction. It then becomes incumbent on us as social scientists to discover the 'frictions' that operate in the real world just as it is incumbent on the student of fluid dynamics or aeronautics to understand how the density and currents of air interact with irregularly shaped objects.

6.2 Two Special Voting Rules and Peculiar Results

Jupiter, Saturn and Condorcet Winners: In 1977 the California Institute of Technology's Jet Propulsion Lab, in concert with America's National Aeronautics and Space Administration (NASA) launched two Voyager space probes intended to fly by Jupiter and Saturn and then to move on to Uranus and Neptune. Both probes were loaded with a variety of experiments and measurement instruments designed by and under the direction of a number of distinct science teams, each with its own interests. Some teams were focused on planetary atmospheres, others on the various moons and still others on the composition of Saturn's rings. Needless to say, there were strong preferences among the teams over alternative trajectory pairs. Those teams whose instruments were directed at planetary atmospheres wanted to fly close to the planets themselves, with perhaps one team concerned with Saturn and another with Jupiter. Scientists interested in Jupiter's moons, on the other hand, preferred different trajectories, while anyone interested in Saturn's rings had their own preferred route for the probes. How then to pick a pair of trajectories? The approach taken by two of JPL's (Jet Propulsion Lab) staff, James Dyer and Ralph Miles, was to first identify 35 or so alternative feasible pairs and get the preferences of the 15 science teams involved with each probe's instrumentation (see Dyer and Miles, "An Actual Application of Collective Choice Theory to the Selection of Trajectories for the Mariner Jupiter/Saturn 1977 Project," *Operations Research* 24, March 1976). In addition, efforts were made to get a sense of cardinal preferences (utility functions) via a series of in-depth interviews of the science teams. But now, what to do with this data? The problem was not simply to pick a pair of trajectories but to have the selection process itself seem fair to everyone involved. As a consequence, a variety of somewhat normative measures were calculated, such as the alternative that ranked highest on average, the alternative with the highest Borda Count, the alternative that maximized the sum of cardinal utilities after the utilities were normalized to [0,1] and so on. With all of this, the pairs were narrowed to three, euphemistically labeled A, B and C. After a bit of tweaking of A, Dyer and Miles recommended it to the selection committee, whereupon that committee chose B. While various teams may have grumbled a bit about not getting their first or even second choice, the process was, apparently, deemed fair, but the question is: Why did the committee choose B and not A? We don't know the answer to this, but in a reanalysis of their data it was learned that among the 35 trajectory pairs considered, B was the Condorcet winner. Thus, B could have been realized with any number of voting schemes that would choose such a winner when one exists.

Our interest in JPL's problem and approach does not lie in any professional interest in space probes, but rather in the fact that a Condorcet winner was chosen—and not only chosen but, in combination with the method of selection, was deemed a fair choice. This, perhaps as much as any other thing points to the centrality of such winners as a means of evaluating alternative elections systems. Specifically, if there is a Condorcet winner, there should be good reasons for choosing some other alternative—reasons we cannot formalize here by some set of rigorously stated criteria, but that nevertheless render the procedure employed as being judged by participants to be "fair." To that end, let us consider two voting schemes that we have not yet discussed but that nevertheless receive considerable attention in the literature—the Borda Count and Hare voting.

The Borda Count: People are perhaps never more creative than when inventing rules for voting, and much of the study of voting originated in a dispute between the Marquis de Condorcet and Jean-Charles de Borda over how to elect members to the French Academy of Sciences. Condorcet, objecting to a method proposed earlier by Borda, argued for a system that would necessarily yield a Condorcet winner (not called that at the time, of course) if one existed. Borda's scheme, in its simplest form, required all voters to rank the alternatives from first to last on their ballot. Assuming n alternatives and taking the ballots one at a time, n points would then be awarded to the alternative ranked first on the ballot, $n - 1$ points to the alternative ranked second, and so on. The number of points awarded to each alternative would be summed across all ballots and the alternative with the greatest point total declared the winner. This scheme, then, would, unlike Cordorcet's criterion, necessarily result in an outcome. It is interesting to note, though, that the Borda Count has found preciously few applications in government, and instead its applications have been restricted largely to non-governmental institutions. American sports, in particular, seems enamored of this scheme, using it for example to select the Heisman Trophy winner in college football, to bestow the Most Valuable Player award in professional baseball and, by AP and UPI, to rank college football teams based on the rankings of sportswriters and coaches. The Eurovision Song Competition also employs a version of this method.

Aside from guaranteeing the selection of something even when there is no Condorcet winner, the Borda Count initially at least seemed to solve a somewhat different problem. Note that ordering alternatives by majority rule is but one way we might define what we mean by "society's preferences." But having run afoul of intransitivities, we shouldn't expect the search for an alternative to end, and a seemingly natural one would be to measure people's utilities over the outcomes, add up utility across individuals for each outcome, and then rank the outcomes by their total "utility count." The insurmountable difficulty here, however, is that there is no way to compare and add utility across people. Utility scales are arbitrary and there is no standard, no platinum rod in Paris, with which to calibrate comparisons. Nor, given the purpose of the concept of utility, which is simply to represent a person's attitudes toward risk, will there ever be such a rod that inter-personally standardizes and justifies a comparison. The Borda Count, then, seems to many to be the "next best thing"—something guaranteed to yield an outcome that depends in some way (albeit in a somewhat arbitrary way) on individual preferences.

We suspect that the Borda Count would find wider application were it not for the fact that it exhibits some peculiar properties and can yield some seemingly paradoxical results. First, it is especially susceptible to strategic misrepresentation of preferences. If, for example, an alternative that a person sincerely ranks in the middle of his or her preference is believed to be a serious competitor to those alternatives ranked higher, then there is a clear incentive to rank that alternative near the bottom in one's reported ballot or reported preference ranking. Of course, we already know that any interesting voting scheme is susceptible to manipulation so the incentive to rank one's most preferred alternative first and its viable competitors last is hardly a devastating "defect." More problematical is the fact that the Borda Count might not yield a Condorcet winner if one exists, which necessarily formed the core of Condorcet's objection to this scheme. Consider, for example, the following preferences among 7 voters:

Voters	1	2	3	4	5	6	7
	A	A	A	B	B	B	C
	B	B	B	C	C	A	A
	C	C	C	A	A	C	B

Clearly, A is the Condorcet winner. It defeats B 4 votes to 3, and defeats C 4 to 3. However, the Borda point count for these three alternatives is A with 15 points, B with 16 points and C with 11 points. Now consider these preferences:

Voters	1	2	3	4	5	6	7
	X	X	X	A	A	B	B
	C	C	C	X	X	A	A
	B	B	B	C	C	X	X
	A	A	A	B	B	C	C

Here X receives 22 points, A gets 17 points, B gets 16 points and C gets 15 points. So X is the clear Borda winner. However, suppose X proves to be ineligible after the balloting. It seems only reasonable for the choice to fall to A, which came in second in the count. But suppose instead that X is deleted from all the ballots and new count calculated. Now A gets 13 points, B gets 14 points and C gets 15 points. Paradoxically, then, the alternative that initially ranked lowest now ranks highest.

If this seems strange, consider what happens if we reverse the preferences in the preceding example so that they become

Voters	1	2	3	4	5	6	7
	A	A	A	B	B	C	C
	B	B	B	C	C	X	X
	C	C	C	X	X	A	A
	X	X	X	A	A	B	B

Here X ranks last: A secures 18 points, B 19, C 20 points and X only 13. Thus, C is the winner. Suppose, though, that the voting rule is a runoff ballot using a Borda Count among the top three competitors. With X eliminated from all ballots, the point counts become A with 15 points, B with 14 and C with 13 points. Thus, with the elimination of a seemingly irrelevant alternative, the winner becomes the loser and a loser becomes the winner. One can be certain, of course, that those in a position to control who is and is not allowed on a ballot will seek to take full advantage of such paradoxical possibilities.

One generalization of the Borda method entails giving positive points to only the top $k < n$ ranked alternatives. That is, if ranked first, an alternative is given k points; if ranked second, $k - 1$ points;. . . if ranked kth, 1 point; and any lower ranking, 0 points. So consider these preferences:

Voters	**1**	**2**	**3**	**4**	**5**	**6**	**7**
	A	A	A	B	B	C	C
	B	B	B	C	C	D	D
	C	C	C	A	D	A	A
	D	D	D	D	A	B	B

For $k = 1$ we have simple plurality rule, in which case A is the winner. For the usual Borda method ($k = 4$), C gets 20 points, A and B each get 19 points, and D gets 12 points, so C is the winner. However, with $k = 2$ the winner is B with 7 points (and 6 points for A and C, and 2 points for D). Thus, not only can final outcomes be manipulated by introducing or eliminating otherwise irrelevant alternatives and by strategic voting, but they can also be manipulated by the details of the Borda Count's implementation.

For our final example, let us consider a situation in which, initially at least, the Condorcet winner is selected by a Borda Count; specifically five voters with the following preferences:

Voters	**1**	**2**	**3**	**4**	**5**
	X	X	X	Y	Y
	Y	Y	Y	X	X
	A	A	A	A	A
	B	B	B	B	B
	C	C	C	C	C

As things stand now, X's Borda Count is 23 ($3 \times 5 + 2 \times 4$) whereas Y's is 22 ($3 \times 4 + 2 \times 5$), so the Condorcet winner, X, is selected. However, to fully assess this method's character, notice that if voter 5 votes strategically by reporting X as his least preferred alternative, Y's Borda Count remains 22, whereas X's count drops from 23 to 20. Thus, voter 5's strategic move secures his most preferred alternative. Of course, any one of voters 1, 2 and 3 can now respond to this maneuver by dropping Y in his or her reported preference so as to defeat Y. Suppose voter 1 does so. This response, though, can be countered by having voter 4 drop X to the bottom of his preference, whereupon another voter who most prefers X, say voter 2, can respond to that move by dropping Y to the bottom of its list. The net result of these sequential moves can, then, yield the following reported preferences, in which case neither X nor Y is chosen but A instead (A's count is now 19, X's is 17 and Y's is 16):

Voters	**1**	**2**	**3**	**4**	**5**
	X	X	X	Y	Y
	A	A	Y	A	A
	B	B	A	B	B
	C	C	B	C	C
	Y	Y	C	X	X

This example, then, underscores the fact that a full understanding of the Borda Count's properties requires an assessment of how voters might strategically choose to misrepresent their preferences. And to see that the situation can become complicated here, suppose we simplify matters by assuming that voters 1, 2 and 3, who share the same preferences, vote as a block and that voters 4 and 5 do the same. For purposes of discussion let us limit these two blocks to two strategies: Voting sincerely and voting strategically by dropping their second choice to the bottom of their preference rankings. In this case we get the following "2-person" game (where, in effect person 1—row chooser—has a weight of 3 and person 2—column chooser—a weight of 2):

	YXABC	YABCX
XYABC	X	Y
XABCY	X	A

What we see here, then, are two pure strategy equilibria: One in which the first block votes strategically and the other in which the second block votes strategically. This simple example, then, raises the possibility that under a Borda Count, there may be an inherent indeterminacy as to what outcomes ultimately prevail, depending on who or whether voters coordinate their strategies.

To see this with another example, consider these three preference orders over three alternatives:

Voters	**1**	**2**	**3**
	A	B	C
	B	C	B
	C	A	A

In this case B is both the Condorcet winner and the Borda Count winner. However, suppose each of our three voters contemplates voting strategically by dropping their second-ranked choice to last. Portrayed in strategic form, then, we have the game in Figure 6.2.

Depending now on the voter's attitudes towards risk, this game has 4 potential equilibria. For example, if voter 3, whose true ranking is C B A, is risk acceptant, he might opt to report the preference C A B in order to bring about a 3-way tie. And if voter 2 acts strategically by casting the ballot B A C because C is B's closet competitor, a risk acceptant voter 1 might prefer to bring about

	CBA		CAB	
	BCA	BAC	BCA	BAC
ABC	B	B	3-way tie	A
ACB	C	3-way tie	C	A

Figure 6.2 Strategic voting with the Borda Count

a 3-way tie by misrepresenting his preferences. Our point here is not that voting sincerely is no longer a Nash equilibrium, but that other equilibria appear when voters try to be strategic, and whether or not there are equilibria in mixed strategies depends not merely on ordinal preferences but on attitudes towards risk as well.

The Single Transferable Vote: If people are inventive when it comes to choosing a single alternative from some list, their inventiveness appears to increase exponentially when they set out to fill a committee. Consider the possibilities: First, we can decide to divide the polity up into geographic districts and elect a single representative from each. Even this simple procedure admits of several variants, where the two most commonly used are simple plurality rule or majority rule in which we require a runoff election between the two top vote getters if no one wins more than 50% of the vote on the first ballot. And here we have some sub-variants wherein the runoff is simultaneous with the first round of voting or, as is more common, conducted at a later date. In the simultaneous case voters rank the candidates and if no one is ranked first on more than 50% of the ballots cast, all but the top two first place vote getters are eliminated from the ballots cast and the votes recounted. Yet another system is that of approval voting whereby voters have the opportunity to "approve" of as many candidates on the ballot as they wish, with the candidate receiving the most approvals being elected. Alternatively, we can make the polity into a single large constituency and require that competition be among parties (or equivalently, among lists of candidates). Seats in the legislature can then be allocated among the lists in proportion to their share of the vote. Here, of course, people worry about how to treat fractional seats, and needless to say any number of alternatives have been proposed and used. There are also two intermediate cases between geographically defined districts and proportional representation by party list. The polity can be divided geographically again, but where it elects several members from each district, either by party list or where the top vote getters are selected (and here we can add the variation wherein voters get a single vote or as many votes as there are seats to fill from a district).

Of course, when designing a legislature, it is often unclear which electoral system advantages whom; in which case, what we often see is a compromise in which a share of the seats are allocated according to party lists and the remainder contested for in single-member districts. This of course opens the door to a debate over what share of the legislature should be elected by each of the two methods, wherein surprises arise among those unfamiliar with democratic practice. Upon the tottering and subsequent collapse of the Soviet empire in Central Europe in the late '80s and early '90s when reformers and incumbent Communist regimes negotiated electoral reforms alongside various power sharing arrangements, the Communist parties in countries like Poland and Hungary assumed that their organizational structure gave them an advantage in single member districts while reformers, largely agreeing with this view, assumed that proportional representation would allow them to secure at least a share of seats. Of course, what the Communists failed to appreciate was that after 40 years of being in power, their electorates had largely decided that a wholesale cleansing was in order, in which case in the long run, at least, it probably mattered little what electoral scheme was adopted (though in the short run, it is possible to imagine voting schemes that would have delayed the fall of Communism). Things are no less interesting in established democratic federal states. Suppose there are natural geographic subdivisions to a polity with unequal populations (e.g., states in the United States, provinces in Canada, and Lander in Germany). In this case we have a variety of options, including dividing each region up into a set of single member districts or, as was practiced in Japan and Taiwan, if a district is entitled on the basis of population to k seats, elect the top k candidates when counting the ballots. And even this system allows for variants whereby voters are given either a single vote or k votes (and here yet another sub-variant is either to allow or disallow voters from casting all k of their votes for a single candidate). Alternatively, we might consider conducting proportional representation (i.e., party list) contests in each region. But consider this variant of simple proportional representation: Suppose each party in each region submits its own separate list, but in determining a party's share of seats in the national legislature, we take its national vote total. Thus, if a party wins X% of the vote nationally and if, of that vote, Y% came from a specific region, then one possibility (see, for instance, Germany) is to fill Y% of the X% of the seats to which the party is entitled from the list submitted by the party's organization in that district.

At this point we suspect the heads of many readers will be spinning with eyes glazing over the confusion of possibilities. This brief (and quite incomplete) survey, though, should give the reader some idea as to the opportunities available to those who might attempt to manipulate things by influencing the choice of electoral procedure. Confounding such things though is the fact that if voters vote strategically, the performance of various procedures can be exceedingly difficult to ascertain. If calculation and the substantive informational requirements of strategic voting are a challenge in a well-defined procedure such as in amendment agendas, imagine what they must be under the various rules for filling the seats of parliaments and legislatures. And indeed, political research to date has only scratched the surface in understanding the nature of strategic voting in but a handful of possibilities. It might be argued, in fact, that if one thinks for whatever reason that voting should elicit sincere preferences, then the best procedure is one that is so complicated that it becomes unlikely that voters will vote strategically by misrepresenting their preferences (as opposed to sincerely) because it is unclear what kind of misrepresentation is in a person's interest.

Faculty Politics and Oscar Nominations: In the 1980s, the Government Department at the University of Texas was highly fractured, split between older tenured members and younger untenured faculty, between those who had been newly recruited from outside the university and paid handsome salaries versus those who had been in the department for years and whose salaries had languished in the process, and split as well between those who sought to infuse the study of politics with the sort of material offered in this volume versus those who argued that formalism was inadequate to develop a full understanding of the complexities of political process. Such disputes are commonplace in academia and hardly unique to Political Science or Government faculties. But what made things especially contentious was the university's decision to turn an important administrative task over to an 8-member executive committee elected by the department. Specifically, while the university would determine the department's overall salary budget increase (if any) for the forthcoming academic year, the executive committee would decide how that budget was to be allocated among the department's 50 or so members. Naturally, the various cleavages within the department all sought some form of representation within the committee, but people were at a loss as to how to organize the vote to achieve that end or even an approximation to it. The procedure that was ultimately selected and that, to our knowledge, is still in place today was Hare voting.

There's little reason to suppose that many people know what Hare voting is, unless, of course, you are an aficionado of Hollywood and know how a movie can become an Oscar nominee for Best Picture of the Year. Hollywood and the world produce thousands of pictures per year, with the vast majority being relegated to after-midnight television. But a multitude do make it to a theatre, and it is from this bag that the approximately 5700 members of America's Academy of Motion Pictures must nominate candidates for Best Picture. And like the government faculty of the University of Texas, there must be more than one nominee (at least 5) but unlike that faculty, the actual number is not fixed (it can be up to 10). Needless to say, not all 5700 members can meet in some room and hold a coherent discussion as to how many and who ought to be nominated, so instead each member mails in a preferential ballot that ranks their five most preferred candidates. The question, then, is: How does the Academy (actually, the accounting firm of Price Waterhouse) aggregate these ballots to construct a list of between 5 and 10 nominees? The answer, as the reader should suspect, is again a version of Hare voting (and it should also be noted that nominations in all other categories are subject to some version of Hare voting as well).

Hare voting, named after its ostensible inventor in 1850, Sir Thomas Hare, but also referred to as the *single transferable vote system*, seeks to ensure proportional representation when a committee of m members must be selected from $n > m$ candidates within a single constituency. Its potential advantage is that it serves as an alternative to artificially constructed divisions within a polity such as when we draw geographically defined district boundaries in a single-member district system. Its other ostensible advantage is that its complexity makes strategic voting especially difficult. In its simplest form this procedure, described as a step-by-step algorithm, operates as follows:

1. Each voter, $j = 1, 2, \ldots, v$, begins with a voting weight, w_j, of 1.
2. Each voter casts a ballot that ranks the n candidates from first to last (we could assume that voters submit partial lists, but we will ignore this complication here).
3. Letting

$$W_i = \sum_{j=1}^{v} \delta_j w_j$$

where $\delta_j = 1$ if voter j's ballot currently ranks candidate i first, and equals 0 otherwise, then candidate i is elected if

$$W_i \geq q = It\left[\frac{v}{m+1}\right] + 1$$

where "*It*" represents "the integer portion of."
4. There are now two possibilities:
   a. If $W_i \geq q$ fails to hold for any candidate, strike from all ballots the names of those candidates who receive no first-place votes. Then delete from the ballots the names of those candidates who receive the fewest first-place votes. (If the number of candidates surviving at any stage in this process equals the number of seats to be filled, then those candidates are elected.) In the event of ties, use a fair lottery to eliminate candidates. Return to step 3.
   b. If $W_i \geq q$ holds for one or more candidates, elect all candidates for whom $W_i \geq q$, and delete their names from all ballots.
5. If candidate i is elected, then, after deleting i from all ballots, set

$$w_j = (W_i - q)/K$$

for voters who had i ranked first on their ballots at the time of i's deletion where K is the number of ballots that rank i first. Return to step 3 unless all committee positions are filled.

There is no reason to suppose that any of this makes intuitive sense and indeed one of the objections to its use is the supposition that it is far too complicated a system for voters to understand. At a minimum, though, the reader should see why the quota q requires that we divide v by $m + 1$ rather than, say, m. If, for instance, we are filling a single seat ($m = 1$), the definition of q requires that the selected candidate secure a majority of the vote as opposed to all of the vote if $m = 1$. And if two candidates are to be elected, than dividing by $m + 1$ simply requires, logically enough, that the quota be a fraction more than one third of the vote. In any event, to illustrate this voting method's character, let us consider a few numerical examples.

Example: Suppose nine voters, who must choose two candidates from the set $\{a, b, c\}$, submit the following ballots ranking these candidates:

1	2	3	4	5	6	7	8	9
a	*a*	*a*	*a*	*a*	*b*	*b*	*c*	*c*
b	*b*	*c*	*c*	*c*	*a*	*c*	*a*	*b*
c	*c*	*b*	*b*	*b*	*c*	*a*	*b*	*a*

The quota q, in this instance, is the integer part of $9/(2 + 1) + 1 = 4$, which candidate *a* satisfies. Thus, *a* is elected, and is deleted from all ballots. However, notice that *a* actually receives 5 votes—one more than the quota. In this case, the first five voters—those who ranked *a* first and who can be thought of as having used four fifths of their vote to elect *a* – are each reassigned the voting weight $(5 - 4)/5 = 1/5$. At this point, the vote for *b* totals 12/5 whereas the vote for *c* totals 13/5. Since neither of the remaining two candidates meets the quota, candidate *b*, with the lowest vote total, is eliminated and *c* joins *a* on the committee.

Suppose now that one or more of these voters contemplate voting strategically—that is, let us ask whether sincere voting is a Nash equilibrium. Specifically, consider whether a voter holding the preference "*a* preferred to *b* preferred to *c*" secures any advantage by ranking *b* above *a*. If one such voter does so, then candidate *a*'s vote equals the quota but *a* has no excess votes to transfer to *b* or to *c* after being elected. After *a* is elected and eliminated from all ballots, candidate *b* has four first-place votes while *c* has five. Candidate *c*, then, is again elected. Thus, insincerity in the form of ranking *b* above *a* is fruitless.

Example: Although the preceding example illustrates a situation where sincere voting corresponds to a Nash equilibrium, it is uninteresting in one sense—the Condorcet winner, *a*, is elected regardless of whether any individual defects from sincerity. Generally, we prefer that Condorcet winners be selected if they exist, but there is no guarantee that Hare voting elects such a candidate. To see this, suppose 38, 37, and 24 voters, respectively, hold each of the following three preference orders over five candidates:

1	2	3
a	*b*	*c*
e	*d*	*d*
d	*e*	*e*
b	*c*	*a*
c	*a*	*b*

If three candidates are to be selected, q equals the integer part of $99/(3 + 1) + 1 = 25$ and the final outcome is $\{a, b, e\}$ – *a* and *b* are elected immediately with 38 – 25 = 13 and 37 – 25 = 12 excess votes, respectively, which, after eliminating *a* and *b* from all ballots, gives 13 votes to *e*, 12 to *d*. Since no candidate meets the quota, the candidate with the fewest votes, *d*, is eliminated, electing *e* with 25 first-place votes. But *d* is the Condorcet winner and since it is never the case that any individual is pivotal, no one has an incentive to defect from sincerity by casting an insincere ballot.

This example, however, is reminiscent of those voting situations in which "bogus" equilibria exist whenever no individual voter is pivotal and where those bogus equilibria can be eliminated only by applying the notions of subgame perfection or weak domination (e.g., as when everyone is unanimous in their preference orders but where, for whatever reason, everyone lists the least preferred alternative first, in which case no voter is pivotal and thus no voter can improve the outcome by unilaterally altering their revealed preference). Thus, ascertaining whether sincere voting is a reasonable equilibrium in this example and learning whether the Condorcet winner is unlikely to be elected requires the application of these ideas. Instead, though, let us consider another example that shows that Condorcet winners need not be elected in equilibrium even when individual voters are pivotal.

Example: Suppose 25, 25, 25, 8, 8, and 8 voters, respectively, hold the following six preference orders over four candidates:

25 votes	25 votes	25 votes	8 votes	8 votes	8 votes
a	*b*	*c*	*d*	*d*	*d*
d	*d*	*d*	*a*	*b*	*c*
b	*c*	*a*	*b*	*c*	*a*
c	*a*	*b*	*c*	*a*	*b*

If three candidates are to be elected, then $q = 25$, and the sincere voting outcome is $\{a, b, c\}$ even though *d* is a Condorcet winner. Notice now that if one of the voters whose preferences correspond to the first order above (i.e., $a > d > b > c$) moves *d* up into first place on his ballot, the outcome becomes $\{d, b, c\}$. But under any reasonable assumption as to how preferences over individual candidates relates to preferences over sets of candidates, $\{d, b, c\}$ is the voter's least preferred possibility (e.g., if a voter's preference for $\{d, b, c\}$ is given by a separable utility function $u(\{d, b, c\}) = u(d) + u(b) + u(c)$). Thus, this voter (and by similar reasoning, all others), has no incentive to shift unilaterally to an insincere ballot in order to secure the Condorcet winning candidate.

After admitting that Condorcet winners need not be selected by Hare voting, there is one final possibility that we might consider—namely, that the strategic misrepresentation of preferences, whenever it occurs, renders the selection of a Condorcet winner more likely. However, as our final example shows, insincere voting here can actually lead away from the selection of such a winner.

Example: Ignoring the preferences in parentheses for the moment, suppose 18, 17, 32, and 32 voters respectively hold the following four preference orders (ignore the alternatives in parentheses):

18 voters	17 voters	32 voters	32 voters
a	a (c)	c	b
b	c (a)	b	a
c	b (b)	a	c

If everyone votes sincerely and if two candidates are to be selected, a is elected first because its vote, 35, exceeds the quota of 34, at which point b is elected because its share of a's excess vote exceeds c's share, breaking the tie between b and c in favor of b. So the Condorcet winner, b, is elected. Now consider the incentives to be strategic. In particular, suppose a voter of the second type casts a ballot that corresponds to the preference in parenthesis. As before, a is elected with 34 votes, but now c's vote exceeds b's so b is eliminated on the second round. The reader should confirm that this single insincere ballot is an equilibrium strategy.

Hare voting and its variants are not widely used, in part because of its complexity. In fact, what we have summarized here with respect to strategic voting and the selection of Condorcet winners pretty much exhausts our theoretical knowledge of this voting scheme. But Hare voting is also not widely used for purely political reasons. In the United States, for example, its sole use is for the city council and school committee of the city of Cambridge, Massachusetts. The city council of Cincinnati, Ohio, employed it from 1924 to 1957, but was subsequently abandoned in favor of a district system. And its reasons for abandonment seem evident: As the city's black population increased, under Hare voting, that population's representation on the council increased as well. Any reasonable population projections past 1957 suggested to the majority white population that black representation on the council would continue to increase were Hare voting maintained. Hence its discontinuance.

Oscar Reconsidered: As noted earlier, Oscar nominations for Best Picture of the Year employs a variant of Hare voting, wherein the number of movies nominated can vary between 5 and 10. This variability in the final number of nominees derives from the particular way in which Hare voting is implemented. Suppose 5500 of the 5700+ members of the Academy of Motion Pictures cast a ballot. To be even in the running for a nomination, a movie must rank first on at least 5% of all ballots cast. Otherwise it is eliminated from all ballots. Now, since 10 is the maximum number of nominees, the quota is calculated by letting $q = 10 + 1 = 11$, so that any movie appearing at the top of $5500/11 = 500$ ballots is nominated.

The Borda Count and Hare voting hardly exhaust the list of voting schemes that might be employed for electing individual candidates or filling a committee. In addition to plurality rule and majority rule with a runoff, which we will examine shortly, two other schemes have been proposed and employed in limited contexts. One is *approval voting* whereby instead of simply indicating a first choice or having one's ballot rank all candidates, a voter indicates all candidates of which he or she "approves"—is willing to see elected or serve on a committee. The difficulty with analyzing this scheme, however, so as to establish its formal properties is that we have no natural way of identifying where any one voter would draw the line between approving and not approving of a candidate. Thus, if given a voter's preference order even with cardinal utility numbers attached, we have no idea where he or she would "draw the line," and two voters with identical preferences might draw their lines differently. Thus, while any number of claims have been made by proponents of this scheme, including that it would act to increase turnout and be more likely to elect Condorcet winners than, say, plurality rule or the Borda Count, these claims are difficult if not impossible to validate. Another scheme is *cumulative voting*, which perhaps can best be viewed as a competitor to Hare voting. Suppose voters are filling an n-member committee and are given n votes. The usual method here is to have voters vote for n or fewer candidates, with the candidates receiving the most votes elected. A variant on this, though, is to allow voters to cast more than one of their n votes for the same candidate, including casting all n votes for that candidate. However, rather than pursue a seemingly endless list of alternative voting schemes and their many variants, we turn next to the two most common methods for filling a single office by election, plurality rule and majority rule with a runoff.

6.3 Two Alternative Rules for Electing Presidents

Plurality Rule: Choosing a method of electing a president is often a contentious constitutional issue for newly emerging democracies, with debate commonly focusing on two alternatives: plurality and majority rule with a runoff. Runoff schemes seem to have gained the ascendancy, with objections against simple plurality rule being of two sorts. First, there is the concern that someone will win with significantly less than a majority of the vote and thereby preclude that person from being viewed as "legitimate." Second, there is the concern that the plurality winner cannot defeat any of his or her challengers in a head to head vote—that the winner will not be a Condorcet winner and may in fact be a Condorcet loser. To explore this second issue, consider a plurality contest among three candidates, x, y and z. There are then six possible strict preference orders within the electorate as follows, where n_i denotes the number of voters holding preferences of the ith type:

n_1	n_2	n_3	n_4	n_5	n_6
x	x	y	y	z	z
y	z	x	z	x	y
z	y	z	x	y	x

To simplify our discussion, however, we will suppose that the election also concerns a single issue and that voter preferences are single peaked over that issue, in which case we cannot have n_i be non-zero for all six preferences—there is no ordering along a single dimensions of x, y and z such that more than four preferences are single peaked. Suppose, for example, that preferences are ordered on the issue (from left to right), x, y and then z, in which case we cannot portray the preferences of voter types 2 and 5 as single peaked. The reader is free, now, to confirm that no ordering of the alternatives on a line yields all six preference types simultaneously single-peaked. It is also useful at this point to establish the following fact:

> *If voters have single peaked preferences in a one-dimensional election and if, among three candidates, there exists a Condorcet winner with no candidate ranked first by a majority, the candidate positioned between the other two candidates is the Condorcet winner.*

The precondition of this result that no candidate is ranked first by a majority of the electorate is intended to preclude the case in which all candidates are to the left or right of the median preference. Now let candidates A, B and C be ordered from left to right alphabetically, and let M_{AB} and M_{BC} be the midpoints, respectively, between A and B and between B and C. Simplifying matters by assuming that preferences are symmetric about their ideas, and letting N(a,b) denote the number of voters with ideal points in the interval [a,b], then the number of voters who prefer B over A is $N(M_{AB},\infty)$, which must exceed 50% of the vote since, by assumption, $N(-\infty,M_{AB})$ is less than 50%. By the same reasoning B is preferred by a majority over C. Candidate B, then, is a Condorcet winner.

Limiting our attention now to situations in which no candidate gains a majority under sincere voting since if there is such a candidate, it is a Condorcet winner and there are no incentives for strategic voting of any sort. Taking account of the arbitrariness in the labeling of the alternatives, if we let candidate x be the plurality winner, there are four cases to be considered with respect to the majority preference relation:

Case 1: x > y > z, so the plurality winner, x, is also the Condorcet winner
Case 2: y > x > z
Case 3: y > z > x, so the plurality winner, x, is the Condorcet loser
Case 4: x > y > z > x, so a Condorcet winner does not exist. This case, however, can be ignored since with a single dimension and single-peaked preferences, a Condorcet winner necessarily exists.

Case 1: If x is the Condorcet winner, neither the preference order n_4 nor n_6 can arise—for the one-dimensional case with single peaked preferences: If no one wins 50% of the vote, then x must lie between y and z, thereby precluding the orders y > z > x and z > y > x. Clearly neither those with type 1 or 2 preferences has an incentive to vote for their second choice, those with type 3 preferences cannot do anything to make y more likely to prevail, and those with type 5 preferences have no incentive to take advantage of the support given to y by type 3 voters to bring about a victory for y by voting for it.

Case 2: Here x is the plurality winner, but y is the Condorcet winner. In this case, with single peaked preferences, y must lie between x and z, in which case type 2 and 5 preference orders cannot exist. Clearly, those with type 1 preferences have no incentive to vote strategically since their candidate is already winning, nor do those with type 3 preferences have an incentive to strategically vote for x since x already has a plurality. Type 4 voters can try to make z a plurality winner by voting for it instead of y, but this move can be countered by type 3 voters voting for x instead of y. Thus, z being the plurality winner cannot be a Nash equilibrium. On the other hand, if type 6 voters vote strategically for y instead of z, y becomes the plurality winner and no one can counter this move (since $n_3 + n_4 + n_6 > n_1$, otherwise x is necessarily both the plurality and Condorcet winner). Thus, the emergence of the Condorcet winner from strategic voting is a Nash equilibrium.

Case 3: Once again, if y is the Condorcet winner, it must lie between x and z so that preference order types 2 and 5 again cannot exist. Now, however, unlike in Case 2, if type 4 voters strategically vote for z instead of y, they give z a plurality, whereas if type 6 voters unilaterally vote strategically for y instead of z, they give y a plurality. And if both types vote strategically simultaneously, they cancel each other out and x again wins. Looking only at these two types, then, we see that unless n_4 or n_6 is great enough to render voting strategically for one's second choice a dominant strategy, types 4 and 6 are in effect playing a "battle of the sexes" game with two alternative and distinct equilibria such that an equilibrium is guaranteed to prevail only if the players coordinate their choices. Type 1 voters, however, can forestall such possibilities by voting for y instead of x, in which case y wins a plurality with certainty. Whether type 1 voters prefer the certainty of y to a probability that types 4 and 6 inadvertently yield x will depend, then, on the degree of risk aversion among type 1 voters. And although there are other conditions to be considered such as when types 4 or 6 have dominant strategies, what we have established here is a case in which there is a Nash equilibrium with sophisticated voting, but the Condorcet winner need not prevail in that equilibrium.

What we have learned here, then, is that unlike in, say, amendment agendas (which, of course, involves an entirely different substantive scenario than multi-candidate elections), there is no guarantee that Condorcet winners emerge under plurality rule even if voters are strategic. The problematical case is Case 3 and thus we might ask how likely this case is to pertain to plurality elections. We suspect, in fact, that it is not altogether uncommon. For example, suppose preferences are distributed uniformly over the interval [0,1], that x is at 0.2, y at 0.6 and z at 0.7. Then, in a straight plurality vote, x gets 40% of the vote, y gets 25% and z gets 35% so x is the plurality winner. However, y clearly is the Condorcet winner since it is closer to 0.5 than x or z, and z defeats x since it is closer than x to 0.5 as well. We note, in fact, that this situation most likely corresponds closely to Chile's 1970 election in which Salvador Allende won the plurality vote with a 36.2% share while the centrist Radomiro Tomic received 27.8% and the right-leaning Jorge Alessandri won 34.9%. The fit with our example comes from the supposition that Tomic's Christian Democrats were most likely ideologically closer to Alessandri's Nationals than to Allende's Socialists and Communists. Allende's victory, then, has as much to do with the center and right parties being unable to coordinate and compromise. Other similar examples include the 1987 South Korean presidential contest and the 2000 Taiwain presidential election.

Majority Rule with a Runoff: If simple plurality rule has a competitor, that competitor is electing presidents under majority rule with a runoff. Here, as well as with simple plurality rule, there are a number of issues to be considered before proclaiming one rule superior to another. One concern with a runoff system, for example, is the question of what incentives it creates for a contentious multiparty system or for a splitting up of existing parties into smaller factions. In this context one can also debate whether requiring a strict majority to forestall a runoff is superior to setting some lower threshold. If, for example, a threshold of say 40% is set, we perhaps lose the legitimacy that comes with the winner ultimately securing a majority of the vote in a runoff, but it would also seem to lessen the incentives for the formation of small special interest parties whose sole purpose is to keep candidates from securing a majority on the first ballot. Here, however, we will look at the same issue we addressed with respect to plurality rule; namely, the likelihood that a Condorcet winner is selected if such a winner exists. We lay out our analysis using the same four cases as before with the precondition that no candidate wins 50% of the vote under sincere voting and voters have single-peaked preferences (see Niou, "Strategic Voting under Plurality and Runoff Rules," *Journal of Theoretical Politics* 13,2, April 2001).

Case 1: If the Condorcet winner is also the plurality winner under sincere voting, then regardless of whether y or z survives the first round along with x, x will prevail in the runoff. So voting strategically for z or y is inconsequential.

Case 2: If y > x > z, the plurality winner, x, will lose to y, the Condorcet winner, in the runoff if no one votes strategically in the first round. The sole hope for defeating y is for z to enter the second round alongside x. If z and x are paired in the runoff, x wins, so only those who prefer x to y have an incentive to defect unilaterally from sincere voting—those with type 1, 2 or 5 preferences. However, since y is the Condorcet winner, if the election is one dimensional, then as before preference types 2 and 5 cannot exist and only type 1 exists to vote strategically—in this case for z. Such a unilateral move, however, cannot be an equilibrium since then type 6 voters would act to counter it by voting for y. And with types 3, 4 and 6 voting for y, no one thereafter has an incentive to shift unilaterally to any other strategy.

Case 3: Here y > z > x, and the plurality winner x is the Condorcet loser. Here, unlike in Case 2, type 1 voters have no incentive to initiate strategic voting since their first choice cannot win in the second round regardless of whether x is pitted against y or z. Type 2 and 5 voters again cannot exist in the one-dimensional context, which leaves type 6 voters, who hold the preference z > y > x, a potential candidate for strategic voting. But voters of this type cannot bring about a victory for z and thus sincere voting by everyone is the Nash equilibrium.

Aside from illustrating the combination of the ideas of unidimensional spatial elections, Nash equilibria and strategic voting, the preceding discussion also tells us something about candidate entry in these election systems. Specifically, if two candidates compete initially and a third candidate must consider whether or not to enter the contest, we know now that in simple plurality rule contests, strategic voting cannot preclude a Condorcet loser from winning (Case 3 in our assessment of plurality rule). This opens the door to entry by a third candidate even if a pre-existing candidate is a Condorcet winner. In other words, the theoretical possibility of strategic voting under plurality rule does not necessarily deter the third candidates from entering the contest even when a Condorcet winner exists. Under the runoff rule, however, the Condorcet winner always wins when one exists. Consequently, there is less of an incentive for a third candidate who is not a Condorcet winner to enter the competition if the candidate's objective is to win the election. We appreciate that this assessment ignores other political considerations such as the incentives of third parties to simply begin building an organization for the future or, by their mere existence, to effect the relative salience of issues. Such possibilities, of course, open the door to other possible modeling exercises.

It is, of course, true that simple plurality rule and majority rule with a runoff are but two of the multitude of rules that might be employed to fill a single elected office. There are those, for instance, who advocate an instant runoff scheme wherein voters submit a preferential ballot and if no candidate crosses the required threshold of first place votes, all candidates except the two with the most first place votes are eliminated from the ballots and the votes recounted. But at this point the reader should note that our discussion of the Borda Count and Hare voting makes no use of spatial preferences while our comparison of plurality and runoff systems does so but only by way of assuming single peakedness so as to eliminate possible ordinal preferences. And quite frankly, it remains to be seen what sorts of general results can be established if a comparative analysis of election systems were to be imbedded somehow in a system that imposes a specific Euclidean topology on preference. Sadly, that research does not exist for us to review, which, we suppose, merely points to the fact that it is not yet the case that everything is known about everything.

6.4 Controlling the Issues Voted On

In 1787, the delegates who drafted the U.S. Constitution moved back and forth among the various issues that concerned them—the powers of the presidency, selection or election of a chief executive, the structure of representation of the states in the national legislature, the powers of the new federal government with respect to the individual states, methods of appointing federal judges, and so on. Moreover, as we noted earlier in this volume, they often voted so as to seemingly decide an issue, only to return to it days or weeks later to vote on the matter again as if their earlier vote hadn't occurred. One can readily imagine, then, a convention that proceeded differently. The delegates were hardly opposed to forming subcommittees such as the Committee of Style and the Committee of Unresolved Matters, so it's conceivable that, as they had done in 1776 when debating independence, they might have first formed various subcommittees where each would address one of the core issues that concerned them. The constitution could then be constructed by combining the decisions of the several subcommittees, perhaps using a Committee to Resolve Inconsistencies. The delegates also well understood that allowing amendments to the various subcommittee reports could yield interminable debate. They did after all report their final product as a take-it-or-leave-it proposition to the individual state ratifying assemblies since they knew their efforts would be for naught in terms of achieving closure if each state made ratification contingent on the adoption of amendments. Thus, we can also imagine the convention operating under a "closed rule" whereby the delegates would have to consider each subcommittee's report with a take it or leave it vote. How different might things have been, then, if the Constitutional Convention had operated in what might seem a more consistent way—allowing for debate and amendment in the ratification process just as it had in its deliberations, or, alternatively, operating with a closed rule that precluded amendments to the decisions of their issue-specific subcommittees?

The point we want to make is that deciding how to discuss and debate complex things is very much a part of the agenda setting process and often involves more than merely choosing which alternatives will be placed on a formally defined agenda. And so the question here then is ascertaining how the theoretical perspectives taken in this volume can contribute to our understanding of this somewhat less precise agenda process. To that end we begin with Figure 6.3, which once again portrays the spatial preferences of three voters over two issues and a status quo at the point x_o. But now, in lieu of supposing that any potential move away from x_o is to be decided by, say, a two-candidate election or by a mechanism wherein 2 or 3 alternative proposals are made and placed on a formal agenda, suppose this 3-member committee operates somewhat less formally with a round-table discussion. Here, however, let us consider two alternatives as to how they might proceed. One alternative is where the committee operates in a "free-wheeling" way—no formal votes are taken until two or more of its members vote to end debate with a specific alternative identified as the alternative to the status quo. Another procedure has the committee deciding beforehand to consider the two issues separately, perhaps debating one in the morning and the other in the afternoon, and that when any two alternatives are discussed they are allowed to differ on only one of the two issues. The first scenario, then, might correspond to a situation in which our three members act as a "committee of the whole" and where quite possibly it is mere exhaustion that terminates discussion. Since the tools for analyzing such loosely structured situations must await a later chapter we instead consider the second method, that of issue-by-issue deliberation and voting.

Figure 6.3 also displays a dashed line h_1 through the status quo parallel to the horizontal axis. If we suppose that our committee takes up issue 1 first,

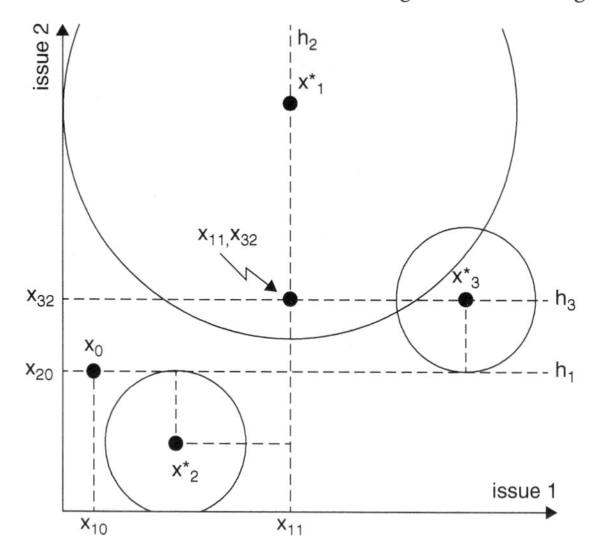

Figure 6.3 Issue-by-issue voting, circular indifference contours

then they are in essence limiting themselves to debating alternatives along h_1. And here, if spatial preferences are of the sort considered thus far in this volume—specifically, represented as drawn in the figure by circular indifference contours—then preferences along h_1 will each be single peaked and where each committee member's ideal preference on h_1 is given by dropping a perpendicular line from their ideal point in the 2-dimensional space to it (since that point corresponds to the point of tangency of that voter's indifference contours to h_1). In our example, then, voter 1's preference is the median preference and thus the issue-constrained winner. Now suppose the committee takes up issue 2, and, with the status quo now changed to (x_{11},x_{20}), it must vote on alternatives along the dashed vertical line h_2. Dropping perpendiculars to that line tells us that voter 3's ideal along h_2 is the median preference, in which case the point (x_{11},x_{32}) is now the issue-constrained winner.

Now, however, suppose our committee debates and votes instead on issue 2 first. In our example, the reader should be able to confirm that x_{32} would still be the value arrived at on this issue and that after considering issue 1, the committee, as before, would arrive at the outcome (x_{11},x_{32}). Indeed, the reader should also confirm that regardless of which issue is voted on first, and even if issues are reconsidered, the point (x_{11},x_{32}) remains the final outcome. This is not to say, of course, that (x_{11},x_{32}) would prevail if the committee allowed itself to consider alternatives that proposed a change on both issues simultaneously since we already know that with three voters portrayed as they are here that there is no overall Condorcet winner—the social preference under majority rule is wholly intransitive. But by limiting themselves to discussions and votes that consider only one issue at a time, the committee has, in effect, guaranteed itself a determinate outcome. In this context, then, we can anticipate either of two possibilities: Either the committee will choose (x_{11},x_{32}) or a majority of the committee, seeing alternatives that are better for it than (x_{11},x_{32}), will attempt somehow to rid itself of its issue-by-issue constraint.

Our example, however, is of a very special sort; namely, it portrays preferences with circular indifference contours. This means that what a person prefers on one issue is invariant with whatever is the status quo on all other issues. Thus, we should not be surprised to see an invariance with the order with which the issues are considered. Specifically, with circular indifference contours (or their n-dimensional generalization) a person's preference on one issue is invariant with whatever is the status quo on all other issues. We can, however, readily imagine situations where this is a poor representation of preference. Suppose, for instance, that the two issues concern investing in urban highway construction versus commuter rail development. To some extent these two transportation systems are redundant and the value a legislator or city council member might place on one will reasonably be a function of how much is invested in the other. Or, for a more headline-driven example, consider the debate of how, whether or when to fund and implement America's Affordable Care Act ("Obamacare") and the contemporaneous scandal over high level bureaucrats within the Internal Revenue Service targeting Obama's ideological opponents (notably various Tea Party groups). It might seem that these are two separate issues, but they are in fact interconnected in that it is the IRS that is to monitor compliance with the act and the act's opponents have sought to use the IRS scandal as a way of sidetracking its implementation. So instead, consider Figure 6.4a, where the indifference contours of our three committee members are now "tilted ellipses"

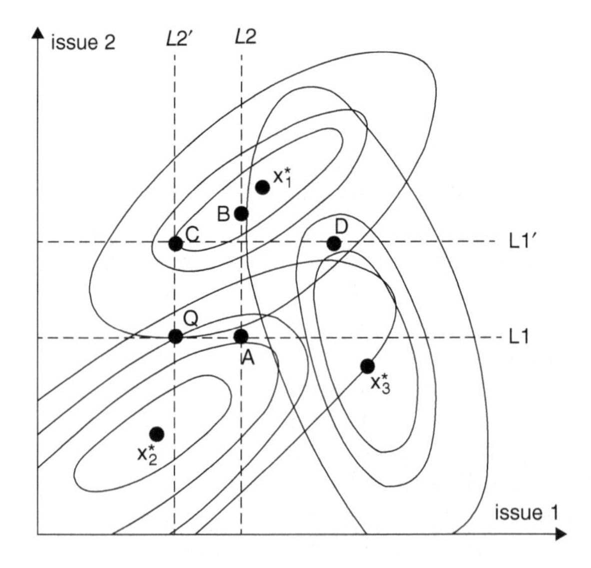

Figure 6.4a Issue-by-issue voting, elliptical indifference contours

so that the greater the value on one dimension, the lower is the preferred value on the other for voter 3, but the higher is the value for voters 1 and 2.

To see now how things change with these preferences, suppose once again that we vote on issue 1 first with Q being the status quo. Looking at the line L1 we see that the point A is the median preference on that line (i.e., the tangency of voter 2's indifference contour to that line is the median tangency) so that if we now vote on issue 2 (i.e., consider alternatives along the line L2) the median preference is at B. Suppose, on the other hand, that we first vote on issue 2 and thus consider the alternatives along the line L2'. In this case, we'd presumably move to the point C (the tangency of voter 1's indifference contour to L2'), and if we were then to consider issue 1 (i.e., alternatives along the line L1'), we'd move to D. And unlike the case of circular indifference contours, D is not the same as B. That is, the order of voting on the issues impacts the eventual outcome. There is, moreover, an additional difference between this example and the previous one with separable preferences; namely, revoting on the issues will change the outcome further still. For example, the reader should confirm that if a vertical line, L2", is drawn through the point D in Figure 6.4a, there is a median preference on it that differs from both D and B.

The sensitivity of the final outcome to the order with which the issues are considered as well as to whether or not we allow a reconsideration of issues raises a new question; namely, whether a process of continual re-voting can ever settle down and converge to a unique and well-defined point? Is there, in other words, an equilibrium? To see that the answer to this query is yes, consider Figure 6.4b, where the curve S1 corresponds to the locus of all tangencies of the

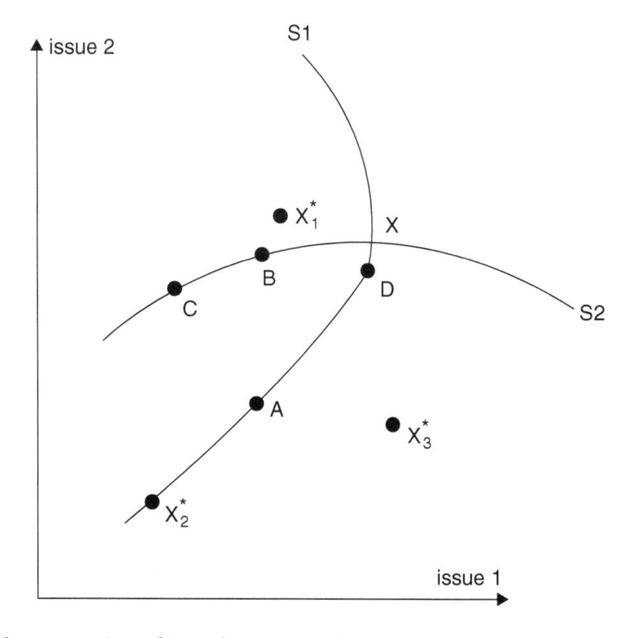

Figure 6.4b Intersection of issue-by-issue median curves

median preference to all the horizontal lines we might draw in Figure 6.4a (the seminal analysis here, upon which most of this section is based, is Gerald Kramer's "Sophisticated Voting Over Multidimensional Choice Spaces," *Journal of Mathematical Sociology* 2, 1972). Thus, if we fix issue 2 at any value, S1 tells us the value on issue 1 that would correspond to the committee's choice as dictated by majority rule. Similarly, the line S2 tells us the value on issue 2 that would be arrived at by the committee if we fix issue 1 at any specific value. Now consider the intersection of these two lines at the point X. Suppose X is the status quo. By construction, there is no issue-by-issue vote that can be taken that would lead us away from X. The point X then is the equilibrium to issue-by-issue voting in our example. The construction of S1 and S2, moreover, tells us that if indifference contours have some specific mathematical properties (e.g., there are no flat spots, they are continuous and they define convex sets), then the intersection of S1 and S2 will be unique. The reader can confirm, moreover, that beginning at any point on S1 or S2, successive votes on alternating issues will converge ultimately to X.

Of course, proof of a proposition using two dimensions does not necessarily mean the result extends to issue spaces of higher dimension. However, we will merely state here that such an extension exists but we skip its proof since there is a more immediate concern that needs to be addressed (and since that proof is more a complex exercise in linear algebra and the geometry of vectors than it is of game theory). Specifically, note that to this point we assume that our committee members vote sincerely. However, if they know that issues can be reconsidered and that the order of voting on the issues can impact outcomes, might they not find it advantageous to vote strategically? We approach this question by supposing that X is the sincere voting equilibrium under issue-by-issue voting and ask the question: Is there an agenda that leads away from X to some new outcome Y that a majority prefers to X? Limiting our illustrative discussion once again to two dimensions, we know that Y cannot have the same value as X on either dimension. Otherwise X is preferred to Y by a majority. So suppose the path to Y from X has the committee voting on issue 1 first and, with at least one voter voting insincerely, forcing a move from X to W, which is then pitted against Y where Y differs from W on issue 2 and is preferred by a majority to W. To see, then, whether such a move is possible we need to analyze the agenda in Figure 6.5 where A is the outcome achieved if the move to W is defeated but a horizontal move away from X is nevertheless voted on in the second round that otherwise would have led to Y if W had passed.

With strategic voting now, there is only one way to realize Y; namely, the committee votes for W over X in the first vote, and then Y defeats W (since, with X being the sincere voting equilibrium, X defeats A in a majority vote). Thus, for strategic voting to move away from X it must be the case that Y defeats both W and X. Whether this is possible or not, however, depends on the precise nature of indifference contours. The first possibility is that preferences are separable, in which case we revert to the special case of circular indifference contours to simplify our presentation. And as a further simplification let us limit discussion to two dimensions. So, suppose that Y is preferred to W by a majority—that is, let $u(\mathrm{Y}) > u(\mathrm{W})$ for all members of this majority. We know, though, that W

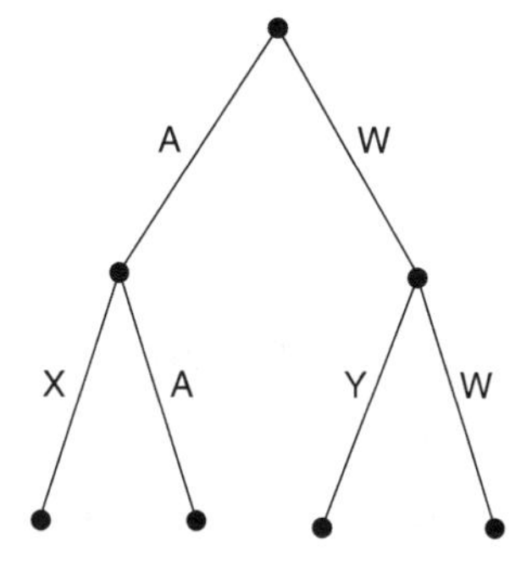

Figure 6.5 Strategic voting agenda under issue-by-issue voting

differs from Y on only one dimension (ostensibly the second) and that if we let W = $(w1,w2)$, we can write Y = $(w1,w2 + b)$, where b is the amount that W and Y differ on that dimension. However, since preferences are separable, we can write $u(Y)$ as the sum of two utility functions $u1(y1) + u2(y2) = u1(w1) + u2(w2 + b)$ and, correspondingly, W as a similar sum, $u1(w1) + u2(w2)$. With Y preferred to W by a majority, if indifference contours are circles (indeed, if they define convex sets generally), a paper and pencil exercise shows that if V is any point on the line between Y and W, then V is preferred to W as well by that same majority. So let $V = tY + (1 - t)W$. Of course, since the line between Y and W is parallel to the second dimension, V can differ from Y and W only on this second dimension. That is,

$$V = (w1,t(w2 + b) + (1 - t)w2) = (w1, w2 + tb).$$

And since $u(V) > u(W)$, with separable preferences this must mean that

$$u2(w2 + tb) > u2(w2).$$

However, recall that X, the current status quo, differs from W only on issue 1 and not 2. That is, $x2 = w2$, so the previous inequality can be rewritten as

$$u2(x2 + tb) > u2(x2)$$

for members of the majority that prefers Y to W, in which case it must be that $u(x1, x2 + tb) > u(x1, x2)$ for this same majority. This, however, violates the assumption that the point X is the sincere voting equilibrium . . . that no majority prefers an alternative to X if we limit the voting to votes over only one issue. Hence, X must also be the strategic voting equilibrium.

A great deal of fuss has been made in the literature over the preceding result, with some writers proclaiming that issue-by-issue voting is the solution to some of the instabilities inherent in situations without Condorcet winners. Labeled *institutionally induced equilibria*, the argument is that the committee system in, for example, the U.S. Congress or parliamentary procedures wherein debate is constrained by a rule requiring that discussion be germane to the issue at hand are viewed as ways in which deliberative bodies can ensure a stable outcome when otherwise there is none. Notice, however, that the preceding proof of stability depends critically on the assumption of separable preferences. Indeed, while there may be a strategic voting equilibrium under issue-by-issue voting without separability, there is no guarantee that such an equilibrium exists. Issue-by-issue voting, then, can induce stability, but it is guaranteed to do so only under the restrictive condition that each voter's preference over one issue is invariant with preferences over all other issues that might be considered.

Example: To see the effect of non-separability on issue-by-issue voting, consider the three-member committee in Figure 6.6, with ideal points at x_1, x_2, and x_3, each of whom perceives some relationship between the issues. Suppose the committee votes first on issue 2, then issue 1, and consider what eventually prevails as a function of what the committee agrees to on issue 2. For instance, if they choose x^* on issue 2, they must then vote on the line h_1, in which case C ultimately prevails, because C is the median preference on h_1. On the other hand, if the committee chooses x^{**}, then, B is the median on the line h_2. And if the committee initially agrees to x^{***} on issue 2, then the final outcome is A. The curve S, then, maps all such outcomes and tells us what prevails as a function of what is chosen on the first ballot over issue 2. To this point, then, matters seem unexceptional, but consider the significance of S and its shape. S tells us what eventually prevails on both issues, given a specific decision on issue 2. Thus, when evaluating how to vote on issue 2, committee members should look ahead and evaluate the outcome on S implied by their initial decision. However, because S is not a straight line (which it is with circular indifference contours), committee members can have preferences that are not single-peaked along it. For example, voter 2's ideal preference on S is the point A, but notice that the point C would lie on a higher indifference curve for 2 than would B. Thus, it is not the case that preferences decline monotonically as we move away from a person's ideal—voter 2's utility declines as we move from A until we get to B, and then it increases until C is reached. That is, preferences are not single peaked on S.

The fact that at least one committee member does not have a single-peaked preference means, of course, that a cycle can develop when evaluating what to do on issue 2. Indeed, A, B, and C cycle in our example. (Voter 3 most prefers B and least prefers C, voter 1 most prefers C and least prefers A, and voter 2 most prefers A and least prefers B—which occasions the usual Condorcet cycle.) Thus, issue-by-issue voting fails to induce a stable outcome.

In this example, then, although the committee's procedure seeks to enforce a separate consideration of the issues, this separation does not exist in the minds of the committee's members. So if those members anticipate the consequences

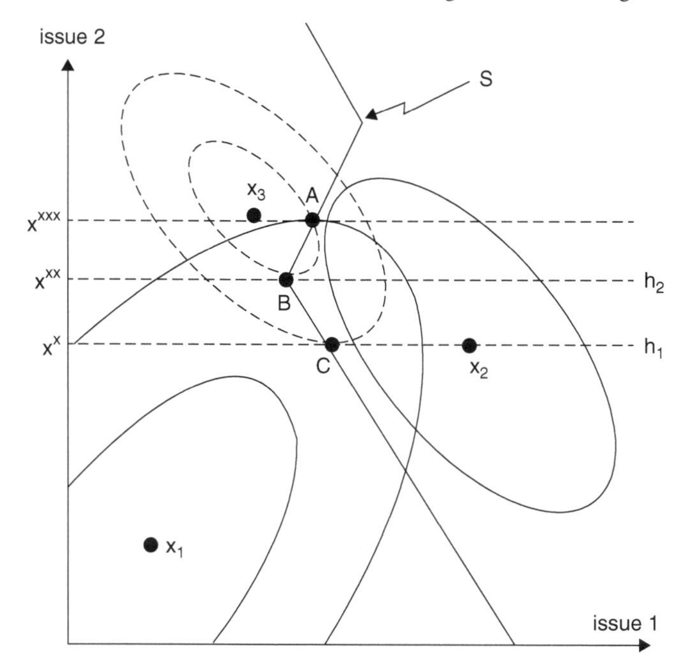

Figure 6.6 Disappearance of stability without separability

of their actions on one issue, procedures alone cannot ensure stability. Hence, the extent to which rules induce stability depends on many things, including, as our examples show, the extent to which the rules themselves are subject to revision and whether issues take a special, separable form. This conclusion does not mean, however, that rules are unimportant. Indeed, if we take the previous example and suppose that the committee votes first on issue 1, rather than issue 2, then the reader should be able to confirm that an equilibrium exists (provided that the committee does not allow a reconsideration of the issues). Thus, even a simple rule such as one that designates the voting order has a profound effect. We are, then, emphasizing two things. First, in determining the influence of a rule or institution, we should suppose that people react strategically to its imposition, with the understanding that all others are trying to do the same thing. And second, to reiterate a lesson learned from the analysis of agendas, those who wish to study legislative outcomes must understand legislative procedures and the opportunities for individuals and majorities to manipulate those procedures for their own ends.

6.5 Referenda and Separability of Preferences

One alternative, of course, to the vagaries of legislative or parliamentary representation and voting is to do away with politicians in their entirety and to institute a system of direct democracy by referenda. The resurrection of direct democracy through referenda is one of the clear trends of democratic politics. Nevertheless, opponents of this trend raise a variety of concerns, the most notable being an overall distrust of the abilities of the average citizen to process information on complex public issues and to make political decisions that serve their interests. We've already seen in Chapter 4, however, how voters might use indirect and relatively costless sources of information to guide their vote. Nevertheless, there are other objections to referenda, including a concern about the susceptibility of the public to well-rehearsed advertisements and well-financed campaigns by special interests and the impact of the initiative and referendum mechanism on other political institutions such as parties and legislatures. Here, however, as a further illustration of the application of the perspectives of this volume, we can offer a different critique of using referenda to measure the preferences of voters. Briefly, our critique does not apply if there is a single issue on the ballot. But if, as was the case in California in 2012, for example, there are upwards of eleven state-wide propositions to be voted on, referenda as currently practiced compel people to separate their votes on issues that may in fact be linked. That is, when voters have *non-separable preferences* across the issues being voted on, the common practice of tallying votes one issue at a time can fail to select an overall Condorcet winner if one exists. Worse yet, as we show shortly, referendum voting can select an overall Condorcet loser or an outcome that is Pareto-dominated by every other possible outcome.

We have, of course, made reference to the concept of separable and non-separable preferences earlier, especially with reference to spatial preferences in the preceding section wherein we note that circular indifference contours correspond to separable preferences. For a somewhat more homey example, consider the food connoisseur's view that red wine should only be eaten with meat dishes and white wine with seafood. Thus, one's preferences for wine, so we are told, are not or should not be separable from our dinner entre. In a domain more relevant to this text, it is not difficult now to identify real instances of referenda that failed to yield separable preferences. For example, a 1978 Swiss national referendum contained one issue on the age of retirement and another on a revision of the old-age pension. Similarly, another Swiss referendum in 1990 contained one proposal to establish a 10-year moratorium on nuclear plant construction, while a separate proposal called for an end to the use of nuclear energy. And in November 1988, California voters confronted five separate ballot proposals dealing with automobile insurance reform. Since the five proposals addressed essentially identical issues, many voters probably had non-separable preferences across the alternatives.

To illustrate the problem non-separability occasions with these voters, suppose a referendum is held on two issues, where, for example, YN indicates approval of issue 1 and defeat of issue 2. Voters can vote YES or NO on each issue, and their (non-separable) preference ranking is given for outcomes on both issues as follows (the analysis that follows relies heavily on Dean Lacy and Niou, "A Problem with Referendums," *Journal of Theoretical Politics* 12,1, January 2000):

1	2	3
YN	NY	NN
YY	YY	YY
NY	YN	NY
NN	NN	YN

Thus, if each voter votes according to his first preference, the outcome is NN: Voters 2 and 3 vote N on issue 1 and Voters 1 and 3 vote N on issue 2. This is the behavior we might expect in a simultaneous vote where voters do not know each other's preferences, and the problem is apparent: YY is the Condorcet winner whereas NN is the Condorcet loser. In this case, then, it would seem that the referenda provide the wrong answer to the question, "What do the people want?"

The outcome of a referendum can be even worse than our example suggests. Consider the case of a bond referendum where three voters must vote on three bonds. Suppose all three voters, in their preferences, impose a budget constraint of two bonds; they want two to pass, but they disagree on which two. All voters prefer the passage of any two bonds to the passage of any one bond, and all voters rank last the passage of all three. The voters may believe that passage of three will increase state debt and raise taxes while passage of one or two will not create an unreasonable financial burden on the state. Looking then at the preferences below, we can surmise that if voters act myopically in the absence of information about each other's preferences, all three bond issues pass even though no voter wants such an outcome. Voter 1 votes to approve of issues 1 and 2, voter three votes to approve of issues 1 and 3, and voter 3 votes to approve of issues 2 and 3, thus resulting in majority approval of all three bond referenda. In other words, the referendum selects an outcome that is Pareto-dominated by every other possible outcome.

1	2	3
YYN	YNY	NYY
YNY	NYY	YYN
NYY	YYN	YNY
NNY	NYN	YNN
YNN	YNN	NYN
NYN	NNY	NNY
NNN	NNN	NNN
YYY	YYY	YYY

Strategic Voting: Under the assumption that voters are uninformed about each other's preferences, we have thus far assumed that voters vote for their most preferred outcome. But if voters know the preferences of others, strategic interaction may change their vote choice and thus the outcome. Nevertheless, as the preferences below show, in a simultaneous vote, strategic voting does not guarantee the selection of a Condorcet winner.

1	2	3
NY	YN	NN
YY	YY	YY
NN	NN	NY
YN	NY	YN

Since voters 1 and 2 have separable preferences, their dominant strategy is to vote for their most preferred alternative. To see this, suppose voter *i*'s most preferred outcome is $(x_1, x_2, \ldots, x_m)$, and suppose *i* is pivotal on issue *k*. From the definition of separability, *i* prefers $(x_1, \ldots, x_k, \ldots, x_m)$ to $(x_1, \ldots, x_k', \ldots, x_m)$ and it must also be the case that he prefers $(x_1, \ldots, x_j, x_k', \ldots, x_m)$ to $(x_1, \ldots x_j', x_k', \ldots, x_m)$. Thus, by transitivity *i* prefers $(x_1, \ldots, x_k, \ldots, x_m)$ to $(x_1, \ldots x_j', x_k', \ldots, x_m)$. Voter 3 has non-separable preferences and is pivotal for each issue. His best response to the votes of 1 and 2 is to vote NN, in which case NN is the social choice and a Nash equilibrium outcome even though YY is the Condorcet winner. To see this notice that if 1 unilaterally changes his vote to YY, the outcome YN prevails—voter 1's least preferred outcome; and if 2 changes his vote to YY, the outcome NY prevails, which is his least preferred outcome. In short, strategic voting alone proves insufficient to resolve the problem created by non-separable preferences.

Sequential Voting: There is one final thing we might consider insofar as finding a method whereby referenda avoid undesirable outcomes; namely, voting on the issues sequentially. Sequential voting, of course, allows people to avoid the problem observed in our example of three bond referenda: Once bond issues 1 and 2 pass, the voters are unanimous in defeating issue 3. We can begin by noting that sequential voting offers a solution to the problem observed in our first example, where NN is majority dominated by every other pair of outcomes in a simultaneous vote. Suppose, however, that each voter begins by voting for her or his most preferred outcome on the first issue, which produces N as the outcome. Once N is revealed, voters 1 and 2 will choose Y on the second issue, yielding NY as the outcome. If the order of the issues in the voting sequence is reversed, then on the second issue (voted on first), the outcome is N and on the first issue (voted on second), the outcome will be Y, resulting in the outcome YN. Thus, sequential voting here prevents the selection of Condorcet losers.

We can, in fact, say more; namely, if voters are sophisticated, then sequential voting will select the Condorcet winner if one exists. Suppose all voters reason through the possible outcomes of sequential voting in our first example. If issue 1 is voted on first and the outcome is seen to be Y, voters 2 and 3 will vote Y on issue 2, and the outcome will be YY. If the outcome of issue 1 is N, then voters 1 and 2 will vote Y on issue 2 and the outcome will be NY. All voters can look ahead to see that if Y wins on issue 1, the outcome will be YY; if N wins, the outcome will be NY. Voters 1 and 3 prefer YY to NY, so they will vote Y on issue 1, then Y on issue 2, and the outcome is YY. Similarly, suppose issue 2 is voted on first with Y as the outcome. Voters 1 and 3 will vote Y on issue 1 and the outcome will be YY. If the outcome on issue 2 is N, then voters 1 and 2 will vote

Y on issue 1 and the outcome will be YN. Since all voters can reason through this sequence, they know that a Y vote on issue 2 will produce YY while an N vote will lead to YN. Since voters 2 and 3 prefer YY to YN, they will vote Y on both issues in the sequence, producing YY as the outcome. This result can in fact be generalized by a formalization of this argument so as to establish that *sophisticated sequential voting on multiple binary issues will produce an overall Condorcet winner when one exists, regardless of whether voters have separable or non-separable preferences.*

6.6 Key Ideas and Concepts

uncovered set
undominated set
amendment agenda
endogeneous agenda
Borda Count
Hare voting
institutionally induced equilibria
single transferable vote
approval voting
cumulative voting
plurality rule
majority rule
runoff elections
issue by issue voting
referenda
sequential voting

Exercises for Chapter 6

1. You are a member of a three-person committee that must choose one outcome from the list (A, B, C, D). Suppose the following preference orders describe the committee (from most to least preferred):

you:	*A*	*B*	*C*	*D*
member 2:	*D*	*C*	*A*	*B*
member 3:	*C*	*B*	*D*	*A*

   Which of the following procedures would you prefer to see implemented if you believed that the other two members of the committee were sophisticated: (1) an agenda that first paired B against C, the winner against A, the winner against D; (2) an agenda that first paired C against A, the winner against D, and the winner against B; (3) an agenda that first paired B against D, the winner against C, the winner against A; or (4) you should not care which is chosen.

2. Assume the following preferences by a five-member committee:

1	2	3	4	5
A	B	C	C	B
B	C	A	A	C
C	A	B	B	A

What is the outcome of the agenda (A, B, C) if only voter 2 is strategic? If voter 2 can educate at most one other person to be sophisticated, whom should he or she educate, assuming that voter 1 is too dumb to ever catch on to what is required?

3. Suppose a majority of the legislature prefers A to B. If you are opposed to A, if A and B must be voted on first regardless of what amendments are introduced, and if everyone is a sophisticated voter, which alternative would you prefer to introduce: C or D? C creates the majority rule cycle, "A preferred to B preferred to C preferred to A" while D defeats both A and B. Your preferences are "B preferred to D preferred to C preferred to A."

4. You are a legislative aid advising a committee chair, who must choose between reporting bill A or B out of committee. If A is reported, it is certain to lose to the status quo Q. If B is reported out, it will be amended on the floor (alternative C), and you will be able to offer a substitute bill, D. Suppose the remaining members of the legislature fall into one of three equally numerous groups, with preferences as follows:

group 1:	*B*	*D*	*A*	*C*	*Q*
group 2:	*C*	*Q*	*A*	*D*	*B*
group 3:	*Q*	*B*	*C*	*A*	*D*

You are a member of the second group. Suppose the legislature votes using the agenda: "substitutes against the amendment, the winner against the bill, the winner against the status quo." You are certain that as things stand, everyone but you votes sincerely. Suppose the dollar value to you of each alternative is: C = $2,000, Q = $1,500, A = $500, D = $0, and B = −$3,000. What is the upper limit on how much you would be willing to pay to have someone educate the legislature so you and everyone else votes sophisticatedly?

5. A legislature (which we assume has three members) can consider four motions, where each motion affects the amount of money going to a legislator's district. Let the amounts (in thousands) to each district from each motion be as follows:

	District 1	District 2	District 3
A (status quo)	300	0	−400
B (committee bill)	500	−600	0
C (possible amend.)	0	800	−900
D (possible amend.)	−900	400	450

Motions A and B are on the floor as proposals and you must decide whether to propose an amended bill. If you propose C, the agenda will be C versus B the winner against A; if you propose D the agenda will be D versus B the winner against A. If D and C are both proposed, the agenda is D versus C, the winner against B, the winner against A.

a. Suppose you are the representative from district 2. Which amendment should you propose: C, which pays your district $800,000 or D, which pays $400,000?
b. Suppose you are chairman (and dictator) of the relevant legislative subcommittee, and that you can report out of your subcommittee either B or C or D as the bill that the legislature must consider. But you are also certain that whatever alternatives you fail to report out will be introduced on the floor as amendments. Thus,

   If you report B, the agenda is "C versus D, winner versus B, winner versus A."
   If you report C, the agenda is "B versus D, winner versus C, winner versus A."
   If you report D, the agenda is "C versus B, winner versus D, winner versus A."

   What would you choose as your bill if you were the representative from district 2?
c. With respect to part (b), if majority rule is used by the subcommittee to choose whether to report B, C, or D, what will it report?

6. Consider the following preferences within two chambers (lower and upper houses) of a legislature:

Upper house legislator			Lower house legislator				
1	2	3	1	2	3	4	5
B	C	D	A	B	B	D	C
C	D	A	C	O	D	A	B
D	O	C	B	C	A	O	D
O	A	O	O	D	O	C	O
A	B	B	D	A	C	B	A

Suppose the president cares only that some bill pass to upset the status quo. To achieve this, he can submit a single agenda to both houses of the legislature. All members of each house of the legislature are strategic and the lower house must vote before the upper house. If both houses choose the same bill for the president to sign, he signs it. If the houses choose different bills, the outcome is "No New Law" (O). If the president is strategic, which of the following four agendas should he submit?

i. (A, B, C, D) ii. (D, C, B, A)
iii. (C, D, A, B) iv. (B, D, C, A)

where (x, y, z, w) means "x versus y, the winner against z, the winner against w."

7. A nine-member legislature using majority rule faces a budget that allows them to pass two of three proposed programs (A, B, C). The legislature has the following preference orders (ranked from most to least preferred):

Legislator

1	*2*	*3*	*4*	*5*	*6*	*7*	*8*	*9*
B	A	B	A	C	C	B	A	C
C	B	A	C	A	B	A	C	B
A	C	C	B	B	A	C	B	A

You chair the legislature and are legislator 1. The voting procedure is as follows:
First, vote whether to keep or veto alternative B. If B is vetoed, the voting ends and alternatives A and C are implemented. If B is kept, then vote whether to keep or veto alternative C. If C is vetoed, the voting ends and alternatives A and B are implemented. If C is kept, then vote whether to keep or veto alternative A. If A is vetoed, the voting ends and alternatives B and C are implemented. If A is kept, one member of the legislature must choose which alternative should be vetoed.
   a. As chairman, should you choose yourself to make this veto decision or legislator 4?
   b. Assuming that legislator 4 makes the veto decision, design an agenda of the type illustrated in which B and C are nevertheless passed.

8. Suppose three candidates, A, B, and C, are competing in a plurality-rule election and that an initial poll of the electorate's preferences reveals the following information:

   31% prefer A to B to C with $u(\text{A}) = 10, u(\text{B}) = 9, u(\text{C}) = 0$,

   29% prefer B to A to C with $u(\text{B}) = 10, u(\text{A}) = 9, u(\text{C}) = 0$,

   40% prefer C to A to B with $u(\text{C}) = 10, u(\text{A}) = 1, u(\text{B}) = 0$.

   a. If the pollster accurately reports the poll result, and if all voters assume that those who hold the same preferences vote in the same strategic way, who wins the election in equilibrium?
   b. Suppose the pollster misrepresents the poll and announces that 40% prefer A, 31% prefer B, and 29% prefer C. Who wins?
   c. Suppose the pollster instead announces that 40% prefer B, 31% prefer A, and 29% prefer C. Who wins?

7 Games with Incomplete Information

7.1 Incomplete Information

Thus far, we have assumed that any uncertainty we choose to incorporate into our models has nature as its source. More importantly, if nature's moves are revealed to one person, we have thus far assumed that they are revealed to everyone, so that there are no informational asymmetries—no one has any private information, aside, possibly, from the choices they make as the situation unfolds. More generally, though, many important political processes can be modeled only if we assume that decision makers have private information, such as the details of their own preferences or their capabilities. A great many examples and subsidiary questions come to mind:

> What costs are terrorists willing to incur after hijacking a plane, and how willing should a government be to make concessions or to risk sacrificing hostages?
>
> What are a weapon system's capabilities that might not be observable or measurable by other countries, and how does this asymmetry in information affect the willingness of countries to engage in arms control negotiations?
>
> How can a congressional committee monitor and regulate an executive agency when it knows that the agency will have better information about the program's performance than Congress once that program goes into effect?
>
> How should we approach a negotiation if we don't know an adversary's willingness to compromise or how it values the time spent negotiating?
>
> When some voters are informed about a legislator's actions and others are not, what weight will the legislator give to the informed versus the uninformed members of his constituency?
>
> With respect to the important issue of strategic deterrence, what value does a country's leadership place on exacting revenge with a counter-strike after being attacked even though it knows that such a strike can only invite its own counter-strike?

The Pripyat Marshes: The Pripyat marshes, stretching across northern Ukraine and southern Belarus, were, during WWII, the bane of the Nazis' drive into the Soviet Union. Largely impenetrable by mechanized armor, they were the ideal refuge of partisans, who could launch their attacks against the Wermacht and then disappear. But in 1986, they achieved notoriety for a different reason: Following the Chernobyl nuclear disaster, they glowed in the dark. One can, in fact, date the inexorable dissolution of the USSR by that disaster—when the failings of the regime became apparent to not only the powers that be in the Kremlin, but to ordinary Soviet citizens as well. In the present game theoretic context, though, it is interesting to look at events that bracketed Chernobyl, beginning in 1983 when President Reagan branded the Soviet Union the "Evil Empire" and to recall that six years later that empire, in the form of the Warsaw Pact, dissolved. In the interim, we find not only Chernobyl, but the ex-KGB head, Yuri Andropov, being replaced as Soviet Premier by Gorbachev, who in turn sought to implement a series of domestic reforms under the labels *glasnost* (openness) and *perestroika* (restructuring). And we should also recall that throughout this decade the USSR's ostensible allies—Hungary, Czechoslovakia, Romania, Bulgaria, East Germany and especially Poland—had been edging toward reform of their own economies and, for a variety of nascent "revolutionaries," of their political systems as well. Reformers in the past, however (e.g., in Hungary in 1956 and Czechoslovakia in 1968), had met a bitter end, so needless to say, political leaders and their reformist opposition in Central Europe were, to say the least, cautious. But Gorbachev promised a new regime, so what to make of his admonitions to the heads of the USSR's satellite states that they should do what they think best without interference from the Kremlin? What were they to make of his telling them that the Brezhnev Doctrine had been replaced by the "Sinatra Doctrine"—do it your way? They'd been told such thing in the past, only to learn those were empty words. What ensued thereafter were a series of actions by the satellite states that, explicitly intended or otherwise, caused the Kremlin to signal its true intentions. Perhaps the first such signal, aside from words, came in March 1989 when Hungary's Prime Minister, Miklos Nemeth, told Gorbachev that Hungary intended to tear down the 200 kilometer electronic fence between it and Austria—and Gorbachev did nothing. A second signal came when Moscow merely encouraged a "political settlement" between Solidarity and the Communists after Solidarity won a landslide election victory in June. Again in June there was Moscow's silence after Hungary, following a reburial of the leader of its 1956 uprising, Imre Nagy, labeled that uprising a popular one as opposed to a "counter revolution." Then, when Hungary in August inquired as to Moscow's view of letting the East Germans, who were flooding their country, move on into Austria (and thence into West Germany), Russian Foreign Minister Shevardnadze replied that the issue didn't concern Moscow. In October, while visiting Berlin, Gorbachev signaled the green light for the Communists to remove the sclerotic Communist Honecker. In November, perhaps by accident, the

> Berlin Wall came, metaphorically speaking, crashing down, and once again the Soviets did nothing. The rest, as people are wont to say, is history.

These examples have one common element: One person (or entity, in the case of the Kremlin) knows something another does not know—costs, capabilities, policy position, program performance, and so on. One need not revert, however, to situations that rarely if ever concern us mere mortals as opposed to government decision makers or military planners in order to identify parallel situations of asymmetric information. When one buys a used car, the dealer is generally at an advantage in terms of knowing whether a car has serious mechanical issues that almost any dealer will attempt to disguise with a thorough cleaning of the interior, a removal of dents, and a good coat of wax. Alternatively, when buying a new car, the most common advice given to the consumer is to not reveal beforehand one's willingness to pay since following that advice increases one's chances of a better deal. Or consider the interaction between a person interviewing for a job and the potential employer. The employer's task is to somehow assess the interviewee's capabilities and shortcomings, while the person being interviewed has every incentive to hide the things he or she thinks will keep them from being hired. And then there are those situations that concern the reputations of the players. A high school bully, for instance, may know that there are people who, if they challenge him, could show him to be a phony, so instead he picks fights with lesser foes in the hopes of establishing a reputation for toughness and physical prowess so that he will not be challenged by those who might overcome him.

Treating situations in which there are parts of an extensive form that are unknown to one person or the other will require some additional tools if only because asymmetric information opens the door to new and more complicated forms of "he-thinks-that-I-think" regresses and more complex forms of strategic interaction. To see what we mean, consider this restatement of the "he-thinks" regress: *I believe that you believe that the game we are playing is ___, in which case you will conclude that ___ is an appropriate strategy for me to adopt. And if you believe that I believe that the game we are playing is in fact ___, then you should conclude that I will indeed choose___ as my strategy.* At this point we can label the game "solved," because our beliefs, by assumption, are identical. Now, however, suppose our beliefs are not identical because we each hold some private information. Of course, each of us might be expected to begin the "play of the game" with some initial guess about each other's private information, but since these are only guesses, as the situation unfolds we might find that each of us is making choices that are inconsistent with initial suppositions. At this point both of us might have reason to revise our initial guesses about the situation, but such reasoning opens the door to thoughts of the following variety:

> *If I know something that you don't know and if you are aware of your own uncertainty, then I may try to deceive you. However, based on what you know*

> *about the situation's strategic structure, you might try to infer what I know from what I say or do as the situation unfolds. Of course, I know you will try to make such inferences, and therefore I will choose my strategy carefully so that it is not only a best response to your actions, but leads you to believe things that induce actions on your part that are advantageous for me. However, you know that I am trying to do this, and therefore you have reason to interpret my actions carefully—possibly as part of an attempted deception.*

To bring matters closer to home, recall those glorious years in which dating led to endless periods of introspective agony. "If I try to kiss her, will she humiliate me with rejection or will she respond in kind?" "Will he ever try to kiss me, or is he dating me merely to pass the time until he can get a date with ___?" "How can I signal my desire for a deeper relationship?" "If I ask her out again, will she say no?" "Did he not ask me for a date because he's interested in someone else, or is he too shy to make the first move?" The agony here, of course, is that at least one person knows something the other doesn't know—in general, in fact, each knows something the other doesn't know—in which case, neither person can be certain they understand completely the "game" being played. What follows thereafter, then, is each person, paying special attention to every nuanced word or action, attempts to infer what the other wants or doesn't want, but where both sides to the relationship are aware that the other is doing this and adjusting their beliefs in some manner or another.

Such situations reveal that our current typology of games—which includes games of perfect and imperfect information—is incomplete. To this typology, we must also add **games of incomplete information** in which characteristics of the situation are revealed only to some subset of players so that one or more of them knows something about the extensive or strategic form representation of the situation that others don't know. But even this category of games has some distinct sub-categories that are worth noting. Consider what happens when we match information with the order of moves. If the player who has the first move possesses private information, that first move becomes a signal for the other players, in which case we refer to the situation as a *signaling game*. Games of reputation are typically signaling games, and in our discussion of events leading to the USSR's empire's dissolution it was the Kremlin that was attempting to signal that its reputation, earned under Stalin and Brezhnev, was no longer relevant. Alternatively, if it's the player who lacks private information who must move first, then we call the situation a *screening game*. Our example of a person being interviewed for a job illustrates such a game wherein an employer might offer a job candidate various benefits such as stock options or signing bonuses without knowing what the potential employee knows about their qualifications or what it would take for the potential employee to agree to be hired. Screening also characterizes the task of insurance companies wherein they must decide whether or not to issue life insurance policies to people who perhaps possess private information about the status of their health. There are, of course, other forms of games with asymmetric incomplete information but rather than concoct any new typologies, we begin in the next section with a simple numerical example that illustrates more precisely the complexity asymmetric incomplete information adds to the analysis.

7.2 A Simple Game of Incomplete Information

Example: A lobbyist (l) who receives $20 if a particular bill passes must choose between

C: Offer a campaign contribution of $15 to a specific legislator (L) who will be pivotal on the bill's final vote for passage, conditional on the legislator (L) voting for the bill, and

~C: Take the chance that the legislator will vote for the bill without the contribution

Appreciating that the magnitudes of the numbers we associate with outcomes are not stated with realism in mind, the lobbyist's payoff from each outcome is

$20 If the bill passes without a bribe being paid

$5 If the bill passes with a bribe

$0 If the bill fails without a bribe

The problem for the lobbyist, however, is that he is uncertain about how the legislator's constituents feel about the bill. For purposes of a numerical example, suppose that, depending on the preferences of constituents, the legislator in question derives either $10 or −$10 from the bill's passage (and nothing if it fails). Keeping in mind that the vote of the legislator in question is pivotal, the legislator's payoffs, then, are

$25 If constituents favor the bill and if the legislator is paid the bribe by the lobbyist after voting for it

$10 If constituents favor the bill but no bribe rewards the legislator's vote for the bill

$5 If constituents oppose the bill but the legislator is bribed so as to vote for the bill

$0 If the legislator votes against the bill, there is no bribe and constituents oppose the bill

−$10 If there is no bribe, constituents oppose the bill, and the legislator stupidly votes for the bill's passage

To make this example more interesting, we let the legislator have the first move in this game, which is either to ask the lobbyist for a contribution (A) or to pass on his move and to let the lobbyist approach him (~A). Figure 7.1 portrays this situation's extensive form if we assume that the legislator's information about his or her constituents is no better than the lobbyist's. In this instance, we give nature an initial move in which it sets constituents' preferences by setting the actual payoffs to the legislator. Thus, the legislator prefers, *ceteris paribus*, to pass the bill (denoted P) with probability p(P) since that is the probability his or her constituents favor the bill, and holds the opposite

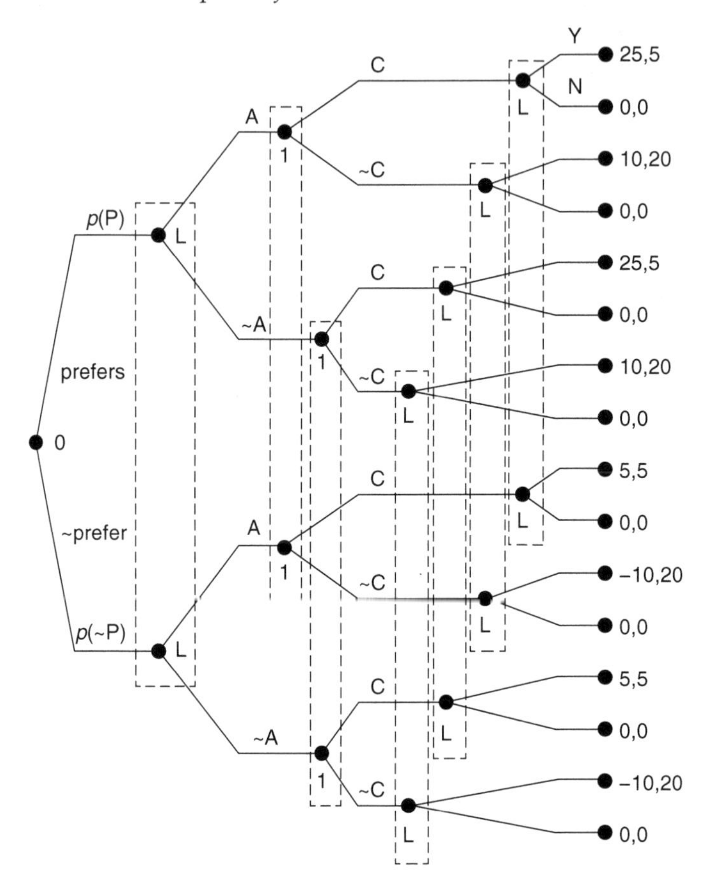

Figure 7.1 Symmetric incomplete information

preference (denoted ~P) with probability $p(\sim P) = 1 - p(P)$. Notice that the top half of this figure is identical to the lower half except for the preferences the legislator associates with specific outcomes.

To this point our example merely corresponds to an extensive form game of the usual sort, but it seems unreasonable to suppose that a legislator is uninformed about the preferences of the people he or she represents. So suppose the legislator knows something the lobbyist doesn't know, namely his or her constituents' preferences. In this case, the situation's extensive form becomes the one in Figure 7.2, which is identical to Figure 7.1 except that it removes the information sets surrounding the legislator's decision nodes. Thus, we now have a situation in which the legislator learns nature's move and is aware of his or her preferences before it is the legislator's turn to first act. The lobbyist, on the other hand, learns the legislator's first move but is uninformed about nature's move.

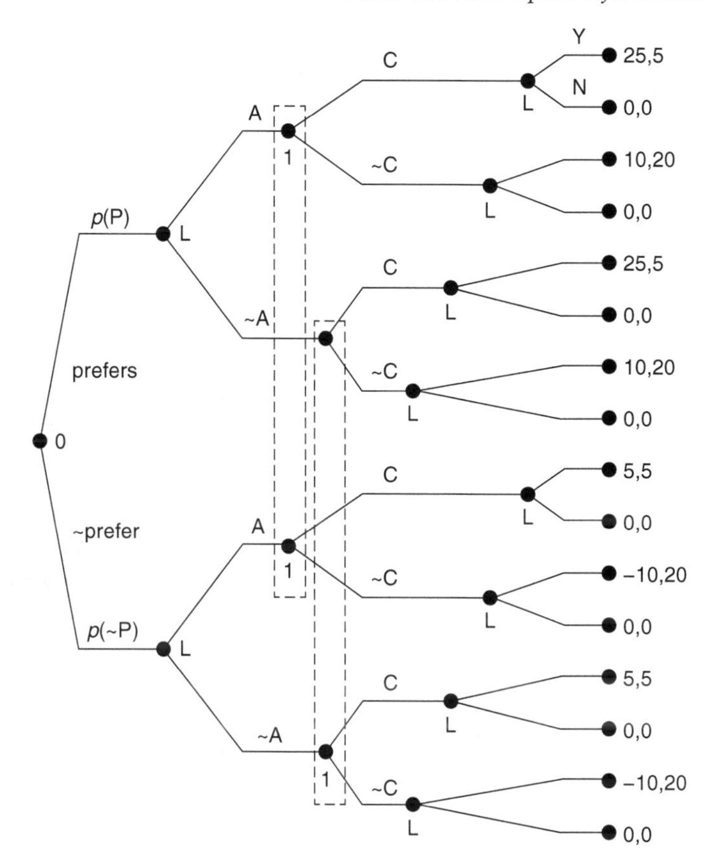

Figure 7.2 Asymmetric incomplete information

With respect now to the issue of common knowledge, notice that there is nothing to stop us from assuming that the probabilities $p(\text{P})$ and $p(\sim\text{P})$, as well as the information conveyed in this figure, are common knowledge. Common knowledge does not mean that one player is precluded from knowing something that another does not know; instead, it simply requires that everyone is aware of the informational uncertainties of others, everyone is aware that everyone else is aware, and so on. Thus, saying that the structure portrayed in Figure 7.2 is common knowledge implies that the lobbyist believes that nature chooses between P and ~P with probabilities $p(\text{P})$ and $p(\sim\text{P})$, the legislator is aware of the lobbyist's beliefs, the lobbyist knows that the legislator is aware of these beliefs . . . and so on. Put differently, both persons are fully informed about the situation's extensive form, each is aware that the other is aware, and so forth.

It might seem, then, that there is nothing unusual in this situation—nothing that we have not encountered before in terms of analytic complexity. However, here we must keep in mind that p(P) and p(~P) are simply the *prior beliefs* of the lobbyist—his or her beliefs prior to the legislator's first move. In deciding what to do and how to approach the situation, one possibility is for the lobbyist to ignore the legislator's initial move and assume simply that the probability that the legislator's constituents favor the bill equals p(P). In this instance, choosing C (offering the bribe) yields the lobbyist a payoff of $5 because under all contingencies the legislator will vote for the bill whereas if no bribe is offered (~C), the lobbyist gets $20p(\text{P}) + 0p(\sim\text{P}) = 20p(\text{P})$. Doing so, though, ignores the fact that before deciding whether to offer a bribe, the lobbyist observes whether the legislator requests the bribe. That is, simply maintaining the probability estimates of p(P) and p(~P) ignores the opportunity for the lobbyist to condition an estimate of these probabilities on the information set reached—on whether the legislator requests a contribution. The lobbyist, in this case, must wonder whether asking or not asking for a bribe reveals something about the information possessed by the legislator—namely, whether or not his or her constituents favor the bill.

At this point in our discussion, we cannot suppose that the method whereby such conditional probabilities are calculated is apparent. Indeed, it is frequently the case that this calculation can be difficult. Nevertheless, we can indicate here its general structure, which makes use of the strategic context of choice. Presumably, the legislator will choose a strategy that specifies whether to choose A or ~A, depending on whether nature chose P or ~P. Thus, if the legislator chooses a strategy that states, in part, "Choose A if nature chooses ~P, but choose ~A if nature chooses P," and if the lobbyist knows this strategy, then, the lobbyist can infer nature's choice after observing the legislator's choice. Of course, we must then contend with the fact that the legislator knows that the lobbyist is making such inferences, in which case the lobbyist must contend with the possibility that the legislator is engaging in some form of deception when choosing A by asking for a bribe. And once the lobbyist asks this question of himself, the he-thinks-that-I-think regress begins anew. So the question remains as to what probabilities the lobbyist should assign to P and ~P **after** observing the legislator's initial move. Notationally, the probabilities of interest are termed *conditional probabilities*, denoted $p(S|e)$, where S is the outcome or state of nature in question and e is the event that is observed and upon which we condition our estimate of p. In the present context, the two probabilities being estimated are denoted $p(\text{P}|\sim\text{A})$ and $p(\sim\text{P}|\text{A})$—respectively, the probability that the legislator's constituents prefer the bill given that the legislator does not ask for a bribe and the probability that those constituents are opposed to the bill given that the legislator request a bribe. The answer to that question can be learned only by solving this regress—only by once again solving for an equilibrium of both strategies and beliefs.

The Taiwan Straits: Before we see how to find such an equilibrium, we should first dissuade the reader from thinking that such situations are of pure academic interest. Recall our brief discussion in Chapter 1 of the situation confronting China, Taiwan and the United States and the uncertainty that confronts the two adversaries across the Taiwan Straits as to the likely response of the United States to any initial hostilities between them. As things stand now, neither China nor Taiwan are certain of that response and our argument is that American ambiguity is deliberate. Were the U.S. to commit unambiguously and unconditionally to Taiwan's defense, Taiwan would most likely declare independence and literally dare China to make a military response. Conversely, were the U.S. to commit unambiguously to a declaration that Mainland-Taiwan relations fall outside of its sphere of interest, China would almost certainly make aggressive moves to absorb Taiwan and negate whatever independence it currently enjoys. The particular difficulty for the U.S. lies in maintaining that ambiguity whenever it reacts to various moves by either Taiwan or China—reactions that might reveal its true intentions. That is, even if the U.S. knows what it would do in the event of a military confrontation between China and Taiwan, it must act in ways that leaves both sides of the conflict unable to infer its intentions. Of course, if there were a deadline of some sort whereby the U.S. had to commit or not commit to Taiwan's defense, the option of maintaining ambiguity about its strategic plans would no longer be available and the situation would more closely match our previous legislative example. The real game, though, is even more complicated because both China and Taiwan also in reality have the option of making choices that might compel the U.S. to reveal its preferences, where their incentives for doing so are also dependent on their prior beliefs and a calculation as to whether or not they in fact want either side to refine their beliefs about the U.S. Similar situations are unfolding now even between China and Japan over islands that both countries claim as well as between China and several other South East Asian states. Nor are strategic scenarios of this sort limited to Asia. A somewhat similar situation exists today in the Middle East. Despite its longstanding commitment to Israeli security, the United States fears an all-out conflict between Israel and Iran—a conflict that could ultimately escalate to involve not only Russia and other states of the region, but also the exchange of nuclear weapons. So here, too, the U.S. has to worry that by being too strong a supporter of Israel, it might encourage it to launch a pre-emptive strike against Iran's nuclear facilities, whereas by being too weak in its support, it will only encourage Iran to take more aggressive action. As with Taiwan and China, then, the U.S. thus far appears to have chosen a policy of deliberate ambiguity, whereby neither Iran nor Israel knows how far the U.S. is willing to let Iran go in the development of nuclear technology. The complication here, though, is that although the U.S. policy appears to be that of avoiding reacting to either state in a way that allows one or the other to infer its intentions with certainty, in the Taiwan straits the players know that that ambiguity is deliberate policy whereas in the Middle East ambiguity can readily engender a reputation for indecision.

The July Crisis of 1914: Although World War I officially began in August 1914, the preceding month provides a clear example of incomplete information in a strategic environment wherein political elites had to update their estimates of the preferences of others as events unfolded. We know, of course, that the assassination of the Habsburg Empire's heir apparent, Franz Ferdinand, led to a series of actions by political leaders in Russia, Austria, Serbia, Germany, Britain and France that resulted in war. If we were to attempt an explanation of that outcome (which, in fact, takes us beyond the confines of this text), suppose, as is often the case when attempting a shorthand explanation for complex events, that we simplify matters by being anthropomorphic about things and treat states as individual decision makers. The difficulty here, though, is that in ascribing preferences, we must take account of the personalities and complex policy disputes within those states. For instance, France's president, Poincare, took a far more aggressive stance against Germany as compared to that country's prime minister, Rene Viviani. Austria's foreign minister, Leopold Berchtold, had to balance between the belligerent position of his country's army chief of staff, Franz Conrad, and a more accommodating Hungarian minister-president, Stefan Tisza. For Russia, an analyst would have to consider the impact of Poincare's summit with the Tsar and the choices made by Russia's normally prevaricating foreign minister, Sergei Sazonov. In Germany, the issue was the Kaiser's willingness to confront Russia in the event that she intervened on Serbia's behalf in combination with the political skills (or lack thereof) of Chancellor Hollweg Bethmann. For Britain, we must assess the strategic skill of His Majesty's foreign secretary, Sir Edward Grey, when dealing with the non-interventionist faction of his own government. And as for Serbia, would its Prime Minister, Nikola Pasic, find a way to balance the demands of his country's military against a desire to appear the innocent victim of Austrian aggression?

The preferences each state would act in accordance with were thus determined by a complex interplay of individual motives, domestic politics and skill at political maneuver and the best anyone at the time could do in choosing a strategy was to assign preferences probabilistically to each state. Would Tisza succeed in moving Berchtold away from the more belligerent Conrad; would Poincare's dislike of Germany dominate an otherwise unstable French polity; would the Tsar, following his summit with Poincare, be swayed wholly to the Frenchman's view or would Sazonov's reputation for indecision dictate Russia's actions; would the crisis over home rule in Ireland and the non-interventionists within Grey's government bring that government down; would Pasic find an accommodating response or would the threat of a military coup dictate Serbia's actions; and once the Kaiser read the text of Austria's ultimatum to Serbia, might he backtrack on his resolve to support the dual monarchy? Anyone who thinks that they, as analysts, can assign certainty to any state's preferences and the choices those preferences would ultimately dictate in explaining the outbreak of hostilities is making too great a use of hindsight. The preferences we might use in any explanation only became clear as events unfolded—as the wording of Austria's ultimatum became widely known, as Serbia mobilized when confronted with that ultimatum, as Russia began doing the same even before the deadline had been reached for Serbia's response, as it became evident that Viviani would leave France's foreign policy to Poincare, and as Germany reacted to Russia's mobilization.

Incompleteness of information and its potential consequences is also well illustrated by events in July of 1914. Although German officials, on the basis of a variety of scattered reports being fed into Berlin, were well informed about the status of Russia's mobilization, Grey arguably believed, on the basis of the report of Britain's ambassador to Russia, Sir George Buchanan, that Russia had not begun to mobilize. Grey, acting as if he believed that mediation was still possible, gave the appearance of taking Russian denials of mobilization at face value, which only served to sustain the hypothesis in German eyes that Britain had taken sides in the dispute. Given Britain's links to France and France's to Russia, a non-neutral Britain yielded Germany's worst strategic nightmare—a Franco-British-Russian military alliance—which thereby required immediate German mobilization and preventive war.

7.3 Bayes's Law and Bayesian Equilibrium

As a first step toward analyzing the strategic complexity that private information admits in such situations, what we need is a model for how people might adjust prior beliefs on the basis of what they observe—a general rule about probability that allows us to calculate conditional probabilities. That rule is *Bayes's Law*, and to illustrate it, it is convenient to step away from purely strategic issues and consider instead a relatively simple one-person decision problem.

Example: A legislator must decide how to vote, believing initially that there is a .75 chance—the legislator's **prior** or initial beliefs—that a majority of his constituents favor its passage. Because the bill is of such profound significance, the legislator will be reelected if and only if his vote on the bill matches his constituent's majority preference. Thus, a pollster is hired to gauge preferences more accurately. The pollster, however, admits that there is only a .95 probability that the poll and the true majority preference will agree, and thus a .05 probability that the poll will indicate a majority in favor (opposed) when in fact a majority opposes (favors) the bill. To represent this situation using more general notation than we employed in Figures 7.1 and 7.2, let

t_1 *denote "a majority favor the bill" (corresponding to P in Figure 7.1),*
t_2 *denote "a majority oppose the bill" (corresponding to ~P in Figure 7.1).*

Similarly, let

> e_1 *denote "the poll indicates that a majority favors the bill,"*
> e_2 *denote "the poll indicates that a majority opposes the bill."*

In accordance with the legislator's prior probabilities,

$p(t_1) = 0.75$ and $p(t_2) = 0.25$.

The pollster's information about the accuracy of the poll, on the other hand, can be expressed as

$$p(e_1|t_1) = p(e_2|t_2) = 0.95,$$
$$p(e_1|t_2) = p(e_2|t_1) = 0.05.$$

where the notation "|" reads "given that." The legislator, however, is not concerned with the accuracy of public opinion polls per se but rather with how he should revise his estimates of $p(t_1)$ and $p(t_2)$ based on these polls. That is, looking at Figure 7.3, he must consider the likelihood that he is at one decision node versus the other, regardless of which information set pertains. That is, a choice must be made, conditional on what is observed (e_1 or e_2), so that the numbers that ought to concern the legislator are the conditional **posterior probabilities** $p(t_1|e_1)$ and $p(t_1|e_2)$. For example, the probability $p(t_1|e_1)$—the probability that a majority favor the bill, given

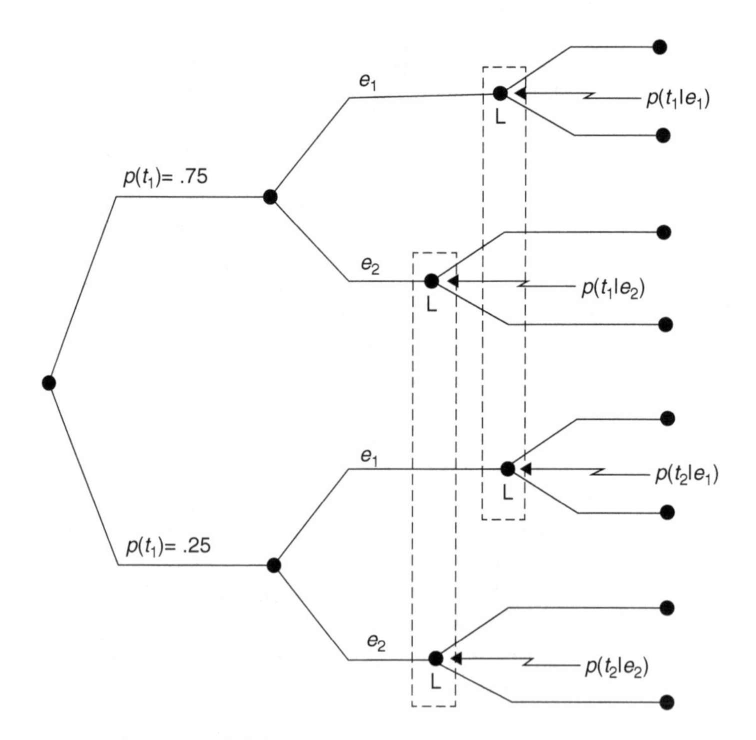

Figure 7.3 Conditional probabilities

that the poll reveals that such a majority exists—corresponds to the probability, given the pollster's report, that the legislator is at the first node of the first information set.

To analyze the problem confronting a decision made in this situation requires, then, a rule as to how beliefs change as a function of observed events. So consider first a simpler situation. First, we require a definition of *conditional probability*, so suppose that O is an observed event, which we believe can occur with probability p(O), and that p(A) is the prior belief that situation A pertains. Thus, the probability that O is observed and A pertains can be expressed in either of two ways: p(A|O)p(O) or p(O|A)p(A). Setting these two terms equal gives

$$p(A|O) = p(O|A)p(A)/p(O)$$

assuming of course that $p(O) \neq 0$. The generalization of this identity, now, using the notation of the preceding example can be written thus: Letting $\{t_1, t_2,\ldots,t_m\}$ be nature's possible choices and $\{e_1, e_2,\ldots, e_n\}$ be the events that a decision maker might observe, and upon which he will condition his guess as to the likelihood that nature makes a particular choice, then ***Bayes's Law*** is

$$p(t_i|e_j) = p(e_j|t_i)p(t_i)/[p(e_j|t_1)p(t_1) + p(e_j|t_2)p(t_2) + \ldots + p(e_j|t_m)p(t_m)]$$

Thus, in our example, $p(t_1|e_1)$, the probability that a majority of constituents prefer the bill, conditional on the fact that the poll indicates that the electorate holds this preference, is given by the equation

$$p(t_i|e_1) = p(e_1|t_i)p(t_1)/[p(e_1|t_1)p(t_1) + p(e_1|t_2)p(t_2)]$$
$$= 0.95 \times 0.75/[0.95 \times 0.75 + 0.05 \times 0.25] = .9828$$

To put such calculations in the context of the USSR's Warsaw Pact allies following Gorbachev's rise to power, suppose those allies believe the following initially:

p(nothing has changed) = 0.7
p(USSR has changed) = 0.3
p(Kremlin reacts negatively | nothing has changed) = 0.9
p(Kremlin reacts negatively | USSR has changed) = 0.1
p(Kremlin does not react | nothing has changed) = 0.1
p(Kremlin does not react | USSR has changed) = 0.9

So, initially, the USSR's allies here begin with the supposition that there's only a 0.3 chance that there's been a fundamental shift in Soviet policy toward its allies that match Gorbachev's words. But now suppose something occurs, such as Hungary opening its border to Austria, and there's no Soviet reaction. Bayes's Law dictates that those allies should then update their belief that nothing has changed as follows:

p(nothing has changed | Kremlin does not react) equals
P(Kremlin does not react | nothing has changed) × p(nothing has changed)
divided by
P(Kremlin does not react | nothing has changed) × p(nothing has changed) plus
(Kremlin does not react | USSR has changed) × p(USSR has changed)
which equals
$0.1 \times 0.7 / [0.1 \times 0.7 + 0.9 \times 0.3] \approx 0.21$

Thus, given how unlikely it is that an unchanged Kremlin would fail to react to a provocation (0.1), observing precisely that—a failure to react—yields a significant drop in the belief that the Kremlin's policies do not match what Gorbachev has been saying. And although it would certainly be a stretch to say that Communist party elites in Central Europe were familiar with Bayes's Law, it is evident that they and the reformers within those countries acted as if they were thus familiar.

For another example of the application of Bayes's law, which illustrates its use in subsequent sections, suppose you are trying to ascertain someone's preferences over three alternatives, A, B and C. Based on your experience with similar situations, suppose the prior probabilities you associate with each of the six possible preference orders over these alternatives are as shown in Table 7.1.

Letting e_1 correspond to "A preferred to B" and letting e_2 correspond to "A not preferred to B," clearly the preceding orders require that

$$p(e_j|t_i) = 0 \text{ if } j = 1 \text{ and } i = 3, 4, 6, \text{ or if } j = 2 \text{ and } i = 1, 2, 5,$$
$$p(e_j|t_i) = 1 \text{ if } j = 1 \text{ and } i = 1, 2, 5, \text{ or if } j = 2 \text{ and } i = 3, 4, 6.$$

Now, however, suppose you observe this person committing an act that cannot be rationalized if that person prefers A to B. So, excluding the first, second, and fifth orders as possibilities, and substituting the preceding values for $p(e_j|t_i)$, we can use Bayes's Law for computing your posterior probability on, say, the third order, $p(t_3|e_2)$ = "Probability that s_3 pertains, given that A is not preferred to B." In this instance we get

$$p(t_3|e_2) = p(t_3)/[p(t_3) + p(t_4) + p(t_6)]$$

Table 7.1 Prior Beliefs About Possible Preference Orders

probabilities:					
$p(t1)$	$p(t2)$	$p(t3)$	$p(t4)$	$p(t5)$	$p(t6)$
A	A	B	B	C	C
B	C	A	C	A	B
C	B	C	A	B	A

To see now how we use Bayes's Law in game theory, let us return to our legislator-lobbyist game and suppose that one or the other player is attempting to establish whether a strategy pair, say (a, b), is an equilibrium. Keeping in mind that any pair of strategies constitutes an equilibrium if and only if each strategy is a best response to the other, consider Figure 7.2 again, which, in accordance with the notion of subgame perfection, we simplify in Figure 7.4 by supposing that the legislator, L, only makes dominant choices at his second move (the game's last move). Notice that in this instance, the legislator's strategies are of the form, "If my preferences are __, then choose __, but if my preferences are __, then choose __ instead," whereas the lobbyist's strategies are of the form, "If my opponent chooses __, then choose __, but if my opponent chooses __, then choose__."

The particular strategic complication confronting the lobbyist, player l, is that to evaluate alternative strategies, he or she must somehow evaluate the probability that nature has selected "prefer" (P = t_1) as against "don't prefer" (~P = t_2). We assume that p(P) and p(~P) are common knowledge, and thus known to the lobbyist. But, as already noted, a lobbyist who simply uses these probabilities when computing a strategy's expected value is failing to make full

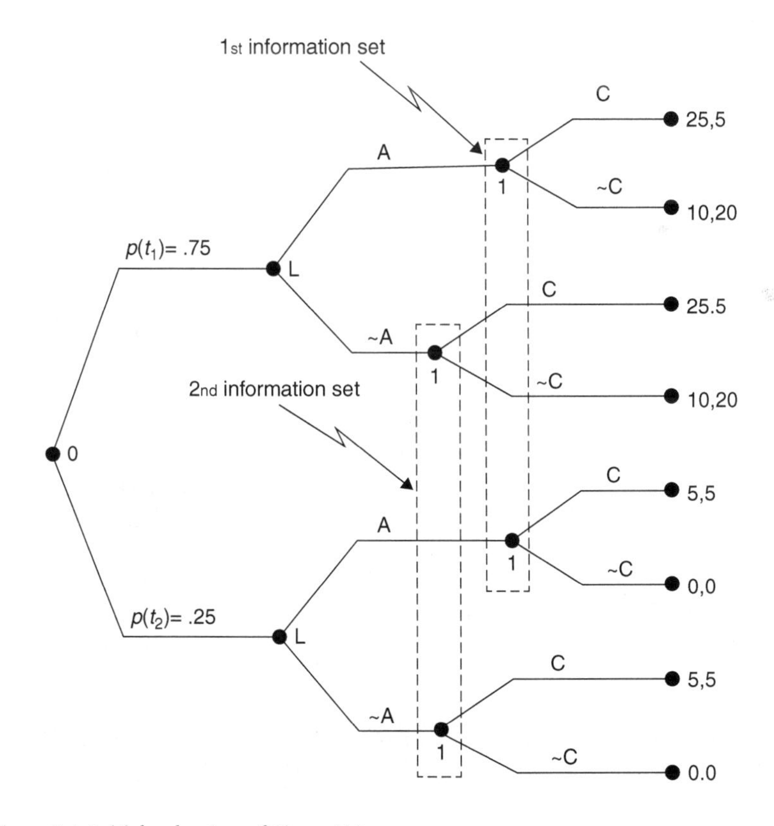

Figure 7.4 Initial reduction of Figure 7.2

use of the available information. Specifically, the lobbyist observes the legislator's choice, which is itself conditioned on the legislator's preference. So in evaluating its strategies, the lobbyist can update the estimate of the likelihood that nature has chosen one state or the other as a function of the legislator's observed action. And this updating, we assume, satisfies Bayes's Law.

To calculate the appropriate conditional probabilities in our example and, subsequently, an equilibrium, suppose you are the lobbyist, and consider the pair of strategies (a, b), where for the legislator

> a = "if P, choose A; but if ~P, choose ~A" [i.e., if a majority of the legislator's constituency favors the bill then ask for a contribution; otherwise do not ask]

and for the lobbyist

> b = "if A, choose ~C; but if ~A, choose C" [i.e., if asked for a contribution then do not offer a bribe, but if not asked, then attempt to bribe]

The first step in determining whether (a, b) is an equilibrium is to determine whether b is a best response to a, which requires first that we calculate the expected value to the lobbyist (l) of b when the legislator (L) chooses a. Owing to the simplicity of our game, we can compute this expected return, denoted $E_{lobbyist}(a, b)$, quite easily by noting that, given a's specification, L chooses A only if nature chooses P, which occurs with probability .75—in which case, given that your strategy is b, you get 20—whereas L chooses ~A only if natures chooses ~P, which occurs with probability .25—in which case you get 5. Thus,

$$E_{lobbyist}(a, b) = .75(20) + .25(5) = 16.25.$$

To see that this calculation is consistent with Bayes's Law, consider the following reasoning: Suppose you find yourself at your first information set (you observe that L chooses A). Then, given that your strategy is b, you choose ~C and get 20 with probability $p(\text{P}|\text{A})$ and 0 with probability $p(\sim\text{P}|\text{A})$. If, on the other hand, you find yourself at your second information set (if you observe L choosing ~A), then, given b, you choose C and get 5 with probability $p(\text{P}|\sim\text{A})$ and 5 again with probability $p(\sim\text{P}|\sim\text{A})$. Thus, your expected payoff from b, given that the legislator acts in accordance with the strategy a, is

$$E_{lobbyist}(a, b) = p(\text{A})[20p(\text{P}|\text{A}) + 0p(\sim\text{P}|\text{A})] + p(\sim\text{A})[5p(\text{P}|\sim\text{A}) + 5p(\sim\text{P}|\sim\text{A})].$$

From L's strategy (i.e., choose A if P and ~A if ~P), however, we know that $p(\text{A}) = 0.75$ and $p(\sim\text{A}) = 0.25$. From Bayes's Law, the probability that the constituency favors the bill when the legislator asks for a contribution is

$$p(\mathrm{P}|\mathrm{A}) = \frac{p(\mathrm{A}|\mathrm{P})p(\mathrm{P})}{p(\mathrm{A}|\mathrm{P})p(\mathrm{P}) + p(\mathrm{A}|\sim\mathrm{P})p(\sim\mathrm{P})}$$

which equals 1 in this instance because L's strategy dictates that $p(\mathrm{A}|\sim\mathrm{P}) = 0$. Similarly, $p(\sim\mathrm{P}|\mathrm{A}) = 0$, $p(\mathrm{P}|\sim\mathrm{A}) = 0$, and $p(\sim\mathrm{P}|\sim\mathrm{A}) = 1$, which yields $E_{lobbyist}(a, b) = 0.75(20) + 0.25(5) = 16.25$.

Of course, determining whether *b* is a best response to *a* requires calculating the expected return of each of the lobbyist's other strategies. And ascertaining whether *(a, b)* is an equilibrium requires that we determine whether *a* is a best response to *b*. However, before we determine the actual equilibrium for our example, let us first review its structure so we can generalize the analysis. Once again, the important feature of the example is that the legislator knows something about his constituency and, thus, about his preferences, which the lobbyist does not know. Although we might have constructed a more elaborate scenario by allowing the interest group to have some private information, the example's essential features are that

1. Each person is characterized by a parameter that is distributed according to a common-knowledge probability density function.
2. The value of this parameter is "chosen" by nature as the first move in the extensive form.
3. The realization of this parameter is private information. (Since the interest group has no private information, we can think of the density that characterizes its "parameter" as allowing only a single value.)

In the example, the parameter in question is the preference of the legislator's constituency and, by extension, the legislator's preference. The probability that the constituency prefers the bill, however, is common knowledge. That is, although both the legislator and the lobbyist have different information about the constituency, the legislator knows the type of information that the lobbyist possesses, the lobbyist knows that the legislator knows the extent of this information, and so forth.

We can generalize the representation of such situations now by supposing that person *i* is a particular "type," which is determined probabilistically from some set of possibilities. The notion of "type" can refer to almost anything, including even whether the player has meaningful choices and is a relevant participant in the game. Generally, though, we let a person's type manifest itself, as in our example, as a particular utility function. Not only does this interpretation allow us to model players who occasionally make mistakes (by maximizing the "wrong" utility function), but it also allows us to consider the possibility that a person is "irrational," by which we mean that the person acts as if he or she is maximizing some improbable or self-destructive utility function. The people we are modeling are then characterized by a common-knowledge density function $p(t^1, t^2,\ldots, t^n)$ that specifies the probability that person 1 is of type t^1, person

2 is of type t^2, and so forth. Finally, we suppose that after observing their own type, each person i computes the conditional density $p(t^1,\ldots, t^{i-1}, t^{i+1},\ldots, t^n|t^i)$ according to Bayes's Law. The simplest possibility, however, is that each person's type is determined independently of every other person, in which case we can characterize the population by a set of densities $\{p_1(t^1),\ldots, p_n(t^n)\}$, where $p_i(t^i)$ is the common-knowledge density governing *i's* type. To keep our discussion of examples as simple as possible, we restrict our attention to this special case.

A *Bayesian equilibrium* to any game in strategic form is defined, now, in a straightforward way. Each person should, of course, condition their strategy on their type, because it is this type that, by assumption, determines preferences over outcomes. Thus, person i knows that at each of *j's* information sets, *j's* strategies will be of the form, "If I am of type t^j_1 then I will choose ___, if I am of type t^j_2, then I will choose ___, and so forth." Holding everything else constant (including the choices of others if there are more than two relevant persons), the payoff i associates with particular strategies on his and *j's* part should be computed as follows: the utility of the outcome that follows if j is of type t^j_1 —the outcome that follows from my choice in combination with the choice j makes when he or she is of type t^j_1—times the probability = $[p_j(t^j_1)$ that j is of this type, plus the utility of the outcome that follows if j is of type t^j_2 times the probability, $p_j(t^j_2)$, that j is of this type, . . . and so forth. A Bayesian equilibrium is then just like a Nash equilibrium—a strategy n-tuple such that, given their types, and given the probabilities that determine types, no person has any incentive to move unilaterally to some other strategy. Similarly, a Bayesian equilibrium is subgame perfect if it is a Bayesian equilibrium for every subgame.

To complete our analysis of our legislator-lobbyist example using our general notation (and here specifically that means letting t^l_1 correspond to P, a majority in the legislator's district favors the bill, and t^l_2 correspond to ~P, a majority opposes the bill), notice that L has four strategies that we can represent as follows:

a_1: *Choose A regardless of type.*
a_2: *Choose ~A regardless of type.*
a_3: *If* t^l_1, *choose A; but if* t^l_2 *choose ~A.*
a_4: *If* t^l_1, *choose ~A; but if* t^l_2, *choose A.*

Similarly, the lobbyist (l) has these four strategies:

b_1: *Choose C regardless of what L does.*
b_2: *Choose ~C regardless of what L does.*
b_3: *If L chooses A, then choose C; otherwise choose ~C.*
b_4: *If L chooses A, then choose ~C, otherwise choose C.*

We complete the strategic form by entering the expected payoffs into the cells of the 4 × 4 game matrix, where the computation of these payoffs uses the fact that

	b_1	b_2	b_3	b_4
a_1	20, 5	7.5, 15	20, 5	7.5, 15
a_2	20, 5	7.5, 15	7.5, 15	20, 5
a_3	20, 5	7.5, 15	18.75, 3.75	8.75, 16.25
a_4	20, 5	7.5, 15	8.75, 16.25	18.75, 3.75

Figure 7.5 Strategic form of legislator-lobbyist game

$p(\text{P}) = p(t^l_{\ 1}) = 0.75$ and $p(\sim\text{P}) = p(t^l_{\ 2}) = 0.25$. For example, (a_3, b_3) yields, "the legislator chooses A and the lobbyist responds with C with probability 0.75 and the legislator chooses ~A and the lobbyist responds with ~C with probability 0.25." Thus, the expected payoff to L of (a_3, b_3) is $0.75(25) + 0.25(0) = 18.75$, whereas the payoff to the lobbyist is $0.75(5) + 0.25(0) = 3.75$. Figure 7.5 now portrays the full strategic form and shows that there are two pure strategy equilibria, both involving b_2, in which legislator L either does (a_1) or does not (a_2) ask for a contribution regardless of type. Looking back at the actions that a_1 and a_2 imply for L (asking for a bribe regardless of type and not asking for a bribe regardless of type), we see that both equilibria in this instance are what game theorists call *pooling equilibria*—equilibria in which a person chooses the same act regardless of their type or the information that nature reveals to them. Thus, in equilibrium, the lobbyist in our example cannot infer anything about the legislator's type on the basis of the legislator's initial action (of course, we should determine whether there are any mixed strategy equilibria). However, in this instance, a check of possibilities reveals that any mixture over a_1 through a_4 that might reasonably be an equilibrium strategy for the legislator would only induce the lobbyist to choose b_2.

7.4 A Game with Two-Sided Incomplete Information

We should not be surprised to learn that the legislator's initial choice of A or ~A does not provide any useful information to the lobbyist. Because the choice of A is costless to the legislator, it never hurts to request a bribe, regardless of the constituencies' preferences on the bill in question. Our example might have been more interesting if we had assumed that there is a cost associated with choosing A (corresponding, for example, to the chance that constituents learn about the request). We will consider such possibilities later, but first we want to suppose that more than one person in a situation has private information. Fortunately, this possibility introduces no new conceptual issues.

Example: Suppose two people prefer different candidates, and for purposes of an example suppose that if both vote (for their preferred candidates) or if both abstain, the candidate that voter 1 prefers wins (our example here is taken from Thomas Palfrey and Howard Rosenthal, "Voter Participation and Strategic Uncertainty," *American Political Science Review* 79, 1985). If C_i is person *i's* cost of voting, and if a person associates a payoff of 1 with his or her preferred candidate and 0 with the opponent, then Figure 7.6a describes the outcomes that prevail, given the choices of both people (row chooser = person 1, column chooser = person 2). Suppose finally that neither person knows the other's cost of voting. Thus, person 1 knows C_1 but not C_2, and 2 knows C_2, but not C_1. Suppose further, however, that 1 knows that C_2 has the same chance of equaling 1/8, 2/8, and 3/8, and that 2 knows that C_1 has the same chance of equaling 4/8, 5/8, and 6/8. Thus, although C_1 and C_2 are random variables, person 1 faces higher average costs than does person 2 (5/8 versus 1/4).

In addition to assuming that both persons have private information, this example differs from the previous one in that neither person has any opportunity to signal anything about their private information. Thus, deception does not appear to be an issue, which might lead us to believe that we can analyze the situation in a more straightforward manner by merely substituting the expected values of C_1 and C_2 into Figure 7.6a and solving the resulting game for equilibrium mixed strategies (there are no pure strategy equilibria). This assumption leads to the 2 × 2 matrix in Figure 7.6b and the conclusion that person 1's probability of voting is 3/4 and person 2's is 5/8.

	Vote	Abstain
Vote	$1-C_1, -C_2$	$1-C_1, 0$
Abstain	$0, 1-C_2$	1, 0

Figure 7.6a Two voters with incomplete information

	Vote	Abstain
Vote	3/8, –1/4	3/8, 0
Abstain	0, 3/4	1, 0

Figure 7.6b Simple substitution of values into Figure 7.6a

This approach to our example, however, does not allow both persons to make full use of their information—specifically, it does not allow them to condition their decisions on the fact that each knows that the other is conditioning a decision on private information. So to see how this fact might affect our analysis, notice that this situation's extensive form, Figure 7.7, shows that each person has three information sets, and at each information set there are two choices. Thus, the number of strategies is $2^3 = 8$, where a typical strategy for person 1

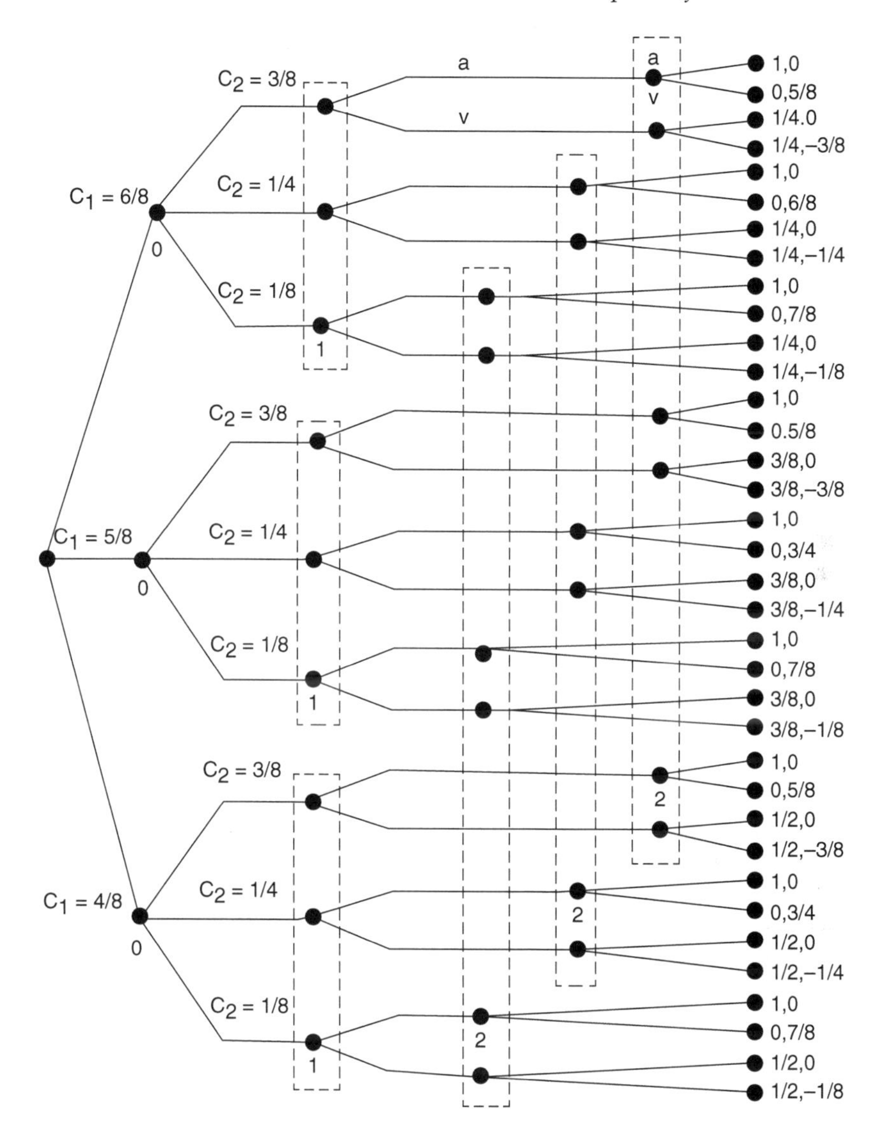

Figure 7.7 Two-person voting decision

reads: If $C_1 = 6/8$, abstain, otherwise vote. We denote such a strategy by (a, v, v) to indicate that the person abstains if the cost of voting assumes its maximum value and votes if this cost takes on a moderate or minimum value. However, four of the eight possible strategies—those that have a person voting if the cost of voting is high but abstaining if the cost is low—are easily seen to be dominated. Restricting our attention, then, to the four strategies (v, v, v), (a, v, v), (a, a, v), and (a, a, a): Figure 7.8 (with all payoffs multiplied by seventy two to eliminate fractions and decimal points) portrays the corresponding strategic

	vvv	avv	aav	aaa
vvv	27, –18	27, –9	27, –3	27, 0
avv	21, 6	29, 7	37, 5	45, 0
aav	12, 30	28, 23	44, 13	60, 0
aaa	0, 54	24, 39	48, 21	72, 0

Figure 7.8 Strategic form of Figure 7.7

form, which establishes that for the parameters of our example, ((a, v, v), (a, v, v)) is the Bayesian equilibrium. In this case, then, both voters vote with probability two thirds (that is, they each vote if either of two circumstances prevails out of the three equally likely possibilities).

To see how the payoffs in these cells are computed, consider the equilibrium cell ((a, v, v), (a, v, v)). Looking at things from person 1's perspective, if $C_1 = 6/8$, then 1 abstains and, given 2's strategy, realizes the payoffs 1, 0, and 0 depending on whether $C_2 = 3/8$, 1/4, or 1/8. Since each of these costs occurs with probability 1/3, when $C_1 = 6/8$, person 1 realizes an expected payoff of 1/3. Similarly, when $C_1 = 5/8$, then 1 always votes and realizes a payoff of 3/8, regardless of 2's cost and subsequent actions, whereas if $C_1 = 4/8$, then 1 again always votes and realizes a payoff of 1/2, regardless of 2's costs and actions. Hence, 1's overall payoff is 1/3[1/3] + 1/3[3/8] + 1/3[1/2] = 29/72.

Because, in this instance, the strategic form disguises the reasons why we have arrived at a different answer than the one we get if we rely simply on a strategic form calculated from simple expected values, let us consider the problem from a different perspective. Notice that each of *i's* strategies can be characterized by a number, C^*_i, such that *i* votes if the cost of voting, C_i, is less than C^*_i, and abstains otherwise. It is straightforward to see, from a simple decision theoretic calculation, that barring other considerations, C^*_i equals the probability that the voter is decisive times the utility difference between the candidates. That is, a citizen votes if

$$(u_1 - u_2)\text{Pr[create or break a tie]} - C > 0,$$

in which case the critical value for C that determines whether a person votes is $(u_1 - u_2)$Pr[create or break a tie]. In large electorates, we might assume that Pr[...] is merely a subjective estimate based on public opinion polls and the like and that it is reasonable to ignore interactive effects of individual decisions. But in small electorates, Pr[...] clearly depends on what a person thinks others will do, which is a function of what they think he will do, and so forth. Suppose then, that we let P_i denote the probability that the citizen is decisive and let p_i be the probability that *i* votes. Then for our two-voter example, P_1 is a function of

p_2, and P_2 is a function of p_1. The critical number C^*_1, then, equals P_1 times the utility difference between the candidates, which is 1 in our example. Thus, the citizen votes if $C_1 < C^*_1 = P_1$. The game theoretic nature of the problem follows from the fact that, since P_1 is a function of p_2, and since p_2 is a function of C^*_2, then C^*_1 is a function of C^*_2, and vice versa—which is to say that 1's optimal strategy depends on 2's strategy (1's strategy should be a best response to 2), and 2's optimal strategy depends on 1's strategy (2's strategy should be a best response to 1). The Bayesian equilibrium identified in Figure 7.8 is merely the pair of strategies that are best responses to each other.

7.5 Agendas Reconsidered

Aside from some additional complexity in the calculation of expected payoffs, the examples of incomplete information considered thus far do not look much different than the games we consider in previous chapters. Also, by setting things into strategic form it does not appear as though we are using Bayes's Law. So instead, let us consider voting agendas again, and recall our conclusion that if an amendment agenda includes an alternative that is a Condorcet winner, it necessarily emerges as the final outcome. However, voting in legislatures and committees assumes a more interesting and strategically complicated character if we suppose that not all legislators know the preferences of all other legislators. Indeed, unless we are willing to suppose that all legislative votes merely to ratify and legitimize outcomes that everyone knows are foregone conclusions, agendas, especially complicated ones, almost certainly occur in an incomplete information environment. We cannot explore all ramifications of this fact, but we can address the question of whether the conclusion about the eventual emergence of Condorcet winners necessarily holds if information about preferences is private while simultaneously illustrating more explicitly the relevance of Bayes's Law. We begin with an especially simple possibility.

Example: Consider a committee of *n* people who can hold one of the following three preferences of the six possibilities when there are three alternatives, where any individual can hold preferences of type *i* with probability p_i, but every person knows their own type:

type 1	type 2	type 3
A	B	C
B	C	A
C	A	B

In addition, assume that all voters assign a utility of 1 to their most-preferred alternative, 0 to their least-preferred alternative, and v $(0 < v < 1)$ to the alternative that ranks second on their preferences. Finally, suppose that the committee uses an agenda that first sets A against B, the winner against C.

What we want to show is something that we suspect is not intuitive—that even if it is almost certainly true that the committee is unanimous in its preferences so that a Condorcet winner is almost certain to exist, that winner will definitely not be selected if the committee is sufficiently large (here our discussion is taken from Thomas Palfrey and Ordeshook, "Agendas, Strategic Voting, and Signaling with Incomplete Information," *American Journal of Political Science* 32, 1988).

To establish this seemingly perverse fact we must proceed a bit differently than before since even with a 3-member committee and three possible preference types for each voter, nature can choose any one of 27 possible committee profiles. Thus, even before we include the branches representing the voters' choices, our extensive form becomes unwieldy. Hence, we must use some shortcuts—specifically, just as we pruned Figure 7.2 by eliminating the legislator's dominated choices (and thus dominated strategies) in order to arrive at Figure 7.4, we should first try to eliminate dominated strategies for different types of voters so as to move as quickly as possible to the identification of subgame perfect Bayesian equilibria.

We begin by noting that as in our earlier analyses of agendas, regardless of the outcome on the initial ballot between A and B, everyone should vote sincerely on the second ballot. Next, notice that we can identify dominant strategies for two of the voter types.

Type 2 voters: With everyone voting sincerely on the last (second) ballot, the uncertainty about preferences means that a victory for A on the first ballot yields a lottery between A and C whereas a victory for B yields a lottery between B and C. The exact nature of this lottery depends, of course, on the probabilities we assign to voters being of one type or the other. However, regardless of these probabilities, a type 2 voter—a voter who prefers B to C to A—prefers any lottery between B and C to any lottery between A and C. From the assumption of common knowledge, then, everyone knows that type 2 voters will vote for B on the first ballot.

Type 3 votes: The analysis of type 3 is a bit more complicated, but once again we can show that a unique strategy is part of any equilibrium for such a type. Briefly, if there are n voters, there are two possibilities:

1. $(n + 1)/2$ or more other voters do not have type 3 preferences.
2. $(n + 1)/2$ or more other voters have type 3 preferences.

No voter knows with certainty which of these possibilities describes the committee, but if case 2 holds, then C prevails regardless of what anyone does on the first ballot. On the other hand, if case 1 holds, then B prevails eventually if it wins on the first ballot, whereas either A or C can prevail if A wins on the first ballot. Since B is a type 3 voter's least-preferred alternative, voting for A on the first ballot is equivalent to voting for a lottery between A and C and is, thereby, a dominant choice for a type 3 voter, a voter who prefers C to A to B.

This discussion means that we can reduce the strategies we must consider for every voter to the following two alternatives:

s_1: *If type 1, then vote sincerely for A.*
If type 2, then vote for B.
If type 3, then vote for A.

s_2: *If type 1, then vote strategically for B.*
If type 2, then vote for B.
If type 3, then vote for A.

As with our previous examples, we might next try to construct a strategic form. However, given the nature of the problem, we encounter too much complexity for our liking. Specifically, if there are n voters, then the corresponding strategic form has 2^n cells that we must check as potential equilibria. At this point, however, we should keep in mind that we merely want to show that something different can occur when there is private information—namely, that a Condorcet winner does not necessarily prevail as the final outcome. Thus, we look only at two specific strategy n-tuples and forgo an analysis of those cells in the strategic form that have different voters choosing different strategies. That is, we restrict our attention to two possibilities: (1) all voters choose s_1, in which case all type 1 voters vote sincerely for A on the first ballot, and (2) all voters choose s_2, in which case all type 1 voters vote strategically for B on the first ballot.

Checking first whether the n-tuple $(s_1, s_1, \ldots, s_1)$ can be a Bayesian equilibrium, what we want to see is whether any voter has an incentive to unilaterally defect to some other strategy. Since we already know that type 2 and type 3 voters have dominant choices on the agenda's first ballot, no voter has an incentive to defect to a strategic choice when it holds either of these types of preferences. The only possible defection is from a type 1 voter who chooses to vote strategically for B. At this point we come to the next trick in our analysis. Notice that there cannot be any **positive** incentive to defect if the voter in question is not pivotal on the first ballot. If, for example, the vast majority of voters have preferences of one particular type, then no type 1 voter can be pivotal, and, thus, no type 1 voter will have an incentive to defect unilaterally from s_1. So in ascertaining whether $(s_1, s_1, \ldots, s_1)$ is an equilibrium, we should examine whether a type 1 voter has an incentive to defect unilaterally from A to B, conditional on that voter being pivotal on the first ballot.

To see whether any specific type 1 voter might defect in this circumstance, notice that the only situation in which such a voter can be pivotal on the first ballot is if precisely $(n - 1)/2$ other voters in the committee have type 2 preferences since, if everyone abides by s_1, it is only these voters who are voting for B. And in this event, a voter must evaluate the two first-ballot vote choices thus:

> *Choice 1*: If the voter in question swings the outcome to B by voting for B, then that voter and the $(n - 1)/2$ type 2 voters who voted for B on the first ballot will join on the second ballot to produce B with certainty.

Choice 2: If the voter in question swings the vote to A by voting for A, then the result is a lottery between A and C. In this lottery, C prevails if there is at least one type 3 voter among those who voted for A on the first ballot and A prevails if all others who chose A have type 1 preferences. Of course, since we are holding constant the strategies of all voters but one, the $(n - 1)/2$ voters who chose A cannot be type 2 voters, and from Bayes' rule, the conditional probability that a particular one of them has type 3 preferences, given that that voter cannot have type 2 preferences, is $p_3/(p_1 + p_3)$ and the conditional probability that a particular one of them has type 1 preferences is $p_1/(p_1 + p_3)$. The probability that all of the $(n - 1)/2$ voters who chose A on the first ballot are of type 1, given that they are not type 2, is

$$p(t_1 | \sim t_2) = \left[\frac{p_1}{p_1 + p_3} \right]^{(n-1)/2}$$

and the probability that one or more of these voters has type 3 preferences is simply

$$p(t_3 | \sim t_2) = 1 - \left[\frac{p_1}{p_1 + p_3} \right]^{(n-1)/2}$$

Thus, whenever the voter in question is pivotal, choosing B yields a payoff of v, whereas choosing A yields an expected payoff of 1 times $P(t_1|\sim t_2)$ plus 0 times $P(t_3|\sim t_2) = 1 - P(t_1|\sim t_2)$, or simply $P(t_1|\sim t_2)$. Thus, this voter prefers to defect unilaterally from the strategy of voting sincerely for A if

$$v > \left[\frac{p_1}{p_1 + p_3} \right]^{(n-1)/2}$$

Since the term in brackets is less than 1, we can make the right side of this inequality as close to 0 as we choose by increasing n. Since $v > 0$, this means that all type 1 voters voting sincerely cannot be an equilibrium if the committee is sufficiently large.

Now consider the second n-tuple, $(s_2, s_2, \ldots, s_2)$ as a possible equilibrium wherein type 1 voters vote strategically for their second choice, B, on the first ballot. If a vote is decisive on the first ballot, there must be precisely $(n - 1)/2$ type 3 voters since it is only voters of this type who vote for A in the conjectured equilibrium. Once again we must consider the consequences of a type 1 voter swinging the first ballot outcome from B to A.

Choice 1: If our type 1 voter in question chooses B, then as before B prevails with certainty since it is only the type 3 voters who will vote for C if B wins on the first ballot and there are only $(n - 1)/2$ such voters.

Choice 2: If the voter in question defects from the presumed equilibrium and chooses A on the first ballot, then again a lottery results in which A prevails if all of the $(n - 1)/2$ voters who originally voted for B are type

1's, and C prevails if only one of these voters has type 2 preferences because such a voter will join with the $(n - 1)/2$ type 3's on the final ballot.

Making a similar calculation as before, the conditional probability that all of the B-voters have type 1 preferences, given that they are not type 3, is

$$p(t_1 | \sim t_3) = \left[\frac{p_1}{p_1 + p_2}\right]^{(n-1)/2}$$

and the probability that one or more of them has type 2 preferences is 1 minus this probability. Thus, for the pivotal voter in question, the expected value of voting for A is 1 times $[p_1/(p_1 + p_2)]^{(n-1)/2}$, plus 0 times 1 minus this probability. The voter prefers to stick to the strategy of voting strategically for B, then, if

$$v > \left[\frac{p_1}{p_1 + p_2}\right]^{(n-1)/2}$$

As before, this inequality is satisfied if n is sufficiently great. That is, if n is sufficiently great, the situation in which all type 1 voters voting strategically for B is part of a Bayesian equilibrium.

To see what we have just done now in strategic form, suppose the voter in question is not pivotal, in which case, depending on whether everyone else is strategic or sincere, he or she receives some fixed payoff. However, when they are pivotal, that voter gets $[p_1/(p_1 + p_3)]^{(n-1)/2}$ or $[p_1/(p_1 + p_2)]^{(n-1)/2}$ depending on whether or not everyone is sincere or strategic. Since we restricted our analysis to looking for symmetric equilibria wherein all voters abide by the same strategy, if we ignore the fixed payoffs when the voter is not pivotal, the situation's strategic form reduces to the one shown in Figure 7.9. It is here that we show, for sufficiently large n, that only the lower right cell can correspond to an equilibrium.

This example illustrates a profound difference between games of complete and games of incomplete information. Notice that we can make p_2 and p_3 quite small, thereby making it nearly certain that A is the unanimous or nearly unanimous committee preference, and still maintain a Bayesian equilibrium in which B prevails. For example, if each voter has type 1 preferences with probability .99, and type 2 and 3 preferences each with probability .005, if the committee has 100 or more members, and if v exceeds .78, then the n-tuple $(s_2, s_2, \ldots, s_2)$ is a Bayesian equilibrium. To see why our conclusions here differ so markedly

	Everyone sincere (s_1)	Everyone strategic (s_2)
Sincere (s_1)	$[p_1/(p_1+p_3)]^{(n-1)/2}$	v
Strategic (s_2)	v	$[p_1/(p_1+p_2)]^{(n-1)/2}$

Figure 7.9 Strategic form of agenda game

from the ones we offer when information is complete, notice that in determining whether a particular strategy n-tuple is an equilibrium, we look only at those situations in which a person is decisive. It is, after all, only in these situations that a person might have an incentive to defect from one strategy to another. With incomplete information, by conditioning on such possibilities, voters are conditioning their decisions on unlikely (but feasible) events. And if people focus on these events, sincere voting becomes an undesirable strategy. When information is complete, on the other hand, such events are no longer feasible—voters know with certainty that voting sincerely is a dominant strategy—and, thus, they cannot create incentives to vote insincerely.

Despite this argument, our result might still seem counterintuitive. We suspect ourselves that it does not apply with full force to most legislatures—that legislatures are not constantly going about passing bills that are otherwise defeated by Condorcet winners that appear simultaneously on agendas. Our reason for supposing that this is true, however, is not because we believe that there is a fundamental flaw in the preceding analysis, but because we suspect that that analysis is incomplete. Actual legislatures are characterized by considerable pre-vote discussions that may include straw votes among subsets of legislators, as when in the U.S. Congress Republican and Democratic members meet in caucus. Prior commitments to vote one way or the other, in conjunction with nonbinding straw votes, may allow the revelation of sufficient information so as to reduce considerably the likelihood of the apparent paradox of our example. However, before we can suppose that prevoting discussions and verbally stated commitments can materially affect our conclusions about agendas, we must consider the fact that such discussion and stated commitments can themselves be part of a person's strategy to deceive others. Thus, we must more explicitly consider the various signals that people can offer prior to acting as part of a situation's strategic character. Another reason for speculating that our example is special is the fact that although we can be nearly certain that a Condorcet winner exists, the probability of its existence is not 1.0. So consider what happens if a Condorcet winner exists with certainty.

Example: Suppose voters have one of the following preferences:

type 1	type 2	type 3	type 4
A	B	B	C
B	A	C	B
C	C	A	A

That a Condorcet winner exists in this instance follows from the fact that if we order the alternatives A-B-C on a single dimension, then all preferences are single-peaked and regardless of the distribution of preferences across types, the Median Voter Theorem applies. Of course, we do not know which alternative is the Condorcet winner since we do not know how many people of the different types actually exist in the committee. Paralleling our previous analysis, and assuming that the agenda is A versus B, the winner against C, notice that both type 2 and type 3 voters have dominant strategies—voting sincerely for B. Holding everyone else's choices constant, let us now look at how type 1 and type 4 voters vote on the first ballot:

If a type 1 voter is pivotal, it gets A with certainty if it chooses A (since such a voter is pivotal if and only if there are precisely $(n + 1)/2$ type 1 voters) and it cannot do better voting for B. Thus, it has no incentive to defect from a sincere strategy,

If a type 4 voter is pivotal, then both choices confront it with a lottery with a clear trade-off. If B is chosen, the voter gets its most preferred alternative, C, only if every type 4 voter constitutes a majority, whereas if it votes for A it gets C if type 3 and type 4 voters are a majority. Thus, choosing between A and B involves a trade-off for a higher probability of a first choice versus accepting some probability of a last versus a middle choice.

We will not derive the exact probability relationship that guarantees sincerity, except to note that it is more easily satisfied, *ceteris paribus*, as n increases. [The exact function is

$$v > \frac{(p_1 + p_3)^n - p_4^{\,n}}{(p_1 + p_2 + p_3)^n - p_4^{\,n}}$$

wherein the right side of this inequality necessarily tends to 0 as n goes to infinity as long as $p_2 > 0$]. On the other hand, the fact that such a condition is necessary indicates that even for the case of single-peaked preferences, there is no guarantee that Condorcet winners will prevail—incomplete information is a qualitatively different situation than complete information, at least for agenda voting.

The preceding discussion suggests that to extend our analysis of agendas we should consider more complicated forms. Unfortunately, such an extension requires a level of analysis inappropriate for this text. In particular, the feature of our examples that renders them tractable (although it might not seem so to all readers) is that they allow only two ballots. With only two opportunities to vote, voters cannot use what they might learn on the first ballot to affect how they act subsequently. Although a voter's initial beliefs about probabilities (the voter's priors) are revised after the first ballot, this information cannot lead to any changes in behavior because everyone votes sincerely on the second (final) ballot. Voting, then, gains an added strategic dimension with three or more ballots—not only must one vote to direct outcomes toward one's preferences, but, as in our legislator-lobbyist example, voters must also be concerned with what others might infer from one's actions. That is, in games of incomplete information, the possibility exists that a person can learn something valuable as the game unfolds—players can begin to refine their estimates of the preferences of others, and if the game allows non-trivial multiple stages, then the "manipulation" of beliefs becomes part of one's strategic concern. Thus, if we allowed multiple ballots in agendas, everyone must be concerned with what interpretation others will give to one's actions, and they must be concerned as well with the interpretation that should be given to the actions of others—all with the understanding that everyone is trying to take account of such matters simultaneously, while making decisions that lead to the best possible outcome under the circumstances. It is perhaps for this reason that we see multistage voting in committees even when, after the fact, there is an evident Condorcet winner.

7.6 Reputation and the Chain-Store Paradox

If we were to consider a formal analysis of games of incomplete information with sequences of moves, the analytic complexity of things would quickly move beyond what we are prepared to present here. Nevertheless, it is evident that such games are an important component of political processes. Consider again the unraveling of the USSR's Warsaw Pact. That unraveling was a process wherein the Kremlin chose to ignore a sequence of actions by its allies as those allies sought to understand Moscow's new policy toward them. There was also, we suspect, a degree of "contagion" whereby Communist officials in one country observed the Kremlin's response to actions elsewhere, in which case the relevant players were able to update their priors even when they themselves did nothing. That updating, moreover, was a complex process for all involved, including political leaders in the West. Innumerable secret and not-so-secret conferences were held in Washington, for instance, in the attempt to assess Moscow's motives, with Gorbachev's visits to New York, London and Bonn all a part of his attempt to shift priors away from what they had been under previous Soviet regimes. And at every stage of this process the players involved, especially those in the West, had to act carefully so as to give Gorbachev the room and support he needed to out-maneuver his potential domestic opposition while allowing for a gradual transition of regimes in Central Europe.

In lieu of delving into analytic complexities that take us far from an introductory treatment of game theory, let us instead outline the general nature of a solution to a specific abstract game that has received considerable attention in the academic literature—the Chain-Store Paradox:

> A large retail store chain with, say, an outlet in each of n markets, enjoys a monopoly in those markets and, accordingly, charges a monopoly price for what it sells. However, a single small competitor "waits in the wings" of each market, and beginning with market 1, each must decide one at a time whether to compete against the chain. If a competitor enters a market, the chain store must then decide whether to appear "weak" by adjusting its prices to the competitive market price or whether to try to establish a reputation for toughness by slashing prices drastically in that market to the point that neither it nor its competitor can earn a positive profit.

The potential advantage of appearing tough, then, is that the chain can try to dissuade future competitors from entering its markets. On the other hand, acting tough can cost the chain considerably in terms of lost profits. Assuming that it is a different potential competitor with whom the chain must content in each market, we have here an "n + 1 person" game. Without incomplete information, we can solve for equilibria in the usual way by working down the extensive form and conclude with reasonable payoffs that competitors should enter in all markets, and the chain should always capitulate. Specifically, in the last (nth) market there is no opportunity for deception or reputation building, in which case a competitor knows that its entry will force the chain to capitulate by lowering its price to meet the competition. Knowing this, the potential competitor in the n-1st market will also enter since it knows that the chain will not attempt to establish a reputation for "toughness" since, regardless of its reputation, a competitor will enter in the nth market. This situation, then, is like the finitely repeated Prisoners' Dilemma in that the chain, preferring to avoid negative profits, will capitulate in the nth (last) market so a competitor enters there; similarly, competitors enter and the chain capitulates in markets $n - 1, n - 2, \ldots 1$.

Such a prediction seems counterintuitive to the extent that we frequently see people in similar circumstances attempting to gain or to maintain reputations for toughness in order to forestall future competition. Anyone familiar with the Robber Baron period of American industrial development in the 19th century and with the actions of such industrialists as Rockefeller, Carnegie, Fisk and Vanderbilt will also be familiar with the strategies they often pursued that, in the short term, were quite costly, but in the long run either eliminated their competition or dissuaded anyone from becoming a competitor. For a more contemporary example, it is convincingly argued that we should not negotiate with terrorists even if failing to do so results in a loss of life among hostages, since displaying a willingness to compromise merely encourages future terrorists. Similarly, mainland China breaks diplomatic relations with those states recognizing Taiwan as an independent entity in order to maintain its reputation for diplomatic toughness, thereby hoping to keep other states from establishing formal diplomatic ties with what it deems to be nothing more than a renegade province.

What allows for the establishment as well as the dissolution of a reputation in each of these situations is some incomplete information about preferences. Terrorists are uncertain about our willingness to abandon hostages, and countries are uncertain about mainland China's willingness to incur the economic costs of breaking diplomatic relations. Thus, if we allow some uncertainty about a chain store's preferences—some uncertainty as to its willingness to incur short-term costs—then the chain may be able to establish a reputation for toughness and a willingness to incur short-term losses in order to realize the longer term gains of keeping others from entering its markets.

The analysis of this game when no competitor knows for certain whether the chain is "tough" or "weak" is challenging. In fact, additional refinement of the Bayesian-Nash equilibrium concept is required in order to restrict predictions and to contend with the possibility that the players must decide what to do if they find themselves at a decision node that is otherwise regarded as being impossible to reach. Nevertheless, if we assume that once the chain fails to "act tough," everyone thereafter knows that the chain is weak, then the general form of the solutions to this game from the perspective of the chain store is as follows: The chain begins by acting tough against any early entrant. Although it incurs a short-term loss by doing so, it gains because one or more potential competitor in the future is scared away. As the game progresses, the chain switches to a mixed strategy between being tough and letting competitors in. Once it acts weak, it acts weak thereafter, and as the last market is approached, the chain, because there are no longer enough markets for it to secure sufficient rewards from its reputation, allows entrants to compete at market prices. We should also note in passing that sometimes it is difficult to rid oneself of a reputation for toughness. Referring again to events in Central Europe, the Soviet Union entered that decade with a reputation for 'toughness'—for not tolerating deviations from an official Party line—based not only on how it treated dissent within its own borders, but how it treated any deviations in its satellite states. Thus, when the Soviet Union attempted to rid itself of what it deemed the economic burden of maintaining Communist regimes in Central Europe, telling the leadership of those countries to follow the "Sinatra Doctrine" of solving their problems, reformers within those countries had a difficult time believing that if reforms went "too far" the Soviet army would not once again make an unwelcome appearance. It took several years and multiple instances of the Kremlin failing to react to significant deviations from the old party line before reformers felt secure in pushing things as far and as fast as they did.

7.7 Signaling, Deception and Mutually Assured Destruction

The Path to WWI: Multiple moves in an extensive form whether in the case of international sanctions or voting agendas open the door to a possibility we cannot study outside of the context of games of incomplete information—signaling and deception—since otherwise, by assumption, the players know all there is to know about each other. If we look again at events leading up to the outbreak of World War I, we find a clear example of the role of signaling in a game of incomplete information wherein one player attempts to get another to act in a way that reveals their type. Briefly, German military planners were suffering from the anxiety occasioned by the belief that the military capabilities of the Entente states, and of Russia in particular, were increasing at a rate as to render Germany ultimately indefensible. This assessment, in turn, supported the argument for a preventive war against Russia and, if need be, France as well. There were, of course, opposing views within the Reich, but if it could be shown that Russia was committed to warring against Germany, then Germany's best course of action was to initiate a preventive war immediately. But how to learn Russia's intentions—its type? The answer arrived at by decision makers in Berlin was to stand firm in support of Austria in its confrontation with Serbia.

If Russia chose to fully mobilize in support of Serbia, this would be taken as signaling its determination to seek a revision in the balance of power on the Continent—a revision wholly to Germany's detriment. In attaching this inference to a decision to mobilize, we should keep in mind that mobilization was anything but a costless action. Trains had to be commandeered and rerouted, manufacturing output shifted, thousands of horses requisitioned and re-shoed, troops called back from their homes and farms (thereby impacting agricultural output) and then soldiers equipped and re-stationed. Put simply, mobilization was not a "cheap talk" move—in the early 20th century it was equivalent to a wholesale restructuring of a national economy. Thus, by supporting Austria and forcing Russia's hand, "the Germans were not taking risks, but testing for threats" (Christopher Clark, *The Sleepwalkers*, Great Britain: Penguin, 2012: 419). And, as we know, Russia did mobilize and strategic planners in Germany did conclude that preventive war was their best option.

It goes virtually without saying that signaling is a critical component of international relations—and indeed of politics generally. Not only do we see it here in events leading to WWI, but we saw it as well when Gorbachev sought to change the USSR's relations with its satellite states; when the United States and the Soviet Union negotiated various arms control treaties; when, in negotiating a budget, legislative and executive branches of a government signal what they are willing to accept and what they are unlikely to agree to; and when labor and management negotiate a contract. But the participants in such situations also must concern themselves with the possibility of artful deception. Leaders of the communist states of Central Europe not only had to concern themselves initially that Gorbachev's utterance were empty words, but also with the possibility that nothing had fundamentally changed in the Kremlin and the powers that be there were merely trying to "smoke out" those political elites who were less than loyal to the Soviet regime. Similarly, in labor-management negotiations, union leaders may have an incentive to disguise the compromises they think their members will accept while management will attempt to make compromises on wages and pensions appear more onerous to profits than they might actually be. It's unreasonable to suppose, however, that the participants in such scenarios are unaware of the incentives that operate within them, in which case we must ask: Is it possible to deceive someone who knows you're trying to deceive them?

To examine signaling and the possibility of deception more closely, then, let us return to our legislator-lobbyist example, except that now suppose that the legislator incurs a cost, C, if it asks for a favor. Because this cost assures the lobbyist that a request for a bribe is no longer *cheap talk*, both the lobbyist and the legislator may want to evaluate their strategies more carefully. Hence, looking back at Figure 7.5, notice that the lobbyist's payoffs (column chooser) are unaffected by the cost, and so we needn't concern ourselves with the strategy b_1

	b_2	b_3	b_4
a_1	7.5–*C*,15	20–*C*,5	7.5–*C*,15
a_2	7.5, 15	7.5, 15	20, 5
a_3	7.5–3*C*/4, 15	18.75–3*C*/4, 3.75	8.75–3*C*/4, 16.25
a_4	7.5–*C*/4, 15	8.75–*C*/4, 16.25	18.75–*C*/4, 3.75

Figure 7.10 Legislator-lobbyist game with costs

because it is dominated by b_2. Thus, constructing the 4×3 strategic form in the usual way, we get Figure 7.10.

The inclusion of these costs has clearly destroyed one equilibrium, (a_1, b_2), but it has left the remaining pooling equilibrium, (a_2, b_2), intact. However, no new pure strategy equilibria are introduced, with the possible exception of (a_2, b_3) if C is sufficiently great, although this equilibrium produces the same outcome as (a_2, b_2). What is interesting about this example, though, is that we now have a mixed strategy equilibrium involving the pure strategies a_1 and a_4 for the legislator and b_2 and b_3 for the lobbyist. Such a mixture does not exist if $C = 0$, since if the lobbyist gives b_4 zero weight, a_1 dominates a_4. But with $C > 0$, some messy algebra establishes that the following is an equilibrium.

$$((1/9, 0, 0, 8/9), (1 - C/15, C/15, 0))$$

In this equilibrium, the legislator will most probably choose a strategy, a_4, that would otherwise reveal its type. However, by placing some probability on a_1, which requires that the legislator ask for a bribe even if he or she doesn't need to be induced to vote for the bill, the lobbyist must give some weight to offering bribes when asked. In this example, though, the mixed strategy (as some additional algebra reveals) is not beneficial to the legislator, who now receives an expected payoff of $7.5 - C/6$. This fact, however, is largely a consequence of parameter values and the fact that the interests of the legislator and the lobbyist are not diametrically opposed—the lobbyist, after all, wants to induce the legislator to vote for the bill when constituents oppose it, and the legislator wants to be thus induced.

So, with situations such as the relationship between Taiwan and mainland China or between competing states in the Middle East in mind, let us consider a second example that begins to address one of the most interesting applications of game theory to strategy in political science. Specifically, we know that for upwards of 45 years, the United States and the Soviet Union ostensibly coexisted with each abiding by a strategy of mutual deterrence—if one country attacked the other, presumably with nuclear weapons, the attacked country would retaliate in kind to inflict losses on the aggressor. The difficulty people had with viewing Mutually Assured Destruction (MAD) an equilibrium, though, was somehow negating the advantage given to a first strike. With each country possessing enough nuclear weapons to destroy the other several times over, a first strike would essentially render its victim impotent, or nearly so. And even if the attacked country possessed some residual retaliatory capability, what incentive would it have to use it if that would merely invite the initial aggressor to launch a second strike as a virtual *coup de gras*? Put simply, theoretically MAD shouldn't have worked, but it did!

> **Example**: There are two countries B and L. B is a big country that can almost certainly defeat L, a little country, in any military confrontation. However, although it probably can't win even if it resists an attack, L can impose a heavy cost on B in the event of an outright conflict. Assume B chooses first (between Attacking and Not Attacking) and that L chooses second (between Resisting and Capitulating). Clearly, if B knows L's preferences, we have described a rather uninteresting situation—one that does not leave much room for what people find interesting about strategic deterrence in international affairs; namely, uncertainty and the possibility of deception. So suppose that B is uncertain as to whether L will resist (if it is "strong") or capitulate (if it is "weak") in the event of an attack, and that it associates the probabilities p and $1 - p$, respectively, with these two possibilities. In addition, suppose country L can make an initial move, which consists of signaling or not signaling its resolve to resist B if attacked. Finally, to make matters more interesting, suppose this signal is costly. Of course, we should not expect this signal to convey information if it is costless, so suppose that if L signals its intention to resist, the cost C is subtracted from its final payoff.

The issue is not simply whether L should send a costly signal, but whether the opportunity to signal can improve L's welfare. We are interested in such matters because it is not uncommon, for example, for people to question the value of some weapons system, especially if that system seems overly costly given its ostensible strategic mission. But consider this possibility: By building such a system, can a country communicate its willingness to fight if attacked? Can a country say, in effect, something like the following to an enemy:

> "Although you may be able to defeat me, I can punish you severely if attacked. And although you might question my resolve, I would certainly punish you; otherwise, I would not have built such a costly weapons system in the first place."

To see whether this argument makes theoretical sense, consider the outcomes and payoffs in Table 7.2, where we assume the following general relationships: First, the big country prefers that L capitulate if it is attacked, i.e.,

$$U_c > 0 > U_r,$$

Table 7.2 Country Payoff Notation

Outcome	*B*	*strong L*	*weak L*
status quo	0	0	0
B attacks, L resists	U_r	u_r^s	u_r^w
B attacks, L capitulates	U_c	u_c^s	u_c^w

and, second, we assume that what differentiates a strong from a weak little country is these two inequalities:

$$0 > -C > u_r^s > u_c^s \text{ and } 0 > -C > u_c^w > u_r^w.$$

Figure 7.11 portrays this situation's extensive form by assuming that L but not B knows L's type. First, since L will not resist if it is weak or capitulate if it is strong, the X'd branches can be deleted. This reduction of the extensive form allows us to simplify our notation by deleting the superscripts on the *u's* (i.e., only a strong L resists, and only a weak one capitulates). To solve for equilibria we next construct a strategic form using the following strategies for B and L. First, for B

B_1: *Attack regardless of whether or not L signals.*
B_2: *Attack if L signals; don't attack if it doesn't.*
B_3: *Don't attack if L signals; attack if it doesn't.*
B_4: *Don't attack, regardless of whether or not L signals.*

and for L

L_1: *Don't signal regardless of type.*
L_2: *Signal if strong; don't signal if weak.*
L_3: *Don't signal if strong; signal if weak.*
L_4: *Signal regardless of type.*

To see how we determine the payoffs in the game's strategic form shown in Figure 7.12, suppose B chooses B_3 and L chooses L_2. So L signals with probability p, B doesn't attack, and the outcome is $-C$ for L and 0 for B; and L doesn't signal with probability $1 - p$, B attacks, and the outcome is u_c for L and U_c for B. Thus, with (L_2, B_3), the expected payoff for L is $-pC + (1 - p)u_c$, and for B it is $(1 - p)U_c$.

Regardless of the numbers we assign to payoffs, we can see that there are no *separating equilibria*—equilibria in which the little country behaves according to its type by signaling when it is strong and not signaling when it is weak, or vice versa. That is, neither the strategy L_2 nor L_3 is involved in any pure strategy equilibrium. This follows from the fact that if L chooses L_2, then B prefers B_3, given that $U_c > 0 > U_r$, but if B chooses B_3, then L prefers L_4. If, on the other hand, L chooses L_3, B's best response is B_2; but (L_3, B_2) cannot be an equilibrium

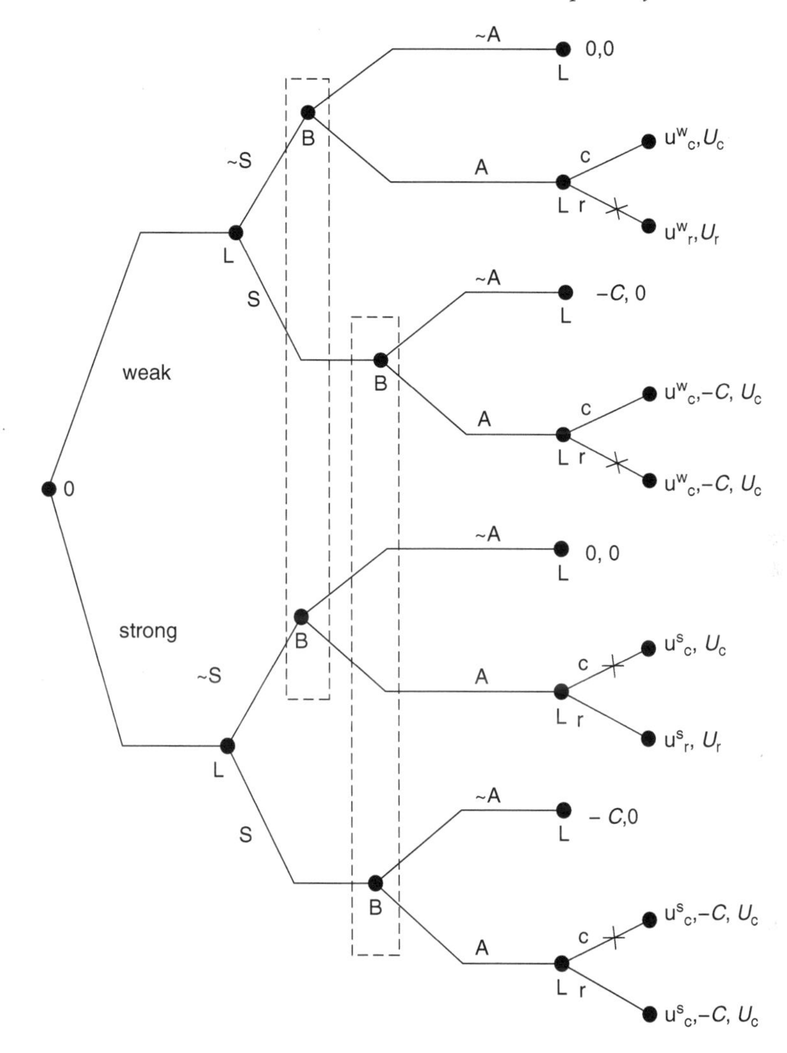

Figure 7.11 A deterrence signaling game

because L's best response to B_2 is L_1. Since there is no *separating equilibrium*, we next look for a *pooling equilibrium* in which L sends the same signal regardless of its type. Ignoring the knife-edged possibility of equality, we have two cases:

case 1: $pU_r + (1 - p)U_c < 0$
case 2: $pU_r + (1 - p)U_c > 0$.

If p is sufficiently large—if B initially believes that L has a sufficiently great probability of being strong and willing to resist aggression, then case 1 applies

	B_1	B_2	B_3	B_4
L_1	$pu_r+(1-p)u_c$ $pU_r+(1-p)U_c$	0 0	$pu_r+(1-p)u_c$ $pU_r+(1-p)U_c$	0 0
L_2	$p(u_r-C)+(1-p)u_c$ $pU_r+(1-p)U_c$	$p(u_r-C)$ pU_r	$-pC+(1-p)u_c$ $(1-p)U_c$	$-pC$ 0
L_3	$pu_r+(1-p)(u_c-C)$ $pU_r+(1-p)U_c$	$(1-p)(u_c-C)$ $(1-p)U_c$	$pu_r-(1-p)C$ pU_r	$-(1-p)C$ 0
L_4	$pu_r+(1-p)u_c-C$ $pU_r+(1-p)U_c$	$pu_r+(1-p)u_c-C$ $pU_r+(1-p)U_c$	$-C$ 0	0 0

Figure 7.12 The strategic form of Figure 7.11

	B_1	B_2	B_3
L_1	−100, 80	0, 0	−100, 80
L_2	−105, 80	−30, −10	−80, 90
L_3	−145, 80	−120, 90	−70, −10
L_4	−150, 80	−150, 80	−50, 0

Figure 7.13 Numerical example of Figure 7.12

and there are at least three (pooling) equilibria: (L_1, B_2), (L_1, B_4) and (L_4, B_4). In addition, if $-C > u_r$, then (L_4, B_3) is an equilibrium as well. All three of these equilibria, however, are equivalent in the sense that B never chooses to attack.

Case 2 is more interesting, because attacking L is a reasonable bet—in the absence of other considerations, B's expected payoff from attacking is greater than the status quo's value, zero, and the unique pure strategy equilibrium is (L_1, B_1). The question is whether the opportunity to signal creates any worthwhile possibilities for L. The answer to this question is, admittedly, sensitive to parameter values. So, in order to merely establish some possibilities, let $p = 0.10$, and

$$U_c = 100,\ U_r = -100,\ u_r = -250, \text{ and } u_c = -75/0.9.$$

Since B_4 is dominated by B_1 when case 2 holds, we can eliminate B_4 from consideration. Finally, for purposes of an example assume that $C = 50$, in which case the relevant strategic form is the one shown in Figure 7.13 (country L = row chooser, country B = column chooser):

Solving for mixed strategies for a 4 × 3 game can be tiresome, and so here we short-circuit some details and, without explanation, focus our attention on the strategies L_2 and L_4 for L and B_1 and B_3 for B. Assuming that both players place zero weight on the excluded strategies, if we compute mixed strategies in the usual way for a 2×2 game, we get a solution of $\boldsymbol{p} = (0, 8/9, 0, 1/9)$ for L and $\boldsymbol{q} = (6/15, 0, 9/15)$ for B. If B chooses between B_1 and B_3 with probabilities q and $1 - q$, then L must be indifferent between the lotteries that L_2 and L_4 offer. In other words,

$$-105q - 80(1 - q) = -150q - 50(1 - q),$$

which solves to establish that $q = 6/15$. Similarly, if L chooses between L_2 and L_4 with probabilities p and $1 - p$, then B is indifferent between B_1 and B_3 only if $80 = 90p$, which yields $p = 8/9$.

Two tasks remain at this stage: (1) showing that this strategy pair improves L's welfare and (2) establishing that it is indeed an equilibrium and B's expected payoff is 80. First, some straightforward algebra establishes that L's expected payoff from $(\boldsymbol{p}, \boldsymbol{q})$ is -90, which is more than what L gets, -100, if L cannot signal. Second, to see that $(\boldsymbol{p}, \boldsymbol{q})$ is in fact an equilibrium, we must show that neither B nor L has an incentive to place positive probabilities on those strategies to which p and q, respectively, assign zero probability. To see that this is indeed the case, notice that B has no incentive to give any weight to B_2 at the expense of the other two pure strategies since B_2 necessarily gives B less than 80 as long as L mixes only between L_2 and L_4. Hence, B has no incentive to shift unilaterally from $(\boldsymbol{p}, \boldsymbol{q})$. And given that B is mixing between the strategies B_1 and B_3, L will not give any weight to L_1 since this strategy earns -100; nor will L give any weight to L_3 since this also earns L an expected payoff of $(-145)6/15 + (-70)9/15 = -100$. Thus, $(\boldsymbol{p}, \boldsymbol{q})$ is an equilibrium.

Using a mixed strategy in a game that has a pure strategy equilibrium might seem strange. Nevertheless, that L can use the opportunity to signal only by implementing a mixed strategy makes sense. The pure strategies L_1 and L_4 convey no information and, thus, they cannot be used to deceive an opponent. On the other hand, using L_2 or L_3 with certainty merely signals one's type to the opponent and invites attack when one is weak. Thus, deception requires a mixed strategy—a strategy that always signals resolve when one is strong, but which leaves something to chance when one is weak.

This fact about deception is useful, at least in the context of the relatively simple game of our example. It reveals that if we observe two persons playing a game that allows for the possibility of deception, we cannot come to any firm conclusion as to whether deception will or will not be attempted. Moreover, since we may only get to observe the game played once—so that we will only have the opportunity to observe a single joint choice of pure strategies—even if a player abides by a mixed strategy, we cannot determine whether that player has played well or poorly on the basis of the final outcome nor can we determine whether that player intended to deceive. Our example also gives us a clue as to how strategic deterrence worked during the years of the U.S.–Soviet standoff. Specifically, there were at least two sources of incomplete information. First, even if both sides knew the capabilities of their own weapons system (which we might question), there was necessarily some uncertainty as to the precise capabilities of the opponent. Thus, the country launching a first strike could not be certain as to the surviving capabilities of the other. Second, in terms of the willingness of the attacked country to invite a second retaliatory attack by retaliating to the first strike, one had to know the attacked country's willingness to incur additional losses. What value did either the U.S. or USSR place on retribution even if the administration of that retribution would almost certainly invite a further degradation of one's country? This, we now know, was a question that neither country wanted to see answered.

7.8 Economic Sanctions in International Affairs

One substantive context in which players have an opportunity to "signal" their types lies in the context of economic sanctions, which have become an increasingly common coercive tool in international disputes. We have, of course, the successful example of sanctions being imposed by a vast array of states on South Africa in the 1980s so as to induce an end to its system of apartheid. On the other hand, we can look back at the sanctions imposed by the U.S. on Japan prior to WWII wherein steel and oil exports were halted in combination with the freezing of Japan's financial assets in the unsuccessful attempt to induce a withdrawal from China. This, and the more recent examples of Iran, Cuba and North Korea leave the impression that sanctions are at best a clumsy tool of international affairs and rarely meet their objectives. This raises the question as to why they are so often threatened when they are unlikely to succeed.

To see how games of incomplete information might help answer this question and to further illustrate the analysis of games of incomplete information, we offer here a 2-person game that involves a dispute on some issue and the potential for one of them to impose sanctions if that issue is unresolved (our analysis here is from Dean Lacy and Niou, "A Theory of Economic Sanctions and Issue Linkage," *Journal of Politics* 66,1, February 2004). Each disputant can be of two types. The coercer—the player that can threaten and impose sanctions—can be either resolute or irresolute, which will dictate whether it prefers to impose sanctions if the target state does not comply with its demands. The target, in turn, can be resilient or compliant, defined by whether it prefers to capitulate to the coercer's demands instead of suffering sanctions. And as in our earlier examples of incomplete information, a player's type is known with certainty only by that player—information is incomplete and asymmetric. Finally, for simplicity, suppose the issue under dispute has a binary outcome, either x = coercer's demand is met and $\sim x$ = coercer's demand is not met. Sanctions also involve a binary outcome, where s = sanctions are applied and $\sim s$ = no sanctions.

The strategic moves in the game begin with a choice by the coercer of whether to threaten sanctions. If there is no threat, the game ends and the status quo, O1, prevails. If the coercer threatens sanctions, the target complies or does not comply. If it does not comply, the coercer then chooses whether to impose the sanctions, after which the target decides whether or not to capitulate to the coercer's demands. The extensive form of this scenario wherein each player is uncertain as to the type of its opponent is then shown in Figure 7.14, where the possible outcomes are as follows:

O1 = (no threat) = the status quo
O2 = (threat, compliance) = $(x,\sim s)$
O3 = (threat, no compliance, sanctions, no capitulation) = $(\sim x,s)$
O4 = (threat, no compliance, sanctions, capitulation) = (x,s)
O5 = (threat, no compliance, no sanctions) = $(\sim x,\sim s)$.

Preferences will depend on a player's identity and type. For a potential target of sanctions, O5 and O1 imply the same outcome, except that O5 entails a reputation gain due to having resisted the coercer's demands. We assume, then, that targets prefer O5 to the status quo since the actions leading to O5 may carry future benefits such as insurance against future sanctions. The worst outcome is O3 for the *compliant* target whereas for the *resilient* target the worst outcome is O2 or O4, and, for purposes of analysis, we will assume that O4 is worse than O2. However, the results that follow hold regardless of the target's relative ranking of O2 and O4. In short, for the compliant target, O5 > O1 > O2 > O4 > O3; for the resilient target, O5 > O1 > O3 > O2 > O4.

We assume that a potential coercer prefers not to impose sanctions if the target complies. If the coercer has the preference ordering O2 > O4 > O1 > O5 > O3, it is *irresolute* since it prefers not to impose economic sanctions regardless of whether the target complies. Alternatively, it might rank the outcomes O2 > O4 > O3 > O1 > O5. If the target complies, the coercer prefers no sanctions to sanctions; if the target does not comply, it prefers sanctions to no sanctions. This coercer is *resolute* since it will incur the cost of sanctions if the target does not comply with its original demand. Both types of coercers prefer the status quo, O1, to O5 because of the loss of reputation and credibility associated with backing down against a target after threatening sanctions.

Normally in this context we would express these preferences as a utility functions, where, say, for the target we might use the notation $u_t(\text{O1/R})$ to denote the utility of the target for outcome O1 if it is resilient (R). However, to avoid such notational complexity, we use the notation provided by Table 7.3.

Suppose the target believes that the coercer is resolute with probability p and irresolute with probability $1 - p$, while the coercer assesses probability q that the target is resilient and probability $1 - q$ that it is compliant. Given the preference rankings and reasoning backward along the game tree, if the coercer is resolute, then it is always in its interest to threaten sanctions instead of accepting the status quo. If the target does not comply, the coercer will impose sanctions if it is resolute and will not impose sanctions if it is irresolute. If the target is resilient, its dominant strategy is not to comply with the coercer's demands. After eliminating dominated strategies, we can now solve the game by constructing its normal form using the remaining type-contingent strategies for target and coercer. For the coercer, the first strategy listed is for a resolute player;

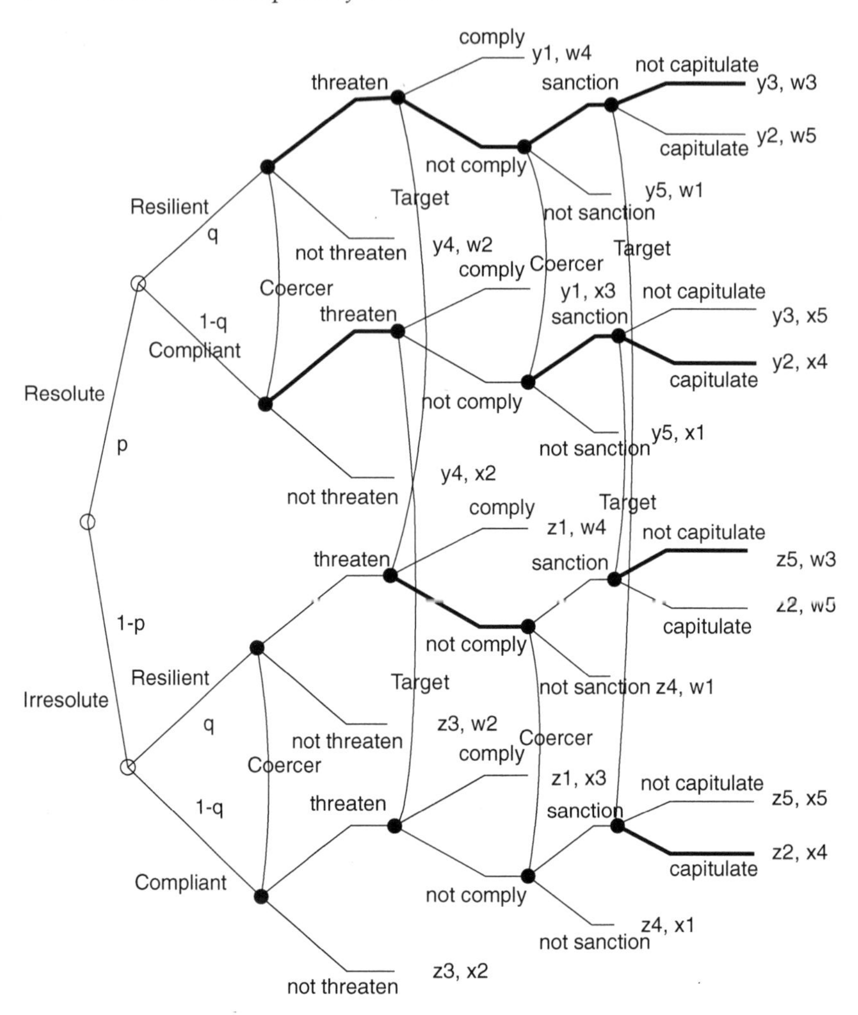

Figure 7.14 Game of economic sanctions in extensive form

Table 7.3 Payoff Notation

Target			*Coercer*	
Outcome	*Resilient*	*Compliant*	*Resolute*	*Irresolute*
O1	w2	x2	y4	z3
O2	w4	x3	y1	z1
O3	w3	x5	y3	z5
O4	w5	x4	y2	z2
O5	w1	x1	y5	z4

the second, for an irresolute player. For the target, the first strategy is played by a resilient player, the second by a compliant one:

Coercer:
(Threaten/Sanction, Threaten/Sanction) = (TS, TS)
(Threaten/Sanction, Threaten/Do Not Sanction) = (TS, T~S)
(Threaten/Sanction, Do Not Threaten/Sanction) = (TS, ~TS)
(Threaten/Sanction, Do Not Threaten/Do Not Sanction) = (TS, ~T~S)

Target:
(Do Not Comply/Do Not Capitulate, Comply/Capitulate) = (~C~C, CC)
(Do Not Comply/Do Not Capitulate, Do Not Comply/Capitulate) = (~C~C, ~CC)

Figure 7.15 gives the reduced strategic form of the extensive form in Figure 7.14, where the payoffs are computed in the usual way.

As algebraically ugly as the entries in this form might appear, we can readily reduce the potential equilibria down to three possibilities (the italicized cells in Figure 7.15).

(Resolute, Irresolute)	(~C~C, CC)	(~C~C, ~CC)
(TS, TS)	pqy3+p(1-q)y1+(1-p)qz5+(1-p)(1-q)z1 pqw3+p(1-q)x3+(1-p)qw3+(1-p)(1-q)x3	pqy3+p(1-q)y2+(1-p)qz5+(1-p)(1-q)z2 pqw3+p(1-q)x4+(1-p)qw3+(1-p)(1-q)x5
(TS, T~S)	*pqy3+p(1-q)y1+(1-p)qz4+(1-p)(1-q)z1* *pqw3+p(1-q)x3+(1-p)qw1+(1-p)(1-q)x3*	pqy3+p(1-q)y2+(1-p)qz4+(1-p)(1-q)z4 pqw3+p(1-q)x4+(1-p)qw1+(1-p)(1-q)x1
(TS, ~TS)	*pqy3+p(1-q)y1+(1-p)qz3+(1-p)(1-q)z3* *pqw3+p(1-q)x3+(1-p)qw2+(1-p)(1-q)x2*	pqy3+p(1-q)y2+(1-p)qz3+(1-p)(1-q)z3 pqw3+p(1-q)x4+(1-p)qw2+(1-p)(1-q)x2
(TS, ~T~S)	*pqy3+p(1-q)y1+(1-p)qz3+(1-p)(1-q)z3* *pqw3+p(1-q)x3+(1-p)qw2+(1-p)(1-q)x2*	pqy3+p(1-q)y2+(1-p)qz3+(1-p)(1-q)z3 pqw3+p(1-q)x4+(1-p)qw2+(1-p)(1-q)x2

Figure 7.15 Sanctions game

(1) (TS, T~S; ~C~C, CC) = (Threaten/Sanction, Threaten/Do Not Sanction; Do Not Comply/Do Not Capitulate, Comply/Capitulate),
(2) (TS, ~TS; ~C~C, CC) = (Threaten/Sanction, Do Not Threaten/Sanction; Do Not Comply/Do Not Capitulate, Comply/Capitulate), and
(3) (TS, ~T~S; ~C~C, CC) = (Threaten/Sanction, Do Not Threaten/Do Not Sanction; Do Not Comply/Do Not Capitulate, Comply/Capitulate).

To illustrate how we can eliminate all other cells from consideration as possible equilibria, consider the cell corresponding to (TS,T~S) and (~C ~C, ~CC). Comparing this cell to the one immediately below it for row chooser, note that these two cells differ only in the last two terms, where the cell in question concerns the utility z4 in those terms whereas the cell below it concerns the utilities z3. But from Table 7.3, z3 > z4. Thus ((TS,T~ S), (~C ~C, ~CC)) cannot be an equilibrium. The remaining un-italicized cells can be eliminated in a similar way. So, turning our attention first to the strategy-pair (TS, T~S; ~C~C, CC), it is an equilibrium when neither the coercer nor the target have an incentive to deviate from the prescribed strategies, which requires that the following weak inequalities be satisfied:

$$qz4 + (1-q)z1 \geq z3 \quad (1)$$

$$x3 \geq (1-p)x1 + px4 \quad (2)$$

Considering inequality (1), *ceteris paribus*, the irresolute coercer is more likely to threaten sanctions if: (a) the target is likely to be compliant (q is low), (b) if the cost of backing down after threatening sanctions compared to the value of accepting the status quo is relatively low (z4 is small), or (c) if the value of the target complying is relatively higher than the status quo (z3 is high). For the compliant target, inequality (2) suggests that the target will comply if threatened and capitulate if sanctioned, (CC), if the coercer is likely to be resolute (p is large), if the cost of complying on X is relatively low, or if the cost of sanctions is relatively high.

When inequality (1) is reversed, (TS, ~TS; ~C~C, CC) and (TS, ~T~S; ~C~C, CC) become equilibria, though both lead to the same outcome and payoffs. If the target is likely to be resilient (q is high), if the cost of backing down after threatening to sanction is relatively high, or if the value of the target complying is not much higher than the value of accepting the status quo, it is more likely that the irresolute coercer will choose not to threaten sanctions in the first place. If inequality (1) is an equality, then all three strategy combinations are Nash equilibria since the coercer has no incentive to deviate from any of the three strategies.

When inequality (1) is satisfied but inequality (2) is not, the game does not have a pure strategy equilibrium, but it has at least one equilibrium in mixed strategies. Inequality (2) is less likely to be satisfied if: (a) the coercer is more likely to be irresolute ($1 - p$ is high); (b) the value of not complying without being sanctioned is high (x1 is high); or (c) the difference between complying after being threatened and capitulating after being sanctioned is small (x3 – x4 is small).

In equilibrium, player types are *separating* in the sanction stage but *pooling* in the threat stage. Both a resolute and an irresolute coercer may threaten sanctions. But if the target state does not comply, then only the resolute coercer will impose sanctions. Two important insights follow from the equilibrium results. First, the threat stage is critical for understanding sanctions. When sanctions are successful, their success will often come at the threat stage. When a target does not comply, then the game proceeds to the sanctions stage. In the sanctions stage, sanctions will be applied only when a resolute coercer meets a resilient target, which highlights the second insight from the model: When sanctions are actually imposed, they often will not succeed.

Based on the equilibrium results, we describe the conditions under which each of the five possible outcomes will arise, giving empirical examples of each.

Outcome 1: *Coercer does not threaten sanctions.*

Conditions: The coercer is irresolute, and the cost of backing down if the target ignores its threat is high while the value of the target complying is not much greater than the value of enduring the status quo.

Example: In the second Clinton administration, the U.S. did not link renewal of China's most favored nation status to improvements in its human rights record. The administration may have been irresolute while perceiving China to be resilient. Furthermore, the reputation cost of backing down against the PRC is high.

Outcome 2: *Coercer threatens economic sanctions; target complies.*

Conditions: Outcome 2 describes the conditions under which sanctions are probably most successful. But note that in this case sanctions are never imposed, only threatened. The conditions are that the target is compliant, the cost of complying is low, and the cost of sanctions is high. To make the conditions sufficient, then (a) the coercer believes the target is likely to be compliant, (b) the cost of backing down if the target does not comply is low, and (c) the value of the target complying is much greater than the value of maintaining the status quo. If we are to judge the success of economic sanctions, then full examination of this category of outcomes is critical. Outcome 2 demonstrates that debates about the success of sanctions that fail to consider the threat stage will miss the point of sanctions.

Outcome 3: *Coercer threatens economic sanctions; target does not comply; coercer imposes sanctions; target does not capitulate.*

Conditions: The necessary and sufficient conditions are that the coercer is resolute and target is resilient. The classic cases of the failure of economic sanctions are examples of this outcome. The Soviet Union, for instance, did not withdraw from Afghanistan after the U.S. grain embargo in 1979; Iraq did not withdraw from Kuwait in 1990 and 1991 despite sanctions by the UN; and the Castro regime in Cuba has been under a general U.S. embargo since the 1960s.

Outcome 4: *Coercer threatens economic sanctions; target does not comply; coercer imposes sanctions; target capitulates.*

Conditions: In either of the perfect Bayesian equilibria in pure strategies described above, the compliant target always complies after the coercer makes a threat. Only the resilient target will hold out. But the resilient target will not capitulate even if the coercer imposes punishment. Therefore, Outcome 4 can occur only in a mixed-strategy equilibrium, that is, when inequality (2), $x3 \geq (1 - p)x1 + px4$, is not satisfied. The conditions for a compliant target to defy the threat of sanctions but to capitulate once sanctions are imposed are: (a) the target believes that the coercer is irresolute (p is small), (b) the value of not complying without being sanctioned is high ($x1$ is high), or (c) the difference in value between complying after being threatened and capitulating after being sanctioned is small ($x3 - x4$ is small). In 1933, for example, the U.K. barred importation of a number of goods from the Soviet Union after the Soviets imprisoned two British citizens. In 1979 several Arab states imposed sanctions on Canada after the Canadian government announced it would move its embassy in Israel from Tel Aviv to Jerusalem. In 1989 and 1990, India imposed economic sanctions on Nepal after Nepal increased military ties with China. In all cases, the target state complied with the coercer's demands after sanctions were imposed. Our model predicts that such outcomes are not often observed since they involve the play of mixed strategies. An irresolute coercer plays a mixed strategy of imposing or not imposing sanctions in order to avoid having target states take advantage of the coercer. In the three cases above, the coercers likely realized that in order to avoid their opponents' taking advantage of them, they must *sometimes* impose sanctions, even if the sanctions may not be successful. Similarly, compliant targets know that in order to avoid being taken advantage of by potential coercers who threaten sanctions, they must *sometimes* avoid backing down to the initial threat of sanctions. The compliant targets then force the hands of irresolute coercers by making them impose sanctions. By assigning some probability to the strategy of not complying with threatened sanctions, the compliant target may deter irresolute coercers from threatening sanctions.

Outcome 5: *Coercer threatens economic sanctions; target does not comply; coercer does not impose sanctions.*

Conditions: The target is resilient and the coercer is irresolute. In addition, the coercer believes the target is likely to be compliant, the cost of backing down if the target does not comply is low, and the value of the target complying is much greater than the value of maintaining the status quo. If the game proceeds to the stage where the coercer must decide whether to impose sanctions when the target has not backed down, it must be that the irresolute coercer miscalculates the target's resilience in the first stage but decides to further test it in the second stage. In the early days of the Clinton administration, for instance, U.S. policymakers apparently wanted to grant China MFN status in exchange for improvements in China's human rights record. But without an improvement in China's human rights status, the U.S. preferred not to grant MFN status. The Clinton administration acted as though it were resolute, but the administration was likely bluffing in order to induce China to improve its human rights record. China did not comply with U.S. demands, and the Clinton administration later pushed for China's MFN status and membership in the World Trade Organization.

The conditions under which each of the outcomes will arise reveal that even if sanctions are threatened, they may not be imposed, and, if imposed, they may not be successful. In only two of these outcomes does a coercer ever impose sanctions. In both cases the target is resilient and unlikely to capitulate. The model explains several puzzles in the literature on economic sanctions and issue linkage. First, some scholars have defined success in narrow terms to mean that a state achieves its most preferred outcome, and they have concluded from case studies that economic sanctions generally do not succeed. The real success of sanctions will not be observable in cases in which sanctions are imposed. Rather, sanctions succeed by convincing potential targets that the coercer is resolute. The model reveals that the threat of sanctions can be as potent a policy tool as the imposition of sanctions. If the target is compliant and if the target believes that the coercer is likely to be resolute, Outcome 2 shows that the threat of imposing punishment can effectively compel targets to comply. Empirical studies that examine only cases in which sanctions were imposed omit a class of cases that represent successful sanctions, though the sanctions were threatened but not imposed. Examining cases of only imposed sanctions generates a selection bias in empirical research on sanctions.

Second, successful sanctions may also be measured as an improvement over the status quo, even if a state's most preferred outcome is not realized. Many scholars consider sanctions a failure if the coercer imposes sanctions and the target does not capitulate. But a resolute coercer prefers the outcome (threat, no compliance, sanctions, no capitulation) to the status quo. Even though the coercer does not achieve its most preferred outcome, it does achieve an outcome it prefers to the status quo. The resolute coercer is certainly no worse off imposing sanctions that are ignored than it is continuing under the status quo.

Third, sanctions are often unsuccessful. In the pure strategy equilibrium, the compliant target always complies after the coercer threatens sanctions. Capitulation by the target after the coercer imposes sanctions is an equilibrium only in mixed strategies: the outcome occurs only when a pure-strategy equilibrium does not exist, which explains why sanctions often fail. However, this does not mean that sanctions should never be imposed if a target does not concede. If a coercer never imposes sanctions, it and other potential targets will always ignore sanction threats. The success of sanctions should be judged not by the cases where sanctions are imposed and fail but by the cases where threatened sanctions change the behavior of targets.

7.9 Rationality Reconsidered

We are now in a position to reconsider the Centipede Game offered in Chapter 2 in Figure 2.16 and the contradiction that subgame perfection appears to imply with respect to the presumed rationality of players there. Actually, though, our interest in this game is not simply that we want to resolve an apparent paradox in the application of the concept of subgame perfection. We should, after all, leave such matters to game theorists. Rather, we want to confront the possibility that "irrationality" itself may be rational.

> **The Cincinnati Kid**: A classic of Hollywood was the 1965 movie *The Cincinnati Kid* in which Steve McQueen, playing the role of the brash, upstart gambler, challenges the "old pro," Edward G. Robinson. After hours of endless play, Robinson's age appears to show as he seemingly wilts under the "Kid's" relentless assault of tactical skill and luck. The dramatic end to their game of five-card stud has McQueen an almost certain winner with a full house against Robinson's flush. The cards, as they are dealt, fall as follows: First, both players are dealt a face down card, which even the movie's audience is not allowed to see. The cards that follow are then dealt face up and Robinson is given the 8♦ while the Kid (McQueen) gets the 10♣. The Kid bets $500 and Robinson calls. Robinson next gets the Q♦ and the Kid the 10♠. The Kid bets $1,000 and Robinson raises $1,000. The Kid calls. Next Robinson is dealt the 10♦ and the Kid gets the A♣. The Kid bets $3,000 and Robinson calls. Robinson's final card is the 9♦; the Kid gets the A♠. With his card in the hole being the A♥, thus giving him a powerful Full House—three aces and two tens—the Kid checks since he doesn't want to signal the strength of his hand and since he is certain that if Robinson guesses he has a Full House, he'd drop out. Seemingly suckered into thinking that the Kid merely has the two pair showing, Robinson bets $1,000. The Kid, thinking he's suckered Robinson into betting an additional $1,000, immediately raises with the remaining money in front of him, $3,500, with the expectation that this bet will lead Robinson to fold. Robinson, however, reaches into his wallet, meets the Kid's bet and raises another $5,000. McQueen hasn't the money, but still convinced that Robinson is merely bluffing and that the best he can be holding is another diamond (and thus a Flush, which his Full House beats), he offers Robinson his "marker"—his IOU. Robinson agrees, the Kid calls the bet and is subsequently stunned into submission as Robinson turns over his face-down card, the J♦ for a Straight Flush. Robinson, of course, wins and McQueen is busted, confused, and humiliated. How, he asks, could Robinson bet as he did with at best an improbable chance of being dealt cards that would defeat a Full House? Robinson's answer is the admonition that poker oftentimes involves making the wrong move at the right time. Robinson's apparent irrationality has suckered McQueen into losing it all.

Our Centipede Game has nowhere near the drama of this scene, especially in its Hollywood portrayal, but it too may require "irrational" action on the part of rational decision makers. To see what we mean, recall that the particular difficulty we confront with this situation centers on this question: Subgame perfection requires that we determine what a player should choose in the initial stages of the game on the basis of what rational players choose subsequently in later stages. But if reaching those stages requires that players choose irrationally in earlier stages, then what is the basis for supposing that players choose rationally in these later stages? Successive pruning of the game's extensive form, then, seems to make sense only if we allow a contradiction of the notion of rationality that underlies the idea of subgame perfection. Put differently, a player at something other than an initial decision node of the extensive form can reason thus: "If everyone is rational, then I should choose . . . But wait a minute! If everyone is rational, then I wouldn't have the opportunity to choose at this node in the first place! So something must be wrong with my assumption that my opponent is rational."

Players can reach successive decision nodes in our example, then, only if one or both of them are irrational, or at least if one or both of them find it rational to *appear* irrational. So consider the possibility that each player selects a strategy under the hypothesis that there is some small probability that the opponent is in fact irrational. We can model this irrationality in any number of ways, including letting people have utility functions that differ from the ones we otherwise specify (for example, we could suppose that irrational people dislike money and that the utility of x dollars equals $-x$). Equivalently, we can suppose that there is some small probability that a player, for whatever reason, simply cannot make a choice that is otherwise optimal.

For the game in Figure 2.16, then, we can assume that there is some probability that, in lieu of picking the money up off the table, any particular player is incapable of such an action and passes at every opportunity. In the Bayesian game format, this means each player knows that he or she is rational, each knows that there is some probability that the opponent is irrational, each knows that the opponent believes that there is some small probability that it is irrational, and so on. The game theoretic question, then, becomes: Is there an equilibrium in which two players, both believing that the other might be irrational, make the "irrational" decision of waiting until the game's final stages before opting to take the money offered? Is it sometimes rational to act "irrationally"? To evaluate this possibility, consider the extensive form in Figure 7.16, which takes the game in Figure 2.16 and assumes that each player has some probability of being unable to do anything except pass. Information sets are drawn, however, to indicate that each player knows its own type but is uncertain about the "rationality" of the opponent.

Because we exclude the possibility that an irrational player can do anything but "pass," each player has three strategies in this extensive form that take account of his or her type:

s_1: If rational, then "take" on the first opportunity (which ends the game), but "pass" if irrational.

s_2: If rational, then "pass" on the first opportunity, "take" on the second (which ends the game), and always "pass" if irrational.

s_3: If rational, then "pass" on both opportunities and "pass" if irrational.

To construct a numerical example, let $q = 0.97$, so each player has a 0.03 chance of being incapable of taking the money. Figure 7.17 shows the corresponding

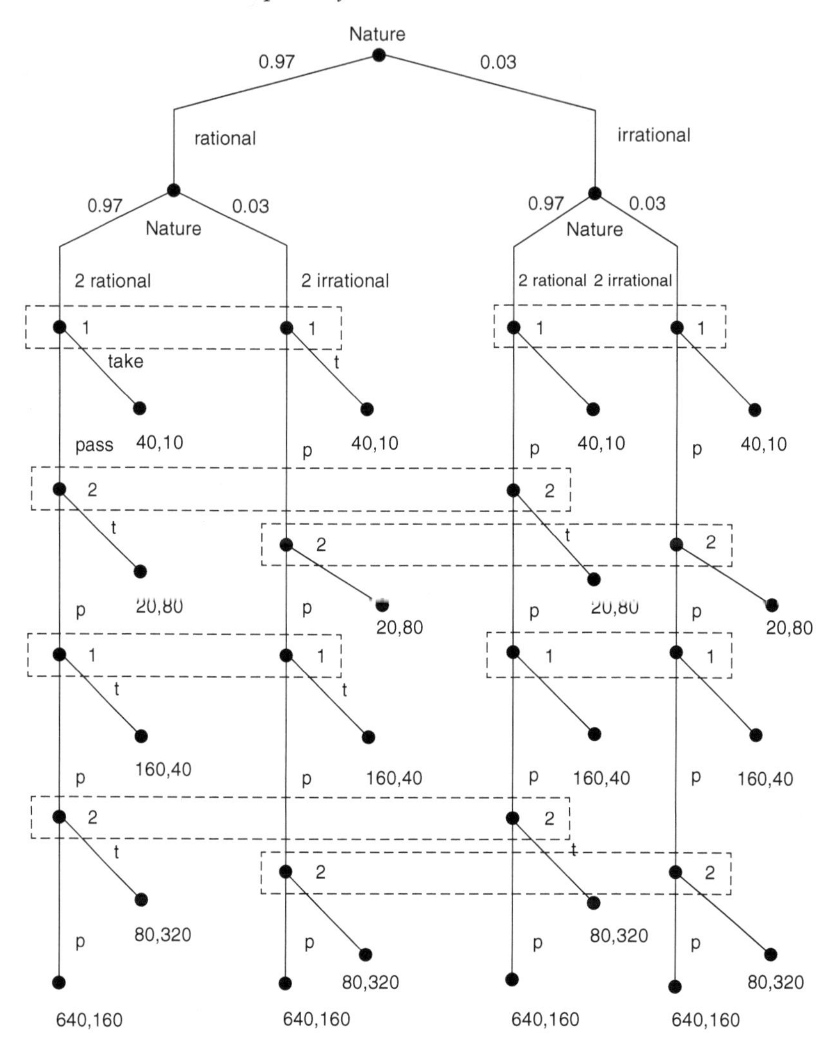

Figure 7.16 Centipede game with possibility of irrationality

strategic form. For example, if both players choose s_1, then 1 receives a payoff of 40 [if 1 and 2 are both rational] with probability $(0.97)^2$, a payoff of 40 [if 1 but not 2 is rational] with probability (0.97)(0.03), a payoff of 20 [if 1 but not 2 is irrational so that 1 passes but 2 takes] with probability (0.03)(0.97), and a payoff of 640 [if 1 and 2 are both irrational and always pass] with probability $(0.03)^2$. Thus, we enter a payoff to 1 of

$$40[(0.97)^2 + (0.97)(0.03)] + 20(0.03)(0.97) + 640[(0.03)^2] = 40$$

	s_1	s_2	s_3
s_1	40, 12.2	41.7, 19.2	58, 14.5
s_2	24.6, 78.9	158.1, 48.3	17.4, 43.6
s_3	38.2, 82.4	96.8, 315.2	640, 160

Figure 7.17 Strategic form of Figure 7.16

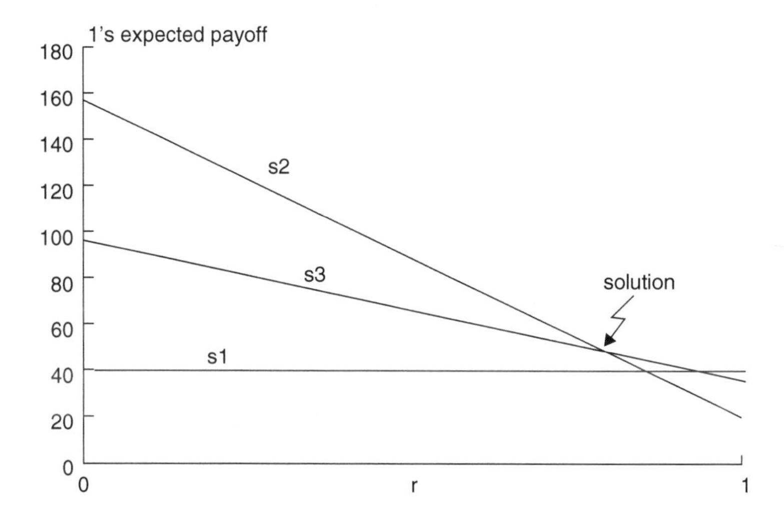

Figure 7.18 Finding a solution to the game in Figure 7.17

in the (s_1, s_1) cell. The remaining cells are computed in a similar fashion.

Notice that s_3 is dominated for player 2 by s_2, which makes sense since s_3 requires that 2 pass at the last branch even if rational, rather than simply take the larger payoff. Checking the other cells reveals that the game does not have a pure strategy equilibrium, so mixed strategies must be considered. At this point, however, we must proceed carefully in order to compute a mixed solution, because we do not yet know whether player 1 should mix over all three pure strategies or over some subset of them. We begin then by supposing that player 2 mixes between s_1 and s_2 with probabilities r and $1 - r$. Then player 1's pure strategies must each yield the following expected payoffs:

$$E(s_1) = 41.7 - 1.7r,$$
$$E(s_2) = 158.1 - 133.5r,$$
$$E(s_3) = 96.8 - 58.6r.$$

We know, of course, that if strategies s_i and s_j each have nonzero probability associated with them in player 1's mixed equilibrium strategy, it must be that

$E(s_i) = E(s_j)$—otherwise player 1 would prefer to unilaterally alter his or her mixed strategy by shifting probability from one pure strategy to the other. However, Figure 7.18, which graphs each of the preceding expected value equations against r, reveals that only two expected values can be equal simultaneously, and that there are three possible pairings. This figure also reveals, however, that player 1's expected payoff is greatest if he or she sets $E(s_2)$ equal to $E(s_3)$—if player 1 chooses a mixed strategy that assigns positive probability only to s_2 and s_3. Hence, setting $E(s_2) = E(s_3)$ gives $r = 0.814$—that is, player 2's mixed equilibrium strategy is (0.814, 0.186, 0).

We can solve now for 1's mixed equilibrium strategy $(0, p, 1 - p)$ by using the fact that 2 must be indifferent between s_1 and s_2 whenever 1 abides by $(0, p, 1 - p)$. For player 2,

$$E(s_1) = 82.4 - 3.5p$$
$$E(s_2) = 315.2 - 266.9p$$

Setting these two equations equal to each other and solving for p gives $p = 0.884$. Thus, player 1's mixed equilibrium strategy is (0, 0.884, 0.116).

Although this particular solution depends on our initial assumption that each player's probability of "irrationality" is .03, what we want to emphasize is that this irrationality is now a self-fulfilling prophecy. Both players, even if they are rational in a traditional sense, have non-zero probabilities of passing rather than taking at any decision node. Appearing to be irrational is now rational and is sustained on the part of both players by the initial assumption that each person has some small probability of actually being (as opposed to appearing to be) irrational.

There are other models that we could use to "explain" any observed irrationality, models that do not require players knowing with such specificity the probability that the other is irrational. For example, we could suppose that each player's probability of irrationality is itself a random variable drawn in accordance with some commonly known probability density. The important point, however, is that we can now begin to study "irrationality" empirically to see if our models can account for it in the framework of the rational choice paradigm. The implication of such models in political science is that barring some complicated measurements that are allowed only if we observe the same game played many times (in which case we must be concerned about the supergame that participants might perceive), there is no way for us to determine whether any observed irrationality is "real" or strategic. Thus, debates in both the academic and popular literature over the meaning of rationality or which pertain to some evidence we assume supports the contention that someone is irrational ought to be seriously scrutinized with our example in mind. If we observe some national leader pursuing policies that make little sense to the rest of us, our first instinct as analysts should not be to presume irrationality; instead, we should examine the nature of the uncertainty that confronts us and try to determine whether those actions are consistent with some strategic imperative. It may be true, of course, that we will learn eventually that these policies were dictated by poor information on someone's part or that a national leader was mentally unbalanced. But an appearance of "irrationality" may also be merely part of the deception that incomplete information allows, and it may follow a logic that is readily understood by supposing that people are rational and strategic.

7.10 Key Ideas and Concepts

incomplete information
asymmetric information
conditional probability
Bayes's Law
Bayes equilibrium
separating equilibrium
pooling equilibrium
beliefs
reputation
chain store paradox
signaling
deception
rational

Exercises for Chapter 7

1. In late 1941 the British were concerned as to how to respond to an anticipated invasion by Japan of Thailand's (and Myanmar's) Kra Isthmus and an eventual invasion of Malaya and the critical British outpost and naval base at Singapore. Thailand was ostensibly neutral and the critical issue that confronted Britain's strategic planners was whether they should attempt to forestall a Japanese occupation of Thailand by invading the country themselves beforehand and thereby block the ports and airfields Japan would most likely use in its move toward Malay and Singapore. As the situation has been described by one historian of the period, "If British forces only entered the Kra Isthmus after a Japanese incursion . . . they would probably set off a war with Japan. If [they moved] . . . before the Japanese entered Thailand . . . Tokyo could use this as a pretext for its own invasion . . . [The British] had been told, possibly accurately, that the Japanese might even attempt to trick Britain into taking the first step . . . [in which case] Britain might lack for the support of the United States" (Evan Mawdsley, *December 1941*, New Haven: Yale University Press, 2011: 54). More specifically, as the British concern was most clearly stated by Churchill, "We should not resist or attempt to forestall a Japanese attack on the Kra Isthmus unless we had a satisfactory assurance from the United States that they would join us should our action cause us to be involved in a war with Japan." Of course, this concern became moot in less than a week with Japan's attack on Pearl Harbor. Nevertheless, without assigning values to the possible outcomes, sketch out a description of this situation in extensive form, treating the

preferences (responses) of the United States as not known with certainty by either Britain or Japan.

2. Consider the following strategic form game:

0, 0	−K, 5
K, 10	5, 0

Suppose only person 1 (row chooser) knows K with certainty, which nature has set equal to 10 and −10 with equal probability (a fact that is known to column chooser, person 2). Solve for this situation's Bayesian equilibrium strategies.

3. In his unsuccessful surgeries (those in which the patient dies) Dr. Ian Competent has only a fifty-fifty chance of not being at fault. With this in mind, the relatives of his latest victim have asked for compensation: $1,000,000. If Competent (who has already had his insurance policy revoked but who as a shareholder in a local savings and loan association is quite wealthy) refuses to settle, the relatives can take the matter to court (or they can forget it). Once in court assume that justice is done. (This is not an exercise that concerns the competence of lawyers.) So if Competent is innocent (and only he knows for sure), he loses nothing and the relatives lose $1,000,000 (attorney's fees being what they are). On the other hand, if he loses, then he loses $3,000,000 and the victim's relatives gain $2,000,000 (again, lawyers take their cut).

    a. Portray this situation's extensive and strategic forms.
    b. Determine the game's equilibrium.
    c. Interpret this equilibrium.

4. Country 1 has secretly approached country 2 with a disarmament proposal that focuses on a new weapons system developed by 2. However, this system has only a 70% chance of being effective, although 2 knows the system's capabilities with certainty. Country 2 can ignore 1's proposal and allow the arms race to continue; alternatively, it can publicly make the same proposal, at which time country 1 must "fish or cut bait" by formally accepting 2's public offer. Associate what you regard as reasonable payoffs with the outcomes, portray this situation's extensive and strategic forms, and find any pure strategy equilibria. Interpret these equilibria, if any exist.

5. Consider the following *sequential elimination* agenda: "Alternatives A and B are first paired. If A wins, it is the outcome; but if B wins, B is paired against C and the winner of this vote is the final outcome." Suppose that only these three preference types are possible (ranked from first preference to last):

    t_1: A B C
    t_2: B A C
    t_3: C B A

Assuming that a person has type t_i preferences with probability p_i, that the p_i's are common knowledge, and that a person knows his own preferences, show that the selection of a Condorcet winner corresponds to an equilibrium.

6. Consider the following two-person, zero-sum game. Nature first chooses player 1's type, L, C, or R, with equal probability. Player 1 knows his type and chooses "Yes" or "No." After player 1 chooses, player 2 chooses "Yes" or "No." Player 2, whose type is common knowledge, does not observe 1's type, but does observe 1's choice of "Yes" and "No." Player payoffs are determined according to the following table:

1's type	1's choice	2's choice	1's payoff
L	Y	Y	1
L	Y	N	2
L	N	Y	4
L	N	N	3
C	Y	Y	3
C	Y	N	7
C	N	Y	9
C	N	N	7
R	Y	Y	3
R	Y	N	4
R	N	Y	2
R	N	N	1

   a. Draw the game's extensive form.
   b. Identify the pure strategies available to each person.
   c. Portray the situation's strategic form.
   d. Find all the pure-strategy equilibria.

7. Refer to section 7.8 and reproduce our analysis of the Centipede Game, assuming that each player's probability of irrationality is .3 rather than .03.
8. The political philosophy—authoritarian or democrat—of President Bulsky of Lower Slobovia is unknown to everyone but himself, including his chief rival, Drinksalotov. So people assume that Bulsky values democratic principles with probability q, $.5 < q < 1$. Seeking favor with the members of NUKE (Nations United for Kapitalist Expansion), Bulsky has pledged to allow a national election for president—an election that can go either way with equal probability if Drinksalotov runs against him. Presently, Bulsky must decide whether to use recent ethnic unrest as an excuse for reneging on his pledge. If he reneges, he retains his position but his wife cannot anticipate an invitation to shop on Rodeo Drive at the next summit whereas Drinksalotov wins international sympathy and an appearance on *Nightline*. If Bulsky commits to an election, Drinksalotov must decide whether to challenge. An unchallenged Bulsky wins by a slim margin in an uncontested election. If Drinksalotov challenges and loses, he is revealed as someone who cannot defeat an incumbent who has steered his country to 40% unemployment

and a −20% GNP growth rate. This outcome's valuation depends also on Bulsky's commitment to democracy, since losing elections has been a "very bad thing" in Lower Slobovia historically. Drinksalotov adheres to democratic principles, but a loss by a like-minded Bulsky is a crushing psychological blow that leads to his semiretirement as assistant director of the Crapski Tractor and Screen Door plant. We have, then, these outcomes:

O1: Bulsky cancels the election.
O2: Bulsky allows the election, but Drinksalotov declines to compete.
O3: Bulsky allows the election; Drinksalotov opposes him but loses.
O4: Bulsky allows the election; Drinksalotov opposes him and wins.

In terms of the world's foremost hard currency—Disney Dollars—suppose valuations, conditional on Bulsky's philosophy, are as follows (with payoffs to Bulsky and Drinksalotov respectively and assuming, as has always been the case in Lower Slobovia, that Disney Dollars and utility are equivalent):

	Authoritarian	Democrat
01	(−1, 1)	(−1, 1)
02	(0, 0)	(0, 0)
03	(2, −6)	(2, 0)
04	(−2, 4)	(−8, 4)

a. Portray this situation's extensive and strategic forms.
b. Describe the players' equilibrium strategies as a function of q.

9. In an upcoming election on insurance rate reform, it is common knowledge that you will cast the decisive vote. You are uncertain about the identity of the reform's sponsor but have (correct) beliefs that there is a 7-in-10 chance that the reform is pro-insurance (INS) and as a consequence will raise your insurance rates (making you poorer) and that there is a 3-in-10 chance that the reform is pro-Consumer (CON), in which case the reform will keep your insurance rates at their present level. A campaigner, who knows whether the reform is the insurance type or the consumer type, and, in either case, is paid only if the bill passes, must decide whether or not to go to your House to tell you to vote for the bill. (Campaigner chooses "House" or "No.") On election day, you must decide whether to vote "Yes" or "No" on the reform. The payoffs are determined as follows: It costs the campaigner $5 to go to your house. The campaigner gets paid $15 if the reform passes and $0 if the reform fails. If either an insurance-type reform passes or a consumer-type reform fails, your rates go up—you lose $10. If either a consumer-type reform passes or an insurance-type reform fails, your rates stay the same—you get $0.

   a. Draw the game's extensive form.
   b. Specify the pure strategies available to each player.
   c. Portray the situation's strategic form.
   d. Find all the pure-strategy equilibria.

8 Cooperation and Coalitions

8.1 The Concept of a Coalition

Of the political processes we have thus far considered or argued can be modeled using game theory, there is one central to politics that we have not yet fully explored—coalitions. Briefly,

> *A* ***coalition*** *corresponds to an agreement on the part of two or more players to coordinate their actions so as to bring about an outcome that is more advantageous to its members than what prevails from uncoordinated action.*

The concept of a coalition encompasses a great many things in politics and studying them includes studying the processes whereby governments are formed and prime ministers chosen, alliances are negotiated and maintained in international affairs, and legislators maneuver to pass mutually beneficial legislation. Indeed, given the generality of our definition, nearly any entity that facilitates the objectives of two or more persons simultaneously can be interpreted as the manifestation of a coalition. Thus, a political party, a legislative caucus, a military alliance, a Soviet cooperative, a labor union, and a citizen's interest group all represent coalitions. More fundamentally, a constitution that establishes and defines a state is the consequence of a coalition among those who choose to give up some degree of individual sovereignty in order to secure the gains that ostensibly flow from collective action.

Of course, no one disputes the fact that coalitions are an important part of politics. However, rather than proceed into a substantive discussion of their content, let us take the more abstract view suggested by our definition and consider at least these four general questions about processes of coalition formation:

1. What coalition structure is likely to prevail—which players will coordinate?
2. What will be the extent of this coordination—will they agree to coordinate on all decisions or on only some subset of decisions?
3. What will be the specific intent of coordination—what outcomes will they seek to realize or avoid?
4. How will the members of a coalition enforce the agreements they reach?

Although we cannot answer any of these questions in isolation from the rest, our answer to the last one sets the stage for how we approach the first three. Taking an extreme possibility, consider a circumstance in which no agreement can be enforced, so that in principle every person is free to renege on any agreement he or she might reach with anyone else. In this event, the concept of a coalition adds little to our understanding of events, because all action is necessarily "non-cooperative" and our explanations for outcomes that "appear cooperative" must be formulated in terms of the ideas set forth in the preceding chapters. Of course, even if no agreements are enforceable, the negotiations that lead to them might signal something about preferences and intentions, in which case they can assist people in coordinating to a particular equilibrium, or they can affect beliefs in such a way as to influence the character of equilibria. But beyond this they cannot affect future actions, because such actions are necessarily dictated by the individual strategic imperatives of the sort we treat earlier. This is, in fact, the character of the Prisoners' Dilemmas wherein even if players agree to coordinate beforehand in order to avoid the mutually disadvantageous outcomes, unless they somehow change or subvert the game's individual motives so as to render it something other than a Dilemma—unless they invent some way to enforce agreements—the game remains what it is and is hardly characteristic of the cooperative processes we normally associate with those things we label "coalitional."

If we view the sale of a house as a coalition between buyer and seller, the "glue" that holds the coalition together and makes such activity worthwhile for both participants is not only a mutual interest, but also the courts and a body of contract law that protect a seller against theft and a buyer against fraud. This glue keeps both buyer and seller from unilaterally reneging on their contract, because without it the buyer prefers to halt mortgage payments after taking possession of the property and the seller prefers to retain control of the property after receiving a down payment. Remove the legal mechanisms for enforcing contracts and one takes away the foundation of competitive markets. Take away or modify the mechanisms of enforcement that define property rights in markets and you change the character of markets, including what it is that will be bought and sold in them, or whether anything will be bought, sold, or produced for sale at all.

To the extent that the various mechanisms of enforcement render some but not all types of agreements workable, identifying which coalitions might form and the agreements they might implement depends on our understanding of these mechanisms and the opportunities for participants to alter them as a game proceeds. At this point, however, the reader might want to interject the observation that, regardless of whether agreements are enforceable, no new theoretical baggage is required to understand the processes whereby coalitions form. If coalitions are agreements to coordinate choices in some extensive or normal form game, then saying that a coalitional agreement is enforceable is equivalent to saying that once we include the enforcement mechanisms as part of our description of the game, the individual components of that agreement correspond to a (Nash, subgame-perfect, perfect, Bayesian, etc.) equilibrium of the extended game.

Earlier, for example, we described the cooperation that emerged during World War I between opposing troops in their trenches—the Christmas Truce. Surely, that cooperation can be described as a coalition—arguably, in this case, a coalition against the opposing countries' high commands. And as we argued then, the mechanism for enforcing that "coalition" was not any external agent or institution, but simply the fact that the troops were in effect playing a repeated Prisoners' Dilemma wherein defections from cooperation can be punished in subsequent plays of the "game." The perspective taken in the major part of this chapter, though, is that, just as we commonly ignore the details of the judicial system when studying the general character of markets, it is sometimes better to focus on some simple possibilities and to postpone seeking answers to all questions. Consider vote trading in legislatures. We know that we are unlikely to witness the signing of legal documents when two legislators agree to support each other's bills, despite the fact that there are incentives for defection—if legislation is considered sequentially, then once one legislator votes for a bill that he otherwise prefers to see defeated, his partner in the trade prefers to renege on the agreement when the second bill (on which he must now vote against his preference) comes up for a vote. Nevertheless, just as was the case in World War I, enforcement mechanisms exist that render abiding by the trade an equilibrium. That mechanism is the shared knowledge that, if such an agreement is broken without mutual consent, the defecting legislator's reputation will be damaged and his or her subsequent participation in advantageous vote trades will at best be problematical.

How we approach matters, then, depends on our research goals and on our beliefs about the enforceability of agreements. If we are not primarily interested in, say, the evolution and maintenance of legislative norms, but if we nevertheless believe that any vote trade can be accomplished, then it is more convenient to ignore those questions pertaining to enforcement, to assume that any agreement is enforceable, and to focus instead on explaining and predicting agreements. If, on the other hand, we believe that the available technology of enforcement is critically important to the determination of what agreements can be reached and implemented—if some agreements are enforceable while others are not—then we must include an analysis of those mechanisms in our models.

In this chapter we will first concentrate on trying to learn what agreements people might reach when any and all agreements are somehow enforceable. Later we will turn our attention to situations in which the subject of enforcement cannot be divorced from the study of potential coalitional agreements. In particular, we will look at an issue that has confounded students of international politics for years, determining whether a balance of power can ensure stability in systems that are otherwise anarchic—that have no exogenous enforcement mechanism other than that countries or alliances with more power can defeat those with less.

The reader, however, may wonder why we might require additional concepts to study coalitional processes, asking the question: Can't we simply approach things as we did with, say, the repeated Prisoners' Dilemma, using the usual notions of strategy and Nash or subgame perfect equilibria? In fact, people asking such questions have good instincts, since a good share of coalitional processes can be studied in this way. Unfortunately, not all such processes lend themselves to an identifiable extensive or strategic form without heroic assumptions. Consider a situation in which three people are seated around a table and, in the context of free and open bargaining, must decide how to divide $100 using majority rule. In this case, there is no self-evident first mover or a predetermined order in which the bargainers can respond to whatever proposals are made. Constructing an extensive or strategic form here to represent bargaining would be an entirely ad hoc enterprise. And what of those bargaining situations that cannot be or are never observed, such as union contract negotiations, electing a Pope, internal Supreme Court discussions, debates among a political party's leadership over strategy and platform? Surely we do not want to relegate such things to the realm of the inexplicable. In the sections that follow, then, we will illustrate both the application of previously introduced concepts to the analysis of coalitions as well as introduce some new concepts intended to treat these less structured and less observable situations.

8.2 Coalitions and Condorcet Winners

The simplest situation we might consider that has political content is a committee that, once again, abides by simple majority rule—where by "abides" we mean that any agreement supported by a majority is assumed to be enforceable. But even if we thus ignore the issue of enforcement, we must consider the institutional structure surrounding the use of any voting system that pertains to whether that system is exogenously or endogenously determined. At one extreme, we can suppose that majorities must operate within well-defined and firmly established rules. Thus, if parliamentary procedures are binding, majorities may have to contend in debate and in voting with rules that disallow joint consideration of more than one issue or topic and that specify an order of speaking among participants. Alternatively, we can view even the rules themselves as the product of some majority decision—as a mechanism that, perhaps, exists by tradition and bureaucratic inertia but can in principle be ignored by that majority. For example, because our analysis of agendas assumes that the extensive form summarized by an agenda tree is the only relevant aspect of the situation, that analysis assumes implicitly that collusive agreements are not enforceable—we do not allow voters the option of coordinating their actions. However, if a majority coalition can coordinate so as to move along any path of the agenda tree, the agenda becomes irrelevant to the final outcome: As long as an outcome appears somewhere on the tree (as long as the outcome is feasible), a majority can decide beforehand to select that outcome and can dictate a path along the agenda that yields it as the final outcome. This fact suggests that if simple majority rule is the final arbitrator of decisions, then the details of institutional structure matter only insofar as there are frictions—transaction costs—impeding the strict enforcement of coalitional agreements or impeding the negotiation of those agreements. Without frictions, if the "institution" of majority rule is otherwise adhered to (if, for example, the participants do not revert to physical coercion), then a majority can over-ride any procedure such as an agenda that tries to direct its actions. Indeed, we might even prefer to view institutional details of such things as agendas as merely a part of some majority's plan to implement a particular outcome.

One example of procedures being part of the package that a majority coalition imposes is the debate that arose within the U.S. Congress in 1981 over President Reagan's first proposed budget. Seeking to cut appropriations for a variety of domestic programs, Reagan's legislative strategists sought to treat his proposed cuts as a package, whereas his opponents sought to consider them on a case-by-case, program-by-program basis. Reagan's side prevailed and the cuts were implemented, with the general understanding that the initial procedural decision was the critical determinant of the final outcome. Specifically, everyone assumed that it would be easier to maintain Reagan's legislative coalition under one procedure and to disrupt it under the other. And because a majority favored the proposed budget cuts, it was able to implement a procedure that was best suited to its ultimate purposes.

This example suggests that if we want to understand the imperatives of coalitions in frictionless majority rule committees, it is oftentimes appropriate to ignore the details of committee procedure and to focus instead on hypotheses about coalitional preference. In order to pursue this suggestion further, however, we require a modest amount of notation and a few new definitions. Thus, consider a committee of n members in which we denote a **coalition** with the notation C and in which the set of all **feasible outcomes** is X. If minority coalitions are powerless to affect matters, we assume that the particular outcome x is feasible if and only if any majority coalition can collude to secure x. Suppose we now allow ourselves the luxury of being anthropomorphic about coalitions by saying that *the coalition C prefers x to x' if and only if the members of C* ***unanimously*** *prefer x to x'*. Thus, if some members of C prefer x while others prefer x', we say that C is indifferent between x and x'. We also say that C is indifferent if, regardless of the preferences of the other members of C, one or more members are indifferent between x and x'.

Having thus extended the notion of preferences from individuals to collectivities, imagine treating coalitions as individuals, where majority coalitions can alter the status quo at will, where minority coalitions have no power, and (if n is even) where blocking coalitions (those controlling half the vote) can only secure the status quo. Consider now the special property of Condorcet winners. Making the bold assumption that such a winner exists, we know that regardless of the status quo's identity, a majority coalition can implement any feasible alternative. The special property of a Condorcet winner, however, is that if it is the status quo, then no majority coalition prefers anything else. Thus, we can think of a Condorcet winner as a Nash equilibrium to the non-cooperative game played among majority (winning) coalitions (where the extensive or strategic form of that game remains obscure). Moreover, since such a winner defeats every other feasible alternative in a majority vote, it must be a unique equilibrium. Thus, quite directly, we can conclude that *frictionless majority rule committees select Condorcet winners as final outcomes.* Our argument loses its relevance if there is no Condorcet winner, but before we generalize our analysis to accommodate this possibility, suppose there is more than one feasible alternative that cannot be defeated in a majority vote. Such outcomes are not Condorcet winners (because individually they cannot defeat everything), but they share the property that no majority coalition has a positive incentive to shift from such an outcome to anything else (because no majority has a unanimous incentive to shift). Clearly, these outcomes should also be thought of as equilibria, except now equilibrium outcomes are not unique.

With but a few new ideas, then, we have extended our game theoretic understanding of politics to include coalitions—albeit for a quite special case. Moreover, we can glean some substantive lessons from this discussion. First, consider the issue of predicting the coalitions that form. Such predictions are important since in many circumstances coalitions are more readily observable than outcomes. In parliamentary bodies, for example, we can easily see which parties coalesce to form a government, but eventual outcomes—agreements among coalition partners to share ministries, patronage, and policy domains, and the policies a government ultimately implements—may be more difficult to detect and measure. Indeed, a governing coalition may collapse before the agreements that ostensibly led to it are implemented, and political scientists have long believed that the preeminent objective of the study of parliamentary systems is an ability to predict what coalition will eventually emerge to form a government. Similarly, journalistic accounts of legislative deliberations explain the passage of some bill using language as "legislators ___, after negotiating an agreement over the bill's content, formed a majority coalition in favor of its passage." However, the preceding discussion implies that *if the set of feasible outcomes and the individual preferences over them yield a Condorcet winner, it may not be possible to predict which majority coalition will form—indeed, we may observe no explicit coalition whatsoever or, with everyone seeing the inevitability of things, a coalition that includes all members of the legislature.* The rationale for this assertion is that the particular coalition that forms to upset a status quo in favor of a Condorcet winner need not be unique. If the status quo is onerous to many voters, then any number of winning coalitions might form to displace it. On the other hand, if the Condorcet winner itself is the status quo, then it should be common knowledge that no alternative outcome can displace it, in which case the futility of forming coalitions should itself be common knowledge. Inaction on the part of a legislature, then, with respect to some issue may indicate merely that the status quo is majority preferred to all other possible policies.

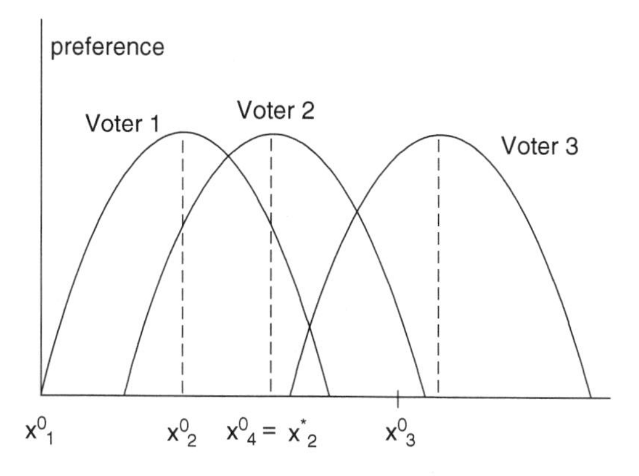

Figure 8.1 Alternative status quos

Example: Consider the single peaked preferences in Figure 8.1 wherein the median ideal, x^*_2, is a Condorcet winner. But consider the four alternative status quos portrayed in that figure: x°_1, x°_2, x°_3, and $x^\circ_4 = x^*_2$. If x°_1 is the status quo, then the members of every two-person (majority) coalition unanimously prefer changing things to x^*_2. Thus, we cannot say whether we will see all three people move unanimously to upset x°_1 or precisely which two-person coalition will do so. If x°_2 is the status quo, then only the coalition {2, 3} prefers a move to the Condorcet winner, whereas if x°_3 is the status quo, then only {1, 2} would vote for a change to x^*_2. Finally, if the status quo is x°_4, which corresponds to the Condorcet winner, then no coalition should form.

This last possibility—that we will see no explicit coalitions—can occur in other contexts.

Example: Consider the vote-trading scenario in Table 8.1, which shows the payoffs that each of five legislators associates with the passage of each of five bills. If the payoff to each legislator from a bill's failure is zero and if payoffs are separable—if each bill has nothing to do with any other bill so that the payoff from the passage of, say, two bills is simply the sum of the payoffs associated with the passage of each—then failing all bills is a Condorcet winner: No combination of pass-fail across the five bills can secure majority approval over the status quo. Notice, though, that legislators 1 and 2 have an incentive to trade votes on A and B, because doing so secures the passage, *ceteris paribus*, of A and B, which increases the payoffs to them from zero to eight. It is the *ceteris paribus* condition, however, that causes difficulty, because legislators 3 and 4 have an incentive to trade in the opposite direction so as to cancel the effect of the trade between 1 and 2 (indeed, 5 has an incentive to bring this possibility to 3 and 4's attention). Because failing all bills is a Condorcet winner, any trade that any

Table 8.1 Vote Trading with Status Quo as Condorcet Winner

Legislator	*Bill A*	*Bill B*	*Bill C*	*Bill D*	*Bill E*
1	10	−2	−5	4	−5
2	−2	10	−5	−5	4
3	4	−8	−5	3	−8
4	−8	5	3	−5	−8
5	−5	−5	4	−10	4

pair of legislators might contemplate will be opposed by the remaining legislators.

If all legislators understand that the status quo is a Condorcet winner and that no set of trades can be sustained by a majority (which is implied by common knowledge), the legislature can avoid time-consuming negotiation and coalition formation by the simple expedient of allowing these bills to remain buried in committee. And as to the issue of whether we might anticipate maneuvers to bring only some subset of bills to the floor, if all legislators are fully cognizant of strategic possibilities, then no such maneuver can succeed since, if such a maneuver threatened to yield an outcome different from the status quo, a majority of legislators would block the attempt. Thus, if a Condorcet winner exists and if that winner corresponds to the status quo, then, at least for the case of complete information environments, using data drawn from actual votes poses a selection problem for empirical research—the only data we will observe will involve situations in which there is no such winner (unless, of course, legislators choose to vote on various measures in order to publicly affirm a position on an issue for the benefit of their reputations among constituents).

Our example also illustrates an interesting general fact about vote trading scenarios known as *Schwartz's Vote Trading Theorem* (see Thomas Schwartz's 1977 article, "Collective Choice, Separation of Issues and Vote Trading," *American Political Science Review* 74, June):

If the payoffs from individual legislation are separable, and if a Condorcet winner exists, then that winner is the outcome that prevails when everyone votes sincerely on each bill.

Thus, if we observe vote trading in a legislature, we should infer that there is no Condorcet winner among the feasible outcomes, that information is incomplete, or that the bills under consideration are not separable. Moreover, this conclusion is not dependent on selecting an example in which the Condorcet outcome is the status quo—our analysis holds regardless of what combination of passage and failure is Condorcet winning. To see the proof of this fact, which consists of showing that any outcome that is not the sincere outcome can be beaten in a majority vote by some other outcome, let **O** denote the set of all possible outcomes—all possible combinations of pass and fail for some set of m bill and let $\boldsymbol{o} = (o_1, o_2, \ldots o_m)$ be an element of $\mathbf{O}$, where $o_j = 1$ if bill j passes and 0 otherwise. Now let $\boldsymbol{o}^* = (o^*_1, o^*_2, \ldots, o^*_j, \ldots o^*_m)$ be the sincere voting outcome and notice that if we change the vote on any one bill, say j, to form the outcome $\boldsymbol{o}' = (o^*_1, o^*_2, \ldots, o'_j, \ldots o^*_m)$, then it must be the case that $\boldsymbol{o}^*$ is majority preferred to $\boldsymbol{o}'$ since preferences are separable by assumption and since a majority prefers o^*_j to o'_j (otherwise o^*_j wouldn't be the sincere outcome on bill j). Now let $\boldsymbol{o} = (o_1, o_2, \ldots o_m)$ be an arbitrary outcome other than the one, $\boldsymbol{o}^*$, that sincere voting produces. Suppose $\boldsymbol{o}$ differs from $\boldsymbol{o}^*$ on at least bill j (since it must differ on one or more bills). However, if a majority prefer $\boldsymbol{o}^* = (o^*_1, o^*_2, \ldots, o^*_j, \ldots o^*_m)$ to $\boldsymbol{o}' = (o^*_1, o^*_2, \ldots, o_j, \ldots o^*_m)$, where $o_j \neq o^*_j$ then from separability, that same majority must yield the social preference $(o_1, o_2, \ldots, o^*_j, \ldots o_m) > (o_1, o_2, \ldots, o_j, \ldots o_m)$, which is to say that if $\boldsymbol{o}$ is any outcome other than what arises from sincere voting on every bill, it can be defeated in a majority vote when preferences are separable by simply taking any bill in $\boldsymbol{o}$ for which the disposition is the opposite of what sincere voting produces and substituting the sincere disposition for it.

8.3 A Generalization—The Core

We know that there are a variety of voting mechanisms that ensure the selection of Condorcet winners under complete information—two-candidate elections, binary agendas, and cooperative majority rule committees with no impediments to coalition formation and the enforcement of agreements. However, although the preceding discussion is useful, it hardly provides a general analysis of coalitional processes. Two questions in particular come to mind: (1) What can we say about processes that are not dictated by majority rule and (2) what predictions about outcomes and coalitions can we offer if there is no Condorcet winner? Answering the first question requires that we restate the theoretical structure of the preceding section in a way that admits generalization. We begin thus:

1. Let the elements of X, the feasible set of outcomes, correspond to vectors that denote the utility each player associates with each substantive outcome, and let $\boldsymbol{u} = (u_1, u_2, \ldots, u_n)$ be a specific vector in X.
2. As before, assume that the coalition C prefers $\boldsymbol{u}$ to $\boldsymbol{u}'$ if and only if $u_i > u'_i$ for all i in C.
3. Define the coalition C to be **effective** for $\boldsymbol{u}$ in X if the members of C can coordinate their actions so as to ensure that each member, i, of C receives a payoff of at least u_i. Let $v(C)$ denote the set of all utility n-tuples for which C is effective, where $v(C)$ is termed *the characteristic function* in the game theory literature.
4. Say that $\boldsymbol{u}$ dominates $\boldsymbol{u}'$ if there exists at least one coalition that is effective for $\boldsymbol{u}$ and which prefers $\boldsymbol{u}$ to $\boldsymbol{u}'$.

Our final step is to define the concept of a cooperative game's core:

> *The* **core** *of a cooperative n-person game is the set of undominated elements of X.*

More formally, let $v^*(C) \subseteq v(C)$ correspond to those outcomes from which we cannot move, so as to make **all** members of C simultaneously better off. Thus,

> The utility n-tuple $\boldsymbol{u}$ is in a game's core if and only if it is in $v^*(C)$ for all coalitions C.

If $\boldsymbol{u}$ is not a member of $v^*(C)$ for some coalition C and if $\boldsymbol{u}$ is feasible, then, by definition, we can find a utility n-tuple in $v^*(C)$ that dominates $\boldsymbol{u}$, thereby prohibiting $\boldsymbol{u}$ from being an element of the core.

At times, especially when dealing with spatial games, it is convenient to define $v^*(C)$ and the core in terms of outcomes rather than utility, provided that we keep in mind that there are specific preferences defined over those outcomes. For example, consider Figure 8.2, which shows the same single-peaked preferences of three voters that Figure 8.1 illustrates. Recall that our discussion of the Median Voter Theorem revealed that such preferences occasioned a Condorcet winner and an election equilibrium, which corresponded to the median voter's ideal—voter 2. However, rather than interpret these voters as an electorate, suppose they are a committee that must choose a policy through negotiation and majority rule. Notice now that all points lying between voter 1 and 2's ideals correspond to the outcomes that are Pareto optimal for the coalition {1, 2}, all points between 1 and 3's ideals are Pareto optimal for {1, 3}, and all points between 2 and 3's ideals are Pareto optimal for {2, 3}. Thus, defined in terms of outcomes as opposed to utility vectors, these sets of Pareto optimals correspond to $v^*(1,2)$, $v^*(1,3)$ and $v^*(2,3)$ respectively, as shown in Figure 8.2. Thus, since voter 2's ideal is common to all three sets of Pareto optimal outcomes, that ideal is the game's core, which illustrates the fact that *if a majority rule cooperative game has a Condorcet winner, that winner corresponds identically to the core.* A Condorcet winner defeats everything in a majority vote, so no other alternative dominates it; and no other alternative can be in the core since it is defeated (dominated) by the Condorcet winner. However, we have introduced the concept of the core to treat situations other than majority rule committees, so consider what happens when we apply this idea to a cooperative Prisoners' Dilemma:

Example: Consider the two-person Prisoners' Dilemma in Figure 8.3a. Notice, first, that although cooperation can yield a player a payoff of -3, a player can be certain that he or she does no worse than 0 by refusing to cooperate. Thus, we say that each player, acting alone, is effective for any outcome that yields that player a payoff of 0 or less. So $v(\{i\}) = \{\boldsymbol{u}\text{:}\ u_i \leq 0\}$, and $v^*(\{i\})$, the utility outcomes that are Pareto optimal for i, corresponds to all feasible 2-tuples in which $u_i \geq 0$. On the other hand, the coalition of both players, {1, 2}, is effective for all feasible outcomes. Hence, the core of the corresponding cooperative

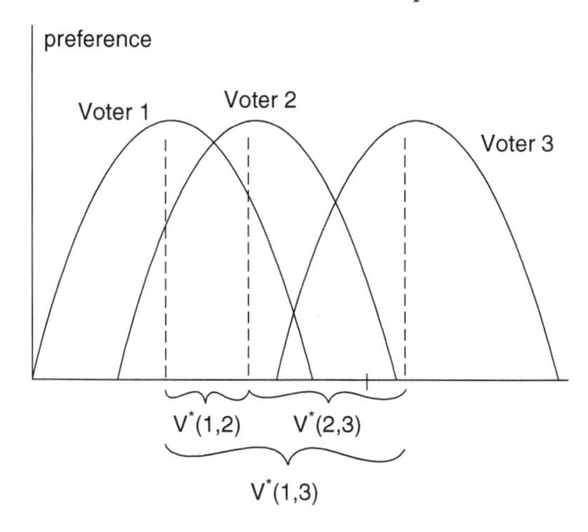

Figure 8.2 Pareto optimals for 2-person coalitions with 3 voters

	Don't cooperate	Cooperate
Don't cooperate	0, 0	7, –3
Cooperate	–3, 7	4, 4

Figure 8.3a Prisoners' Dilemma

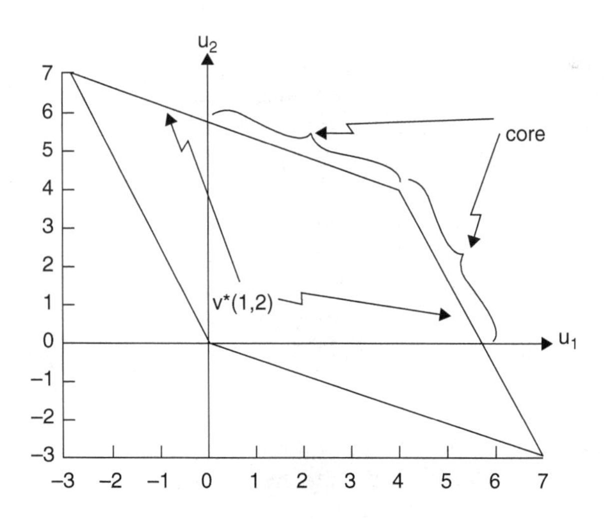

Figure 8.3b The core in the Prisoners' Dilemma

game is any feasible outcome in which both players each receive at least 0, and is Pareto optimal for $\{1, 2\}$. As Figure 8.3b shows, the only such outcome that does not entail lotteries is $(4, 4)$, which is what prevails if both players cooperate.

This example reveals what we can and what we cannot learn from our analysis thus far. Played non-cooperatively, the unique Nash equilibrium yields the outcome (0, 0), whereas played cooperatively, we ostensibly avoid the dilemma. What our analysis fails to specify, however, is the mechanisms these two players might use to enforce cooperation. Thus, we learn what outcomes will prevail if enforcement is feasible, but we do not learn how to ensure that feasibility.

The issue of enforcement is, as we have already indicated, profoundly important theoretically. But it is also important substantively—a fact illustrated by an important argument about the role of the courts in the adjudication of property rights, called the **Coase Theorem** (Ronald Coase, "The Problem of Social Cost," *The Journal of Law and Economics* 3, 1960). This theorem asserts that

> *If transactions costs are zero, if there is a mechanism for enforcing contracts, if there is a freely transferable numeraire (money), and if there is an unambiguous specification of property rights, then everyone involved in a situation in which the actions of some hurt or benefit others can reach mutually beneficial bargains without government intervention. Furthermore, if a unique outcome maximizes social wealth, then the parties will attain that outcome regardless of the prior assignment of property rights and liabilities.*

Recast in the terminology of our discussion of the Prisoners' Dilemma, the Coase Theorem counters the argument that the inefficiencies occasioned by public goods and externalities justify government action, and in particular, government regulation. Instead, governments ought simply to ensure the enforceability of contracts, thereby guaranteeing legally defensible and alienable property rights. If one can receive compensation for the benefits that one's actions bestow on others, and if an institutional arrangement can guarantee such a right, then the public-goods problem dissolves. Political debate would then focus on the allocation of those rights rather than on decisions about what levels of various goods, services, and regulations to supply. For example, an industry might have a property right to pollute, which the "victims" could buy away from it if they value clean air more than the industry values the use of the ambient air as a sink. Alternatively, the victims might have a property right in clean air, which the industry might buy away from them. And if victims (industries) are willing to pay more for clean (dirty) air, then the same outcome, clean (dirty) air, prevails no matter what the prior assignment of property rights has been.

> **Example**: Suppose persons 1 and 2 must choose an outcome from the set $O = \{o_1, \ldots, o_5\}$, where their payoffs are as shown in Table 8.2. Suppose person 1 has the legal right to decide which outcome is chosen—person 1 has control of the relevant property right—and suppose money and utility are equivalent. Because 1 is effective for all outcomes, he can select outcome o_5 and secure a payoff of 12. Thus,

$v(1) = \{\boldsymbol{u}: u_1 < 12\}.$

Person 2, on the other hand, cannot guarantee receiving more than 0, so we set

$v(2) = \{\boldsymbol{u}: u_2 < 0\}.$

Persons 1 and 2 can secure any feasible outcome, including the one that yields a total payoff of 14. Since we have assumed that money and utility are equivalent, exchanging money is equivalent to transferring utility, and since money, presumably, is freely transferable, persons 1 and 2 can thereby secure any utility 2-tuple in which the sum of u_1 and u_2 does not exceed 14. That is, 1 and 2 together can select outcome o_2, realize a combined payoff of 14, and trade money to achieve any outcome in which $u_1 + u_2 \leq 14$. Hence,

$v(1, 2) = \{\boldsymbol{u}: u_1 + u_2 \leq 14\}.$

Since no outcome in which 1 receives at least \$12, 2 receives at least 0, and 1 and 2 together receive at least \$14 can be dominated by {1}, {2}, or {1, 2}, it follows that the core to this game is the set of 2-tuples $\boldsymbol{u} = (u_1, u_2)$ such that

core $= \{\boldsymbol{u}: u_1 \geq 12, u_2 \geq 0, u_1 + u_2 = 14\}.$

Alternatively, suppose 2 has the right to decide the outcome, so $v(1)$, $v(2)$ and $v(1, 2)$ become

$v(1) = \{\boldsymbol{u}: u_1 \leq 0\}; v(2) = \{\boldsymbol{u}: u_2 \leq 12\};$
$v(1, 2) = \{\boldsymbol{u}: u_1 + u_2 \leq 14\},$

in which case the core is

core $= \{\boldsymbol{u}: u_1 \geq 0, u_2 \geq 12, u_1 + u_2 = 14\}.$

Thus, the specification of a property right dictates the eventual distribution of payoffs between 1 and 2, but o_2 prevails regardless of who controls the decision.

Table 8.2 Illustration of the Coase Theorem

outcome	*person 1*	*person 2*	*total*
o_1	\$0	\$12	\$12
o_2	4	10	14
o_3	6	6	12
o_4	7.5	4	11.5
o_5	12	0	12

The implications of this example are viewed as important by those who criticize an expanding role for government. Instead of becoming directly involved in the production of commodities and the delivery of services—instead of establishing specific clean-air and safety standards with which businesses must comply or instead of centralizing decisions about the resources people should allocate to each of these activities—the theoretical argument here suggests that governments ought simply to assist in reducing transaction costs among relevant parties and in establishing and enforcing property rights. That is, if transaction costs can be reduced to zero so that individuals can efficiently negotiate among themselves, then, while ensuring the enforceability of agreements, the state can establish property rights in such a way as to effect whatever redistribution of income is somehow deemed appropriate. Reality, of course, is far more complex than our example, and the presumption that transaction costs can be set to zero is utopian. However, our intent is simply to show how a simple game, in conjunction with the concept of the core, leads to a discussion of important political matters.

8.4 The Politics of Redistribution

Earlier we note that a state's functions can be divided into two broad categories—regulating those Prisoners' Dilemmas that markets cannot resolve (including the implementation of mechanisms for enforcing those contracts that are an essential part of markets) and the redistribution of wealth. Our earlier discussion of the Prisoners' Dilemma played cooperatively reveals that the central problem with respect to the first function is enforcing decisions that yield outcomes in the core. Hence, governments are necessarily coercive entities. Our discussion of the Coase Theorem, on the other hand, suggests that if government can ensure the implementation of effective enforcement, then its primary task should be that of lowering transaction costs and determining the form of redistribution—determining who wins and who loses via the specification of property rights. This argument, however, should not be interpreted to mean that dictating the eventual form of redistribution is a straightforward task for a democratic state. Both Chapters 2 and 3 illustrate electoral instability in redistributive elections, where the source of that instability is the fact that such situations do not have Condorcet winners. To see this problem using the definition of the core, consider a game of pure redistribution in which:

1. Outcomes are defined in terms of a transferable commodity, such as money. Thus, every outcome in the feasible set X is characterized by the distribution of money among the players.
2. Each person's utility for this commodity is linear. So if the commodity is money, then money and utility are equivalent and utility is transferable among people. If, for example, players 1, 2, and 3 can secure the payoffs x_1, x_2, and x_3, then they can also secure the payoffs y_1, y_2, and y_3, provided that $y_1 + y_2 + y_3 = x_1 + x_2 + x_3$.
3. The transferability of utility allows us to summarize the utility outcomes that a coalition can secure, $v(C)$, more conveniently as a single number,

with the understanding that the coalition C, if it forms, can secure any feasible allocation in which the sum of payoffs to its members does not exceed $v(C)$.

4. No resources are created as a function of the way players play the game, so if N denotes a coalition of the whole (of all n players), $v(C) + v(N - C)$ equals a nonnegative constant for all C; nor are resources destroyed by the process of coalition formation, so if C' and C'' are two disjoint coalitions, then

$$v(C' + C'') \geq v(C') + c(C'').$$

5. Strict inequality must hold in the preceding expression for at least two disjoint coalitions; otherwise, no coalition is worthwhile. In particular, if individual players can get as much acting alone as they can get by coalescing, there is no need to form a coalition.

A game of pure redistribution, then, models a legislature contemplating a revision of tax codes under the constraint that the total revenues to be collected are constant. Equivalently, such games correspond to those legislative deliberations in which, after the revenues are collected, the issue is how to allocate them across the population in the form of various subsidies and transfer payments. Notice, moreover, that if the constant to which condition 4 refers is zero—if $v(C) + v(N - C) = 0$—then the pure redistribution game corresponds to a situation in which one coalition expropriates from another, whereas if the constant is positive—if $v(C) + v(N - C) > 0$—then the game is one in which some specific total of resources is to be allocated across the players.

That games of pure redistribution do not have cores—that they share the instability of majority rule games without Condorcet winners—can be established with the following somewhat fanciful example:

> **Valley of the Dump**: Suppose there are n clans, each living in its own part of a valley totally walled in by the surrounding cliffs that tower above it. Everyday, each clan produces precisely one bag of garbage, but the cliffs prevent exporting the valley's garbage. Thus, each clan faces a solid waste disposal problem. The legal code of the valley, strictly enforced, allows clans to dump their garbage in their own back yards or the yards of others, but it also permits coalitions. Assume that each clan evaluates the outcome of the garbage problem solely in terms of the total number of bags dumped in the yards of its members. More precisely, each clan has a linear utility function in bags, preferring fewer bags to more, and that the utility for having one bag dumped on one's yard is -1. To see that this game of redistribution has no core, suppose to the contrary that the payoff vector $\boldsymbol{x} = (x_1, x_2, \ldots, x_n)$ is in the core. Then, since any coalition with $n - 1$ members can dump all of its garbage on the yard of the excluded clan and, presumably, will have one bag of garbage spread across its members' yards, it must be that

$$x_2 + x_3 + \ldots + x_n \geq -1$$
$$x_1 + x_3 + \ldots + x_n \geq -1$$
$$\cdot$$
$$\cdot$$
$$x_1 + x_2 + \ldots + x_{n-1} \geq -1.$$

which is to say that for $\boldsymbol{x}$ to be in the core, it must give every coalition of $n - 1$ members a payoff of no less than -1, since any such coalition can guarantee against having more than one bag of garbage dumped in total across the lawns of its members. Since there are n such equations, if we add them up, we have

$$(n - 1)(x_1 + x_2 + \ldots + x_n) \geq -n.$$

But since there are precisely n bags produced per day and since those n bags cannot be exported out of the valley, it must also be true that $(x_1 + x_2 + \ldots + x_n) = -n$. So $\boldsymbol{x}$ is in the core only if $-(n - 1)n > -n$, which requires that $n \leq 2$. That is, the redistributive game in the valley has a core only if there are not more than two clans.

That games of pure redistribution do not have cores helps explain a great many things. We have already argued in an electoral context that to the extent that the policies with which an incumbent must deal are redistributive, a challenger in the next election can always find some coalition of voters that can be made better off with some alternative policy and pattern of redistribution. The task for the challenger, then, is to find such a coalition and to convince its members that, were he or she in office, an appropriate redistribution would prevail. The incumbent's task, on the other hand, is to convince voters that politics is not purely redistributive or, as in the U.S. Congress, that the incumbent's seniority is critical in keeping the federal government's largesse from being redistributed away from the constituency in question. Incumbent presidents seeking reelection will tend to portray themselves as statesmen who are leading the nation to more Pareto efficient outcomes, whereas especially attractive platforms for challengers will consist of populist appeals identifying the incumbent as an enemy of "us" (a majority) and an ally of "them" (some sinister, easily identified, "overfed," minority). Within Congress itself, redistributive politics is equally unstable since our formal analysis applies to majority rule committees as well as elections. Indeed, some scholars have argued that the U.S. Congress's propensity to form overly large coalitions that grant special benefits to a great many constituencies derives from the inherent risk aversion of legislators with respect to the prospect of finding themselves in a losing coalition. The alternative to a norm of "universal inclusion" is the instability and uncertainty associated with a majority rule game without a core.

8.5 The Core and Spatial Issues

The fact that cores and Condorcet winners are nearly equivalent in majority voting games and the fact that Condorcet winners are rare in multi-dimensional

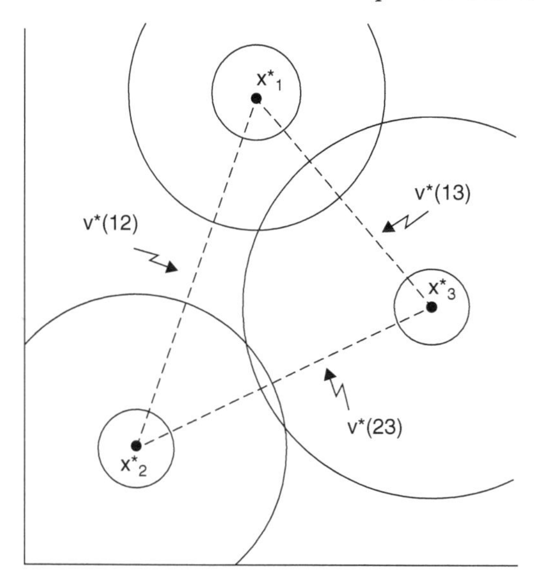

Figure 8.4 An empty core with two issues

spatial voting games tell us that cores are rare in such voting games if played cooperatively. To see this directly, consider again the three-voter configuration of spatial preferences in two dimensions, shown in Figure 8.4. Rather than assume that these preferences refer to voters in some mass electorate, suppose instead that they pertain to members of some committee, and, rather than have the eventual policy correspond to the position chosen by a successful election candidate, suppose that policy is the one agreed to by some majority coalition.

Our earlier discussion of these preferences in an electoral context led to the conclusion that the ideal point configuration in Figure 8.4 does not yield a Condorcet winner. Here the cooperative, majority rule, coalitional game cannot have a core. To see this using the core's definition directly, recall that an outcome is in the core if and only if it is Pareto optimal for every coalition—if and only if it is in $v^*(C)$ for every winning coalition C. In this instance with circular indifference contours, the Pareto optimal policies for any two-person coalition (a majority) corresponds to the line connecting their ideal points. Clearly, there is no policy that is Pareto optimal for all three 2-person coalitions simultaneously—the three lines describing the Pareto optimal outcomes for the three 2-person (majority) coalitions do not have a common intersection. Thus, the disequilibrium that pervades a two-candidate spatial election does not disappear if we assume that the electorate is a committee.

We can extend this argument to assert that when a Condorcet winner exists in an election, it exists also if that electorate acts as a majority rule committee. In this way, then, we can view a two-candidate election as a device for achieving the same outcomes as would be achieved in a "New England town meeting model" of democratic processes, except that the election has the advantage of easier implementation in large electorates whereas the town meeting has the advantage of ensuring that the members of the electorate can inform each other directly about their views.

8.6 Majority Rule Games Without Cores

In Chapter 6 we offer several examples of where institutional "frictions" such as issue-by-issue voting can, if they are unequivocally binding, yield stability in spatial contexts where otherwise there is no Condorcet winner. Before asserting, though, that institutions can induce stability, we need to take into account the fact that the choice of institutions themselves can inherit the instability (social intransitivity) that characterize social preferences over the outcomes with which they deal. For example, if we take the usual three-voter example of the Condorcet cycle over the alternatives A, B, and C (in which A defeats B, B defeats C, and C defeats A), we now know that if voters are strategic, the agenda

> "A versus B, the winner against C," yields B,
> "A versus C, the winner against B," yields A,
> "B versus C, the winner against A," yields C.

So, although a specific agenda yields a determinate outcome because it necessarily constrains what alternatives can be paired in a vote, a cycle reappears if the committee is tasked without constraint with voting on the agenda itself. This example illustrates a principle we might call the unavoidability of cycles: *Unless individual preferences over the basic outcomes themselves occasion a Condorcet winner, we cannot escape the fact that at some point in the process of selecting rules or voting on outcomes, alternatives will cycle, and opportunities will present themselves for people to try to manipulate the agenda selection process to their own advantage.*

This principle tells us that it is imperative that we say something general about majority rule games without cores. Saying something about coalitions and outcomes when there is no core, however, requires an adjustment in our thinking about the nature of prediction. Accustomed as we are to assuming that, except for the chance events of nature, outcomes follow a definitive logic, we commonly overlook the possibility that there may be fundamental indeterminacies in politics. When we observe and attempt to understand the formation of a particular legislative or parliamentary coalition, we commonly approach matters with the implicit supposition that we should attribute an inability to explain why a particular coalition formed as against some other to a failure to measure some relevant parameter such as unobserved interpersonal relationships among the participants. This may be true, but generally such refined measurement is impractical or even impossible; moreover, we have little theoretical guidance in determining what ought to be measured and how measurements and prior observations relate to subsequent events. A similar problem confronts those who wish to predict whether a tossed coin will land heads or tails.

In principle, precise measurement of all physical characteristics of the coin toss will allow a prediction to some refined degree of accuracy. But such measurement is impractical, and, thus, we find it more convenient to treat the outcome as a random event.

In a similar spirit, imagine that three people, who, absent deep psychological testing and observation, seem identical in every respect, must use majority rule to divide $1,000. If utility and money are equivalent, the corresponding pure redistribution game has no core. But because they abide by majority rule, we might be willing to utter the tentative prediction, based on the assumption that each is "sufficiently avaricious" so as to be unconcerned with "fairness," that two people will coalesce to divide the money evenly between themselves, excluding the third from any payment. Notice, though, that we cannot say which coalition will form—from our view, all three 2-person coalitions are equivalent. Indeed, in this circumstance, the only reasonable prediction is that some two-person coalition will form.

To this point nothing seems unusual since we have already encountered indeterminacy in games with multiple equilibria. But recall that whether a particular n-tuple of strategies is or is not an equilibrium depends solely on the properties of that equilibrium's character or the character of the strategies that comprise it. We do not say that some n-tuple is a Nash or subgame perfect equilibrium because some other n-tuple is or is not an equilibrium. In the context of cooperative games without cores, on the other hand, things may be quite different. Specifically, in the cyclic bargaining that seems inevitable in pure redistribution games, we might speculate that the participants will be especially attracted to specific proposals and that these proposals will achieve special significance **as a set**. For the case of three people negotiating to divide $1,000, the set of outcomes $\{(500, 500, 0), (500, 0, 500), (0, 500, 500)\}$ comes to mind. Our attention focuses on this set, we suspect, for several reasons. First, as a set it treats the three otherwise indistinguishable bargainers as equals—each is excluded from precisely one winning coalition and each wins the same amount when in a winning coalition. As a consequence, no proposed outcome in the set advantages a member of the corresponding winning coalition so as to give his or her partner an incentive to switch to another proposal in the set. And second, all other feasible ways of dividing the $1,000 are dominated by at least one proposal in the set—at least one proposal in the set will be preferred by a majority.

Although this reasoning is imprecise, what we want to emphasize is this: we do not focus on the set $\{(500, 500, 0), (500, 0, 500), (0, 500, 500)\}$ because each of the outcomes in it satisfies some property that renders it an equilibrium; rather we focus on these outcomes because they satisfy certain properties *taken together as a set*. Thus, although we do not attempt to predict which of the three outcomes will ultimately prevail, labeling the set as a "solution" to the 3-person bargaining game corresponds to the prediction that some outcome in it will prevail. For example, then, although recognizing that an understanding of why one coalition rather than another forms in a Parliament or a legislature may require a detailed analysis of personalities and chance events, even without such measurements, we may be able to exclude a great many coalitions as possibilities merely on the basis of the participants' general policy preferences. It is in this spirit that we consider the hypothesis that a general abstract characterization of situations allows us to identify sets of feasible outcomes as predictions.

To make this reasoning more precise, we proceed by first trying to generalize the definition of the core. There are several possibilities. First, we could simply play with the component parts of the core's definition, weakening one or more of those parts. Second, using some general ideas about how bargaining might proceed, we could specify "reasonable" properties that a predicted set ought to satisfy. Finally, we could attempt to describe an explicit extensive form model of bargaining and thereby deduce an appropriate solution as a Nash equilibrium to the corresponding non-cooperative model of the coalitional game. Traditionally, game theory focuses on the first two avenues, but some progress has been made with respect to the third. For the moment, though, let us focus on the first.

Recall that the elements of a core share two properties: *The core is **internally stable** in the sense that if there is more than one outcome in it, no outcome in it dominates another outcome in it. And the core is **externally stable** in the sense that outcomes in it are undominated by the feasible outcomes that are not in the core.* The problem in games without cores is that everything is dominated—there is no externally stable set. However, looking back at our example in which three people use majority rule to divide $1,000, one feature of the set of outcomes $V = \{(500, 500, 0), (500, 0, 500), (0, 500, 500)\}$ that we find especially attractive is that coalition partners do not have any apparent incentive to switch from one coalition to another within this set. Maintaining this requirement (which is admittedly arbitrary at this point), we know that feasible outcomes outside the set dominate outcomes in it (otherwise, the set is a core, which doesn't exist). But in this pure redistributive example, we can readily verify that every outcome outside of V is dominated by some outcome in V. Thus, even if bargaining "wanders out of V," there is an outcome in V that takes negotiators directly back to this set. So if we modify only the definition of external stability, we arrive at the following characterization of a stable set:

1. *V is **internally stable** in the sense that no outcome in it dominates any other outcome in it.*
2. *V is **externally stable** in the sense that every feasible outcome not in V is dominated by some outcome in V.*

Sets of feasible outcomes that together satisfy internal and external stability are called *von Neumann solutions*, or *V-sets*. Thus, *V*-sets are like cores except that their definition imposes a weaker definition of external stability. The core requires that every outcome in it be undominated by every outcome outside of it; the *V*-set requires that every outcome outside of it be dominated by something inside of it.

Although we will argue shortly that the V-set is deficient as a general hypothesis, we want to reemphasize the profoundly important and innovative perspective it takes with respect to prediction—a perspective that is not familiar to the general character of political analysis. Specifically, *if we predict outcomes in V, we do so not merely because of the properties of those outcomes taken one at a time, but rather because of the properties of the set, V, to which they belong.* Thus, our explanations for legislative or parliamentary outcomes can no longer take the form "outcome . . . prevailed because the majority . . . preferred it to all other outcomes and because this majority had sufficient skill in forming before any opposition could materialize." To this explanation we must append such statements as "and because that outcome was among a set of outcomes, consisting of . . ., that could prevail in this way." The V-set, then, requires that we diminish somewhat our focus on the particular character of victorious outcomes and coalitions and that we pay greater attention to the general characteristics of such outcomes and to the other outcomes that share these characteristics.

Aside from its innovative perspective, the notion of the V-set is attractive because it seems to be a mathematically straightforward generalization of the core. And historically at least, the V-set served as the focus of research into cooperative games for much of game theory's initial development. However, its definition occasions certain problems, the most important of which is that for a great many games V-sets are not unique. Indeed, the set of outcomes in some V-sets typically include nearly all feasible outcomes. For example, in a 3-person divide-the-thousand-dollar game, the set

$$V' = \{\mathbf{u}\text{: } u_i = c,\ u_j + u_k = 1{,}000 - c,\ c \leq 1{,}000/3\}$$

satisfies both internal and external stability. For example, with $i = 1$, the two payoff vectors (300, 350, 350) and (300, 400, 300) are in V' and in accordance with internal stability neither dominates the other. On the other hand, the vector (350, 300, 350) is not in V', but in accordance with external stability, it is dominated (via players 2 and 3) by the vector (300, 325, 375), which is in V'. A second difficulty is that as a generalization of the core, the V-set's definition is wholly ad hoc and without behavioral justification. For this reason game theorists have sought to narrow its predictions with additional refinements and to associate it with some notions of the bargaining process. One such refinement is the *main-simple V-set*, which treats majority rule games.

> *A V-set is* ***main simple*** *(denoted V_m) if each outcome in it can be associated with a minimum winning coalition so that every member of that coalition prefers that outcome at least as much as any other in V_m and where no such coalition is associated with more than one outcome in V_m.*

The important property of main-simple V-sets is that they are unique for an important class of games:

if utility in a majority rule game is like money and is perfectly transferable among the players, then that game has a unique main-simple V-set.

For example, in the divide-the-thousand-dollars game, V_m is the set {(500, 500, 0), (500, 0, 500), (0, 500, 500)}. More generally, if the game is perfectly symmetric in the sense that each player has one vote and if each minimum winning coalition can win the same amount, X, as any other, then the elements of V_m take the form $(2X/(n + 1),\dots, 2X/(n + 1), 0,\dots, 0)$. Thus, main-simple V-sets rationalize the prediction that, in majority rule pure redistribution games, some minimum-winning coalition will form to divide the spoils evenly among its members.

Universalism in the U.S. Congress: A seemingly disquieting note with respect to the main-simple V-set's predictions is that in the U.S. Congress we often observe nearly unanimous agreement to pass legislation that is redistributive because the component parts of that legislation benefit only specific constituencies. One explanation for this phenomenon is that legislators choose beforehand between being competitive versus cooperative, where being cooperative means to avoid forming a minimal winning coalition in favor of a more universally acceptable result; being competitive means seeking to form a minimal winning coalition that expropriates from a maximal minority. The advantage of being competitive is that it maximizes one's gains whenever one is included in the winning coalition, but the disadvantage is that one takes the chance of being excluded from the majority. Being cooperative ensures against losing, but one cannot win a great deal. For example, in a divide-the-thousand-dollar game with n bargainers, each person in a minimal winning coalition gains $1{,}000/[(n + 1)/2]$ whereas excluded players gain nothing. The probability that any specific person is included in a minimal winning coalition equals the proportion of such coalitions that includes that person, $(n + 1)/2n$, so the expected gain in a non-cooperative legislature is

$$\left[\frac{2{,}000}{n+1}\right]\left[\frac{n+1}{2n}\right]+0\left[1-\frac{n+1}{2n}\right]=\frac{1{,}000}{n},$$

which is what one earns if everyone plays cooperatively and simply divides the thousand dollars among all players. Thus, risk averse legislators or a sense that outcomes ought to be "fair and equitable" is sufficient to explain unanimity within a legislature.

Example: Figure 8.5 reproduces the preferences in Figure 8.4, but now let us focus on the three outcomes A, B, and C formed by the simultaneous tangencies of persons 1 and 2's indifference contours, 1 and 3's contours, and 2 and 3's contours. Notice first that, because the players are assumed to be able to bargain only over policy, utility cannot be transferred among them—one player cannot transfer x units of utility to some other player, because there is no commodity like money in the game. However, this game does have a main-simple *V*-set. First, to see that {A, B, C} is that set, notice that since A and B both lie on the same contour for person 1, 1 is indifferent between these two outcomes. Persons 2 and 3, on the other hand, hold opposite preferences, so A does not dominate B nor does B dominate A. The same argument holds for B and C and for C and A, so the set {A, B, C} is internally stable. To show, then, that {A, B, C} is a *V*-set, notice that the two shaded "petals" in Figure 8.5 correspond to the set of outcomes that defeat A. A simple paper and pencil exercise, though, reveals that B or C defeats any outcome in this set, while those outcomes that defeat B are defeated by A or C and that those that defeat C are defeated by A or B. Hence, {A, B, C} is a *V*-set. The reader should be able to confirm now that this set is a main-simple solution, although as we show later, there are majority rule spatial games without main-simple *V*-sets.

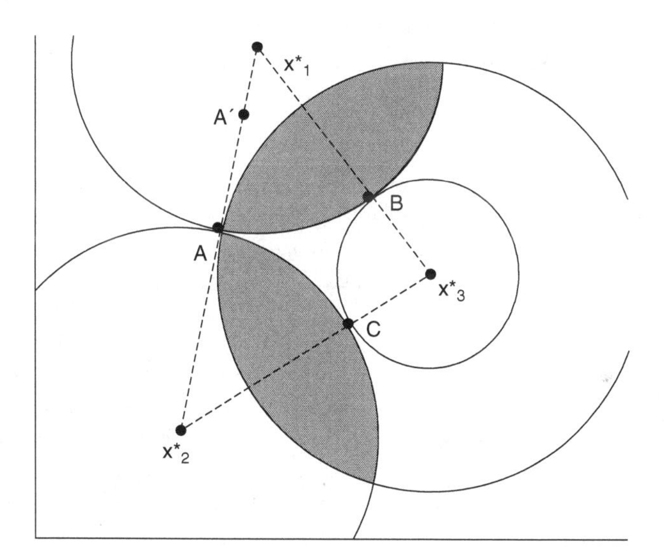

Figure 8.5 A spatial V-set

The impetus for the development of the main-simple *V*-set lies not only with the necessity for refining the *V*-set but also with the view that in a competitive coalitional environment, minimal winning coalitions somehow play a special role. Thus, the definition of a main-simple *V*-set adds some seemingly reasonable conditions to those of internal and external stability in the hope of generating a more refined prediction. Unfortunately, although its predictions seem reasonable within its domain, important problems remain. First, its definition pertains only to majority rule games, and, second, its definition, like that of the *V*-set itself, is ad hoc. Thus, the logical connections of this idea to the general perspectives of game theory are unclear. For this reason two related alternatives to the *V*-set have been offered—the bargaining set and the competitive solution (for formal definitions of the bargaining set, see Robert Aumann and Michael Maschler, “The Bargaining Set for Cooperative Games,” in Dresher et al., eds., *Advances in Game Theory*, Princeton: Princeton University Press, 1964: 443–476 and Richard McKelvey, Peter Ordeshook and Mark Weiner, “The Competitive Solution for N-Person Games Without Transferable Utility,” *American Political Science Review*, June 1978).

The *bargaining set* is based on the idea that if a coalition can form around a specific outcome, then each member of that coalition should be able to defend what he or she receives from the outcome, which the coalition proposes against the possible objections of other members of the coalition. If a person is getting “too much”—if he or she cannot defend a payoff—then we ought not predict that outcome. But if each person can mount a defense against every conceivable objection, then that outcome is “stable” and it should be included in the set of predicted possibilities. Put differently, members of a coalition should be able to offer legitimate *counter-objections* to any *objection* against what they are getting in a coalition; if they cannot formulate such a defense, then either the coalition must adjust the payments to its members or that coalition cannot form.

> **Example:** Consider again the divide-the-thousand-dollars game, and the outcome (500, 500, 0) supported by the coalition {1, 2}—denoted by the *proposal* ((500, 500, 0), {1, 2}). Suppose person 2, seeking a greater share, *objects* against 1 with the proposal ((0, 550, 450), {2, 3}), which he and his proposed new coalition partner prefer to the original proposal. However, 1 can defend his payoff of 500 in the original proposal by *counter-objecting* with ((500, 0, 500), {1, 3}), because this counter gives him 500 and is preferred by 3 to the objection. Indeed, every objection by 2 against 1 (as well as every objection by 1 against 2) can be countered, so ((500, 500, 0), {1, 2}) is in the bargaining set. Alternatively, consider (600, 400, 0). In this instance, 2 can object with ((0, 500, 500), {2, 3}) and 1 cannot counter—person 1 cannot give 3 as much as 500 and at the same time defend the original proposed payoff of 600. So the proposal ((600, 400, 0), {1, 2}) is not in the bargaining set.

The bargaining set clearly provides a more reasonable basis for analyzing political coalitions than does the *V*-set, but like the *V*-set, it is not without difficulties. First, there are several variants to its definition, and there is no clear reason to choose one variant over another. For example, can several persons object simultaneously against more than one person? If an objection can be directed against more than one person, must the targets of the objection all counter simultaneously with the same proposal, or can they counter individually with different proposals? Must those who are included in the counter strictly or weakly prefer the counter to the objection? Must the counter itself be stable (in the bargaining set) and, thus, be believable? (Bargaining sets satisfying this property are called *strong bargaining sets*.) A more serious problem, however, is that this idea is designed for redistributive games and not for spatial policy games. In a pure redistributive game there is presumed to be a commodity (typically money) that can be traded among the players so as to allow for the perfect transfer of utility. Thus, if a player is "getting too little" in a coalition, a transfer is feasible from those who are getting "too much." But such transfers are generally impossible if only policy represented by spatial preferences is considered. If the only way to adjust a coalition partner's payoff is to alter the coalition's policy proposal, then we may be unable to transfer utility to one or more coalition partners without fundamentally undermining the coalition's viability. Nevertheless, the idea of a bargaining set does seem to take us part way to a solution to the bargaining problem since, among other things, it does serve to rationalize main simple *V*-sets. Specifically, it is possible to show that main simple *V*-sets and strong bargaining sets are equivalent. Thus, the bargaining notions behind the bargaining set provide a behavioral rationalization for looking at main simple *V*-sets as predictions.

An alternative to the bargaining set is the *competitive solution, K*. Briefly, this hypothesis is based on the suggestion that coalitions, in order to form, must compete successfully with other coalitions for critical—pivotal—members. If two coalitions, *C* and *C'*, are to be simultaneously viable—if they are both capable of forming in a competitive environment—then it must be the case that the players who pivot between *C* and *C'* (the members who are common to both coalitions) cannot all prefer the proposal of one coalition to the proposal of the other; otherwise, these pivotal players would block one coalition in favor of the other. Referring to the three-person game in Figure 8.5, suppose, for whatever reason, that voters 1 and 2 do not appreciate the potential value of their coalition. Since player 3 is temporarily the sole pivot between the coalitions that might form—{1, 3} and {2, 3}—we can imagine a bargaining process in which voters 1 and 2 begin by offering 3 outcomes such as B and C, but find themselves eventually "bidding up" to 3's ideal point as they each try to secure 3's loyalty. However, unless 1 and 2 are completely dense, one or the other should soon realize that they can avoid compromising in the direction of 3's ideal by forming a coalition of their own. With voter 3 no longer enjoying the position of sole pivot, and finding it necessary to compete with 1 and 2 as potential pivots as well, we can expect 3 to begin proclaiming, "Honest, I was only kidding in asking for my ideal. I'll settle for B or C as originally discussed." The competitive solution tries to formalize this bargaining scenario.

To define the competitive solution, we first introduce the notion of viability. Focusing on proposals that identify a payoff vector and the coalition that forms to realize that outcome (note that *p* does not refer here to a probability),

> *If* $p = (\mathbf{u}, C)$ *and* $p' = (\mathbf{u}', C')$ *are any two feasible proposals, then* p *is* ***viable*** *against* p' *if*
>
> 1. $\mathbf{u}$ *is Pareto optimal for C.*
> 2. *It is not the case that everyone in both C and C' prefer* $\mathbf{u}'$ *to* $\mathbf{u}$ *– at least one common member is indifferent between* $\mathbf{u}$ *and* $\mathbf{u}'$ *or strictly prefers* $\mathbf{u}$ *to* $\mathbf{u}'$.

And p is *strictly viable* against p' if all pivotal coalition members strictly prefer $\boldsymbol{u}$ to $\boldsymbol{u}'$.

The notion of domination used to define the core and the V-set focuses on all the players in a coalition that might enforce a particular outcome. The notion of viability, on the other hand, focuses only on the players that are pivotal between alternative proposals. With this difference, we then say that the set of feasible proposals K is a competitive solution if

1. *No coalition is associated with more than one proposal in K.*
2. *K is* **internally stable** *in the sense that all proposals in it are viable against each other.*
3. *K is* **externally stable** *in the sense that if some p'' not in K is strictly viable against some p in K, then there is a p' in K that is strictly viable against p''.*

Clearly the definition of K parallels the definition of the V-set—like the V-set, it makes use of notions of internal and external stability and it focuses on the properties of a set of outcomes rather than on the properties of outcomes taken one at a time. However, note the important differences.

First, because it is based on the idea of coalitions competing against each other, because a coalition's strategy is its proposal, and because "players" can choose but one strategy in a game, K, like the main simple V-set, allows coalitions to have only one proposal.

Second, as we have already noted, K uses the notion of viability rather than that of domination.

Third, rather than supposing that all feasible proposals not in V are dominated by at least one proposal in V, we require that some member of K be strictly viable against only those feasible proposals not in K that might upset K—that are strictly viable against something in K.

Despite these differences, there are important relationships between the V-set and K, as well as between the core, the bargaining set, and K. Specifically,

An outcome in the core (associated with any winning coalition) is a competitive solution. The outcomes in a main-simple V-set (associated with the appropriate minimum winning coalition) also constitute a competitive solution. Finally, the elements of the strong bargaining set are a competitive solution.

In this way we can think of K as a generalization of the core, the V-set, and the strong bargaining set (at least for majority rule games), so consider this example:

The "Corrupt Bargain" of 1824: In the U.S. Presidential election of 1824 no candidate secured a majority of the Electoral College vote, whereupon John Quincy Adams was elected president over Andrew Jackson by the House of Representatives even though, following the election, Jackson controlled more electoral votes and had secured a majority of popular votes in more states than had Adams. The actual Electoral College vote tabulation following the general election was

Jackson:	*99 votes and a majority in 11 states*
Adams:	*84 votes and a majority in 7 states*
Crawford:	*41 votes and a majority in 3 states*
Clay:	*37 votes and a majority in 3 states*

In the House each state has one vote, but because the Constitution dictated that only three candidates could be considered there, Clay was eliminated. The issue for Clay, then, was: Whom should he support? An intense series of negotiations preceded the final vote, with some of Jackson's support shifting to the other three candidates, and with Clay's support in Missouri shifting to Adams, at which time, with the remainder of Clay's support still uncertain, the candidates' support became

Adams:	*10 states*
Jackson:	*7 states*
Crawford:	*4 states*
Clay:	*3 states*

Suppose for purposes of a numerical example that if C is winning, then $v(C)$ equals the size of the opposition. That is, taking considerable liberties with the meaning of numbers, suppose size measures resources, which a winning coalition can expropriate from the losers. For example, then, $v(\text{Adams, Crawford}) = 10$. Assume that a coalition's value can be divided in any way among its members, and let J, A, Cr, and Cl denote Jackson, Adams, Crawford, and Clay, respectively. There are three facts now that we can use to find this game's competitive solution:

1. The game has a main-simple V-set, because it has only winning and losing coalitions and because utility is transferable among the players.
2. The outcomes in this set correspond to a competitive solution, and each minimal winning coalition is represented by a proposal in K.
3. If K is a competitive solution, then each proposal in it must be Pareto optimal for the respective winning coalition—otherwise that coalition has a proposal that is strictly viable against its own proposal in K and viable against all other proposals.

Consider the set of proposals $K = \{(\mathbf{w}, \{A, J\}), (\mathbf{x}, \{A, Cr\}), (\mathbf{y}, \{A, Cl\}), (\mathbf{z}, \{J, Cr, Cl\})\}$ made by the four minimum winning coalitions. Because the elements of K must be Pareto optimal for the corresponding winning coalition, we must have,

$$w_A + w_J = 7,$$
$$x_A + x_{Cr} = 10,$$
$$y_A + y_{Cl} = 11,$$
$$z_J + z_{Cr} + z_{Cl} = 10.$$

To ensure that those who pivot between the coalitions with proposals in K are indifferent—that no proposal in K be strictly viable against any other proposal in K—we must have

$$w_A = x_A = y_A,$$
$$w_J = z_J,$$
$$x_{Cr} = z_{Cr},$$
$$y_{Cl} = z_{Cl}.$$

These equalities, taken together, solve to establish that Adams receives 6 whenever included in the winning coalition, Jackson 1, Crawford 4, and Clay 5. Thus,

$$K = \{((6, 1, -4, -3), \{A, J\}),$$
$$((6, -7, 4, -3), \{A, Cr\}),$$
$$((6, -7, -4, 5), \{A, Cl\}),$$
$$((-10, 1, 4, 5), \{J, Cr, Cl\})$$

is a competitive solution. Hence, Clay could join either with Adams alone or with Jackson and Crawford.

8.7 Parliamentary Coalitions

It would be foolhardy to claim that the above example and its analysis explains why Clay and Adams coalesced or whether their bargain was indeed corrupt, as was claimed by Jackson's supporters when Adams subsequently nominated Clay as his Secretary of State and heir apparent—an implicit promise that Adams was ultimately unable to deliver on owing to Jackson's victory in the next election. The liberties taken with the meaning of numbers are far too great to lend confidence to any substantive interpretation of the example, which merely illustrates the calculation of a competitive solution using what we know about main simple V-sets when utility is transferable. But questioning the validity of any substantive interpretation here also causes us to examine more closely how we should model legislative and parliamentary coalition processes.

It is of course the case that one of the most widely studied coalitional processes in politics is that of parliamentary coalitions wherein parties maneuver to form governments and the "commodities" most often subject to negotiation being policy and the allocation of ministerial positions. That both policy and ministries are negotiated simultaneously, though, presents some modest conceptual difficulties. Specifically, we should ask: Are ministerial positions valued as badges of prestige and as sources of employment or are they valued because their control also implies control of dimensions of public policy? If the primary motivation of parties is to secure prestige and jobs, then parliamentary maneuvering ought to be conceptualized as a purely redistributive game in which the parties vie for some (imperfectly) divisible commodity—the total of ministries. On the other hand, if control of policy is the primary motivation, then we must conceptualize the preferences of political parties differently.

To take the most interesting possibility, assume that policy is the principal motivation, and to simplify matters, suppose we view a political party in a parliamentary setting as a unitary decision-making entity with a complete and transitive utility function defined over a policy space. Thus, we suppose that the set of all alternative governmental policies can be represented by a simple coordinate system with each dimension of the coordinate system corresponding to a specific issue, that each party has an ideal point (most preferred policy) in that policy space, and that the party's utility declines as we move away from that ideal point in any direction. We know, of course, from Arrow's Impossibility Theorem that this simplification cannot be sustained generally, and, in particular, that it glosses over the policy conflicts within parties that often manifest themselves as competitions for party leadership. Nevertheless, this simplification does allow a "first-pass" at the issue of parliamentary coalition processes.

Example: Consider once again the three policies A, B, and C in Figure 8.5, except in this instance we will associate those ideal points specifically with parties 1, 2, and 3. Recall that the outcomes A, B, and C, which together constitute a main-simple *V*-set, are selected so that party 1 is indifferent between A and B, party 2 is indifferent between A and C, and party 3 is indifferent between B and C. What we want to establish now is that {A, B, C} does in fact correspond to a competitive solution; more precisely,

$$K = \{(A, \{1, 2\}), (B, \{1, 3\}, (C, \{2, 3\})\}.$$

First, no coalition is associated with more than one outcome. Second, having associated the outcome A with {1, 2} and so on, party 1 is pivotal between the two coalitions {1, 2} and {1, 3}—and, as required by *K's* definition, 1 is indifferent between A and B. Similarly, the other pivotal players are indifferent in the required way. Thus, no proposal in *K* is strictly viable against any other proposal in *K*. Finally, to see whether *K* can be upset with a new proposal, consider the possibility that a coalition already represented by a proposal in *K*, say {1, 2}, tries to upset *K* with a new proposal. Referring to Figure 8.5, if {1, 2} proposes A', then although (A', {1, 2}) is viable against (A, {1, 2}) (because the pivots, 1 and 2, do not strictly prefer one outcome to the other) and although (A', {1, 2}) is strictly viable against (B, {1, 3}) (the pivot, 1, strictly prefers A' to B), the proposal (A', {1, 2}) is not strictly viable against (C, {2, 3})—the pivot, party 2, strictly prefers C to A'. Similar reasoning establishes that no larger coalition—in particular, {1, 2, 3}—can upset *K*, so *K* is a competitive solution.

Before we consider parliamentary games with more than three parties, it is useful to pause at this moment to consider the fact that, aside from assuming that no party controls a majority of seats (a majority of the votes in Parliament), our discussion of this example makes no reference to the actual number of seats allocated to each party. Thus, subject only to the constraint that no party is a majority, the identification of {A, B, C} as the predicted set of outcomes is invariant with the actual distribution of seats. If there are 99 seats in the Parliament, then we predict the same set of outcomes regardless of whether those seats are divided (33, 33, 33) or (49, 49, 1)—even if one party's share of the seats is considerably smaller than the other two parties, it is as essential to the formation of a winning coalition as are the two large parties. The suggestion here, then, is that the electoral success of a party should not be measured simply by the number of seats it controls, but rather by whether it is a potentially critical member of a winning coalition.

In addition, we can use the competitive solution and the spatial representation of preferences to demonstrate the importance of the relative positioning of parties in the issue space.

Example: First, consider Figure 8.6, which shows the ideal points of four parties, and suppose that seats are allocated so that only {1, 2}, {1, 3}, {1, 4}, and {2, 3, 4} are minimal winning coalitions. Next, consider the set

$$K = \{(A, \{1, 2\}), (B, \{1, 3\}), (C, \{1, 4\}), (D, \{2, 3, 4\})\}.$$

To see that K is a competitive solution, notice that no coalition has more than one proposal in K. Second, A, B, C, and D are positioned so that the pivots between any two coalitions in K are indifferent between the outcomes associated with those coalitions. Next, suppose {1, 2} tries to upset K with (A',{1, 2}). However, if {1,2} selects an A' that makes (A',{1,2}) viable against, say, (B,{1,3}), that proposal cannot be viable against (D,{2, 3, 4})—if we make any move from A that does not injure party 1, we necessarily make party 2 worse off. Thus, no coalition in K can upset K. And once again, a similar argument establishes that no other coalition can upset K.

Example: Now consider the ideal point configuration in Figure 8.7, which differs from the one shown in Figure 8.6 only in the location of

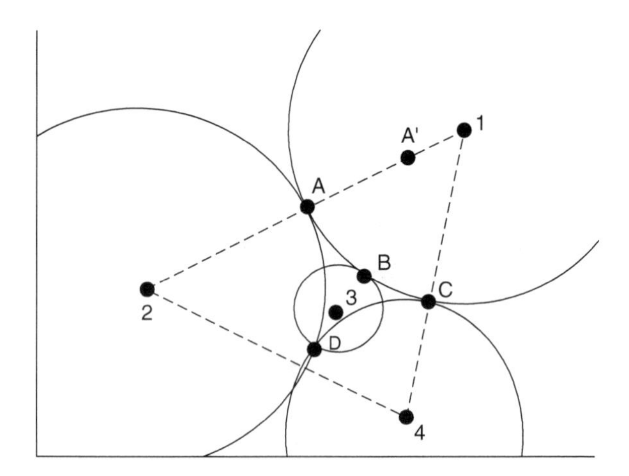

Figure 8.6 A competitive solution

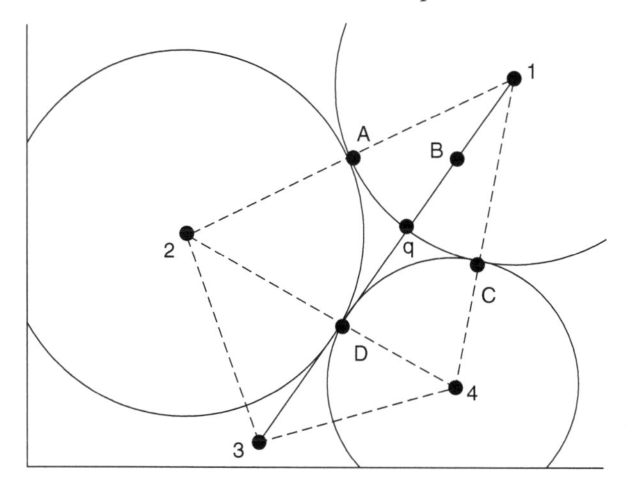

Figure 8.7 A competitive solution

party 3's ideal. In this instance, we can establish that the coalition $\{1, 3\}$ no longer has a proposal in K. Specifically, let

$$K = \{(\text{A}, \{1, 2\}), (\text{C}, \{1, 4\}), (\text{D}, \{2, 3, 4\})\}$$

and suppose {1,3} tries to enter with (B,{1,3}). Notice that to be viable against (A,{1,2})—and Pareto optimal for {1,3}—B must be located on the line between the point q and party 1's ideal, because 1 pivots between {1,3} and {1,2}. However, if we satisfy this constraint, then party 3 strictly prefers D to {1,3}'s proposal, and since 3 pivots between {1,3} and {2,3,4}, no such proposal can be viable against D. Thus, {1,3} cannot upset K. (Since it is not the case that all minimal winning coalitions have proposals in K, one implication of this example is that once we turn to a game in which utility does not act like money and is not perfectly transferable among the players, then that game need not have a main-simple V-set.)

The preceding examples reveal that when predicting parliamentary coalitions we should be concerned not only with what coalitions are winning and losing but also with the parties' policy preferences. This analysis also reveals that the overall policy associated with a parliamentary coalition is not some simple weighted average of the policies preferred by the member parties. Rather, it is the product of bargaining that takes into account the alternatives available to all coalition members. Closer inspection of our example reveals, moreover, that the coalitions with proposals in K are generally those that consist of "contiguous" parties—roughly speaking, parties with diametrically opposite preferences will not coalesce if there are winning coalitions that require less compromise. Finally, the coalitions with proposals in K generally will not contain inessential members—only coalitions just large enough to win will form, and if other parties are allowed to join in to ratify a government coalition, they may be allowed to do so but they cannot affect policy.

8.8 Problems and Some Incomplete Ideas

It might seem from this brief survey that we are well on our way to treating at least a sizeable part of those coalitions that most concern those who study politics—coalitions based on some form of majority rule. Unfortunately, that is not the case, since there is one issue that we have not yet confronted—the existence of these various solution ideas. We know, of course, that the core is often empty and although various existence theorems can be established when utility is freely transferable among the players, for the more general context of bargaining without transferable utility, general existence of these ideas eludes us. Indeed, there are a sufficient number of counter examples to existence in varied contexts to convince us that none of the above-discussed ideas is wholly satisfactory. But we can speculate as to what a more satisfactory approach might look like. We can begin that speculation by imagining a bargaining scenario in which the players each begin with a level of utility in mind, l_i, they believe they will strive for (or, alternatively, settle for). As negotiations proceed, however, some players realize that they can raise their expectations without damaging their prospects while others must lower theirs if they are to have a chance of being included in some winning coalition and realizing a "minimally satisfying" outcome. Now suppose we can formalize this idea and look for a Nash equilibrium of levels—a vector of levels $\boldsymbol{l}^* = (l^*_1, l^*_2, \ldots, l^*_n)$ such that no person has an incentive to raise or lower their level.

This perspective, then, has much in common with Herbert Simon's early idea of *satisficing* and the argument that people do not maximize utility but rather, given the constraints of time and search costs, they satisfice. In this case, rather than assume that the level at which people are willing to be satisfied is exogenously given, we are speculating that such a level is endogenous to the bargaining process. There is, though, a huge chasm between verbal speculation and the precise formalization of an idea. As a start we can introduce the notion of the *Defensible set* which has some of the flavor of the bargaining set. Specifically, let $p = (o, C)$ be an objection against player i at l_i and the proposal $p' = (o',$ $C')$ if $u_i(o) < l_i$ and p is strictly viable against p'. $p'' = (o'', C'')$ in turn is a counter objection by i at l_i to p if $u_i(o'') \geq l_i$ and p'' is strictly viable against p. The idea here, then, is to not have players defend the specific utility they associate with a proposal, but rather to defend only up to some level of utility they are satisfied with. If a person's coalition partners choose an outcome that gives player i more than he or she is asking for, player i need not defend that level utility. To see what we mean here, consider Figure 8.8 where the points $p1$ through $p5$ are the points corresponding to the Competitive Solution. In this instance, we can assume that 2's satisficing level is the indifference curve on which p_1 and p_3 lie. However, suppose players 1 and 3 attempt to coalesce with 2 with p_2. With

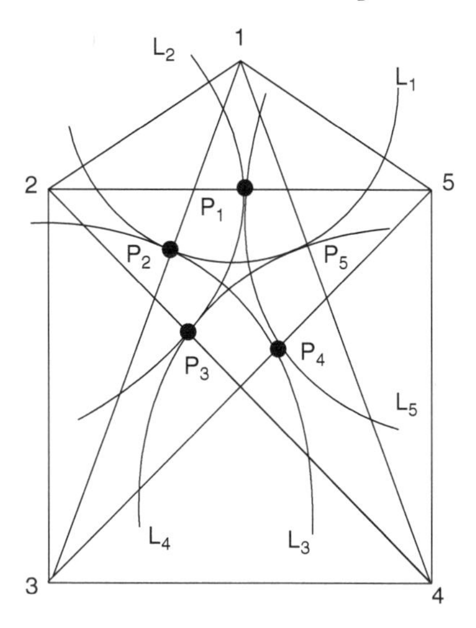

Figure 8.8 Levels for a defensible set

the bargaining set as originally defined, player must defend the level of utility he associated with p_2. With transferable utility this may be possible, but here utility isn't transferable. All that a coalition can adjust is policy (the spatial position of a proposal). So in a sense, if the coalition {1, 2, 3} forms, player 2 enjoys a positive externality, but in defending that externality, he is required only to defend up to the level defined by proposals p_1 and p_3.

To define a solution, now, with this idea, if we restrict the consideration of coalitions to those that are winning, then let

$$S(\boldsymbol{l}) = \{p = (o, C) \text{ such that for all } i \text{ in } C, u_i(o) \geq l_i\}.$$

Thus, $S(\boldsymbol{l})$ consists of all proposals that might be formed that satisfy the levels of the individual bargainers. And the support set for an individual player, $S_i(\boldsymbol{l})$ is simply the set of proposals for which i is a member of the corresponding coalitions. The Defensible Set, $D(\boldsymbol{l})$, in turn, consists of a set of supported proposals at levels $\boldsymbol{l}$ such that each player can defend the proposals in $S(\boldsymbol{l})$ that he supports—that are in $S_i(\boldsymbol{l})$—against all objections and no one can defend any supported set at any higher level, given everyone else's level.

For an explicitly game theoretic formulation of these ideas, let

$$f_i(\boldsymbol{l}) = \max\{l_i \text{ such that } (l_1, l_2, \ldots, l_i, \ldots l_n) \text{ is defensible by } i\}.$$

Thus, $f_i(\boldsymbol{l})$ is the highest level of utility that i can defend, holding constant the levels of all other bargainers. Next, define the payoff

$g_i(\mathbf{l}) = l_i$ if $f_i(\mathbf{l}) \geq l_i$ and 0 otherwise,

assuming that all possible payoffs have been scaled to be non-negative. This expression, then, can be used to define an *n*-person non-cooperative game where the player's strategies are the levels they propose to defend. We have, then, come full circle in a sense of taking a cooperative bargaining game that is most likely impossible to describe in extensive or strategic form and converted it to a non-cooperative game wherein we search for Nash equilibria of satisficing levels. The predicted outcomes of that game are, then, the proposals supported by those levels—the Defensible set. The two questions that remain with respect to $D(\mathbf{l})$, though, are the usual ones—uniqueness and existence. That $D(\mathbf{l})$ need not be unique is established by the ordinal preferences in Table 8.3. Specifically, some complex calculations (aided by a computer search program) that need not concern readers here, we find two "solutions"

{(E,{2,4,5}), (A,{3,4,5}), (O,{1,2,4}), (H,{1,3,5}), (B,{1,2,3,5}), (D,{1,2,4})}
and {(E,{2,4,5}), (A,{3,4,5}), (O,{1,2,4}), (H,{1,3,5}), (F,{1,2,3})}

In the first solution, the players set their levels at B, G (≈ B), A, A and H respectively. In the second, those levels are O, E, F, A and H. Thus, in the first solution, player 3, by committing to a relatively high level, "traps" 1 and 2 into setting their utility values at B and D respectively as the highest levels they can defend. In the second solution, if 3's level is lowered from A to F, players 1 and 2 can raise their levels to establish a different equilibrium. We should not be surprised at this possibility, and its implication; namely, that there most likely exists a third solution in mixed strategies. Nor should we be surprised that some

Table 8.3 Non-Unique Defensible Set

Player 1	*Player 2*	*Player 3*	*Player 4*	*Player 5*
N	J	B	L	B
J	O	H	E	A
F	M	A	D	E
I	F	F	O	H
K	E	I	A	G
H	I	L,N	G	K
O	K,D	P	M	Q
G	G,B	Q	I	D
P,Q	H,C,P	K,M,D	P	M
D,M	L,N,Q	J	K,B	O,C
B	A	E,C	F,C	P
A,C		O	N,J,Q	L,N,J
L			H	F,I
E				

Table 8.4 Empty Defensible Set in Pure Strategies

Player 1	*Player 2*	*Player 3*	*Player 4*	*Player 5*
B	D	H	E	G
H	C	I	C	I
J	A	J	B	B
D	J	C	H	E
A	I	B	J	H
F	F	A	G	D
G	E	E	A	A
C	B	G	F	C
I	H	D	D	J
E	G	F	I	F

games have only mixed strategy solutions. Specifically, the preference orders in Table 8.4 do not give rise to a Defensible set in pure strategies—the players' levels cycle endlessly. However, since the number of players and possible levels is finite, it must have at least one mixed solution.

This brief review of the Defensible set does not, however, solve all problems. First, if preferences are spatial so that each player's strategy space is continuous, the existence of the Defensible set in either pure or mixed strategies is simply unexplored. Moreover, if the Defensible Set exists but the closely related Competitive Solution does not, we do not know the Defensible Set's properties. So clearly, this approach to coalitions and bargaining remains an area for future research. Nevertheless, the ideas behind the Defensible Set—notably, the notion of endogenous satisficing—point the way toward a melding of cooperative and non-cooperative perspectives without recourse to the ad hoc construction of extensive forms when treating various forms of bargaining.

8.9 The Balance of Power Versus Collective Security

Thus far our discussion of cooperative games sweeps aside the issue of enforcement—the *V*-set, Bargaining Set and Competitive Solution all assume that whatever agreements are reached can be enforced. There is, however, at least one circumstance in politics in which the task of understanding how the absence of an exogenous enforcement mechanism affects outcomes is **the** central research issue. Specifically, consider the view of international affairs that argues that stability in international politics, if it arises at all, arises because of a "balance of power." Without delving into the many issues surrounding this argument (such as the definition of "power" and the confusing array of definitions of "balance of power" in the literature), this view sees international affairs as basically anarchic in the sense that whatever cooperation we observe arises and is enforced endogenously by the self-interest of relevant decision. States have interests and they pursue those interests by applying whatever resources they possess toward those ends, including if necessary the threat and use of military force. Institutions and organizations such as the United Nations, then, are deemed either largely irrelevant to the establishment and maintenance of stability because they are themselves endogenous phenomena or merely an instrument whereby states can seek legitimacy for their actions. From this view, then, arguing that international institutions stabilize international politics merely pushes our explanation back a step to where we must explain why countries establish and abide by the edicts of such institutions.

The difficulty with the argument that a balance of power enforces stability is that it does not establish the circumstances under which various alliances but not others—specific subsets of countries—are enforceable in the sense that abiding by an agreement is an equilibrium strategy for the states in question. Why, for instance, prior to the outbreak of WWI did Britain stick by its commitment to France and why did Russia feel compelled to defend Serbia against Austria? It cannot be merely because they made agreements to do so—after all, Germany and Russia signed the Ribbentrop-Molotov pact a few decades later and we know how that fared subsequently. Of course, we have already seen how cooperation can emerge in repeated Prisoners' Dilemmas, but balance of power politics differs from such games in important ways. Cooperation in the repeated Prisoners' Dilemma arises in a game in which all players gain when cooperating as compared to when they are not cooperating. That is, cooperation moves outcomes from Pareto inefficient to Pareto efficient ones. However, power is presumably a relative variable and in constant supply, so unlike a Prisoners' Dilemma, balance-of-power politics is, in its base form, necessarily constant or zero sum. Hence, with games like "divide-the-dollar" in mind, the question that arises in arguments over the viability of a balance of power system and international stability is the following: If power politics is like a majority rule game in which coalitions with a majority of power can defeat those with less, then why doesn't a majority coalition of countries form to eliminate the rest, with this process continuing until there are only two equally powerful countries in the system?

There exists in the literature, though, an alternative view that stability in international affairs can be secured instead via *collective security* agreements wherein countries agree to refrain from hostilities and the remaining members of the agreement punish whoever might choose to defect. Institutions, in this view, play a profound role in that it is through them that agreements are reached and the punishment of defectors applied. This was, after all, the original intent of the United Nations when it was first envisioned in the heat of WWII. This was also the glue that held the Warsaw Pact together—a fact fully on display when it was enforced in 1956 and 1968 by Soviet troops in Hungary and then Czechoslovakia. The problem for proponents of this view, however, is that absent a dominant player such as the USSR in the Warsaw Pact who can punish any and all defectors, it leaves unstated the mechanisms whereby threats or promises of punishment are credible—where potential defectors believe it will in fact be in the interests of others to punish them in the event of a defection.

We have, of course, necessarily greatly simplified the positions of those whose view of international affairs corresponds to either the balance of power paradigm or that of collective security. But what interests us here is that the proponents of these two paradigms often argue as if they were treating different universes—that if one argument is correct then the other must be incorrect. A game theorist is well aware of the fact, however, that most games of interest have multiple equilibria. Thus the question arises as to whether the "game of international politics" has two alternative types of equilibria, and whether the academic debate here is which type is the easier one to achieve and maintain. To examine this possibility, we follow the model of balance of power politics offered by Niou and Ordeshook ("Stability in Anarchic International Systems," *American Political Science Review*, December 1990) that differs from a divide-the-dollar game in that even if we play a "divide-the-dollar game" several times in succession, we can define the game so that the players' voting weights remain constant, with each player controlling $1/n$th of the vote. In a balance of power scenario, on the other hand, a "voting weight" is presumably endogenous and equals that country's proportion of resources ("power"). That is, countries can expropriate resources from others in order to augment their ability to threaten and overcome adversaries. In a balance-of-power game, then, a country must be concerned about the possibility that if it coalesces with someone to eliminate others, it will become the next victim of its coalition partner.

To see how such a concern changes our view of coalitional possibilities, consider a system with three countries, $i = 1, 2$, and 3, and suppose country i controls r^o_i resources, which it values and which can be used to defend against or to overcome an adversary (for convenience, let $r^o_1 \geq r^o_2 \geq r^o_3$). Thus, if $r_j > r_i$, then in principle at least j can defeat i and expropriate all of i's resources. Next, assume that no country controls more than half the resources in the system, so no country can overwhelm the other two. Now consider the following extensive form game, which we denote by Γ:

1. Nature randomly picks one of the countries—say, i.
2. Country i chooses whether to make a threat, $\boldsymbol{r} = (r_1, r_2, r_3)$, or to pass; if it passes we return to step 1.
3. If i threatens $\boldsymbol{r}$ and if $r_j \geq r^o_j$, j is i's coalition partner and must approve or reject i's proposal. If j rejects, we return to step 1; if j accepts, then $\boldsymbol{r}$ becomes the **current threat**.
4. Let k be the country threatened by $\boldsymbol{r}$ (which requires that $r_k < r^o_k$). Country k has three responses: do nothing, in which case $\boldsymbol{r}$ becomes the status quo and we return to step 1; propose an alternative threat $\boldsymbol{r}'$; or transfer resources to i or j in the attempt to "buy them out."
5. If k chooses the second option, its coalition partner (i or j) must approve or reject k's proposal. If this partner rejects, $\boldsymbol{r}$ becomes the new status quo and we return to step 1, whereas if it accepts, $\boldsymbol{r}'$ becomes the new current threat and we return to step 4 with k replaced by the newly threatened player. If k chooses the third option of transferring resources, then if the country who

is a party to the transfer accepts, the transfer is consummated to form a new status quo, and we return to step 1, but if it is rejected, then the current threat ***r*** becomes the new status quo and we return to step 1.

We also impose two assumptions that help define the game's strategic character. First, we suppose that if two countries "war," then a third country that is larger than each of them can take advantage of the situation to realize some unspecified gains. This assumption has the effect of "freezing" systems in which one country controls half the resources. For example, if (150, 100, 50) is the status quo, then 2 will not attack 3, because 1 can use the conflict to become predominant—in which case it will eliminate both 2 and 3. Similarly, if 1 attacks 2, then 3 must come to 2's aid lest it become 1's next victim (which illustrates some of the "glue" that sustains alliances in international politics). Since 1 knows this beforehand, it will not attack 2.

Our second assumption accommodates in a modest way the costs that might reasonably be associated with conflict. Specifically, if given a choice between securing X units of resources by implementing a threat versus securing those resources via a transfer, a country prefers the transfer.

The analytic problem that confronts us now is that, in theory at least, our game can continue forever—1 and 2 could threaten 3, 3 could counter by proposing that it and 2 threaten 1, 1 can counter by proposing that 1 and 3 threaten 2, and so on. This fact creates two problems. First, if we once again define a strategy as a "plan of action" for the entire game, there are infinitely many possible strategies. Thus, we may have to appeal to various intuitive criteria in order to narrow the list of possibilities we consider. Here we will initially consider only stationary strategies whereby countries merely try to form the best available coalition, without regard to the threats or counters others have previously participated in—countries will form coalitions with whomever they can, without regard to what country might have defected from some coalition or agreement in the past. One might say, then, that the strategies considered here correspond to General de Gaulle's dictum that "France has no friends, only interests." The second problem with which we must contend is that it is inappropriate to suppose, as we did in the analysis of infinitely repeated Prisoners' Dilemmas, that future payoffs are discounted or that even payoffs ought to be discounted at all after one threat is replaced by another: the reasoning that our game models may merely occur in the heads of participants as they contemplate strategic possibilities. Because the status quo never changes along those branches of the extensive form in which threats and counterthreats follow each other in an infinite sequence, there need not be any stream of benefits or losses to discount.

However, we can solve this game by pretending that it is finite—by supposing that we know the consequences of all branches in the game's extensive form. After postulating these consequences, a subgame-perfect equilibrium is characterized by strategies in which no one has an incentive to defect unilaterally to any choice not dictated by that player's strategy, and the postulated consequences are consistent in that they are "self-fulfilling prophesies"—the subgame perfect choices they imply must yield those consequences.

To formalize this idea, suppose, in accordance with the assumption that countries consider only stationary strategies, that all countries respond to a threat $\boldsymbol{r}$ without regard to the histories associated with each player. Next, let Γ_r denote the subgame that follows the threat and acceptance of $\boldsymbol{r}$. If each country i associates the value $v_i(\Gamma_r)$ with playing that subgame, then $\boldsymbol{v}(\Gamma_r) = (v_1(\Gamma_r), v_2(\Gamma_r), v_3(\Gamma_r))$—the continuation value of Γ_r summarizes what countries 1, 2, and 3 believe follows from $\boldsymbol{r}$ becoming a standing threat. Thus, $v_i(\Gamma_r)$, when compared against whatever follows if $\boldsymbol{r}$ is rejected, determines *i*'s preference for acceptance or rejection of $\boldsymbol{r}$ or for making this threat in the first place. Once values for all threats are specified we can assume that the acceptance of a threat or counter is a terminal node with its continuation value as the "final outcome." We then analyze Γ like a finite extensive form game of complete information, deducing the subgame perfect equilibrium strategies by working backward from the "terminal nodes" in the same way we treat finite agendas in majority voting games—we deduce what each country ought to do any time it must choose a threat, a counter, or accepting or rejecting a threat or counter. Hence, *an equilibrium is a set of continuation values—one for each threat—and a set of strategies for each country such that these values and strategies are consistent.* That is, in equilibrium, the choices implied by the continuation values—the strategies that are a subgame perfect equilibrium given the continuation values—must, in turn, imply those continuation values.

Example: Suppose $(S, \boldsymbol{r}^o) = (\{1, 2, 3\}, (120, 100, 80))$, and let country 1 propose the threat $(\{1, 3\}, (150, 0, 150))$. Limiting our discussion to threats in this form (i.e., two players threaten a third with elimination) and to transfers, consider the representation of the situation in Figure 8.9, where • denotes a terminal node. After 1 proposes its threat, its proposed coalition partner 3 must decide whether to accept or reject. If 3 accepts, then $(\{1, 3\}, (150, 0, 150))$ becomes a standing threat. At this point, given the limits we have temporarily placed on the threatened player's actions, if 2 hopes to survive it must then offer a counter that is either a coalition with 1 or with 3 to divide 300, or a transfer of thirty units to 1 (which, if offered, 1 accepts since this is the best possible payoff for 1 given that 2 and 3 will not allow an outcome that gives 1 more than 150). Country 2 need not consider a transfer to 3 since, being larger, 1 can be made to prefer the transfer over the original threat with the expenditure of fewer resources. Depending on which counter 2 chooses, either 1 or 3 must decide whether to accept or reject the offer. In the event of a rejection, the threat is implemented and 2 is eliminated. In the event of an acceptance, the counter becomes the new current threat, and the subgame that follows is denoted by Γ with an appropriate subscript. Figure 8.9 portrays the next step in this process with either 1 or 3 offering a counter.

Without concerning ourselves for the moment with how they arrive at such numbers, suppose the countries in this example assume that the following continuation values hold for the three threats that eliminate a country (we need not consider threats that give someone more than 150 since no one will allow it to go unchallenged):

$$\mathbf{v}(\Gamma_{(150,0,150)}) = (150, 70, 80),$$
$$\mathbf{v}(\Gamma_{(0,150,150)}) = (70, 150, 80),$$
$$\mathbf{v}(\Gamma_{(150,150,0)}) = \mathbf{v}(\Gamma_{(150,0,150)})/2 + \mathbf{v}(\Gamma_{(0,150,150)})/2 = (110, 110, 80).$$

The value for $\mathbf{v}(\Gamma_{(150,150,0)})$ is based on the assumption that country 3 counters with (150, 0, 150) and (0, 150, 150) with equal probability whenever it is indifferent between these two choices (or that 3 accepts each counter with equal probability if it is indifferent). Once we have assigned values to each subgame we can then deduce the choices they and subgame perfection imply for the extensive form in Figure 8.9. Referring to this figure, then, after 1 proposes a threat and 3 accepts, country 2 has three choices:

1. Counter with the threat (0, 150, 150).
2. Counter with the threat (150, 150, 0).
3. Transfer resources to 1 so as to make 1 near-predominant and to freeze the system.

Consider the top-right of Figure 8.9, which assumes that 2 counters with (0, 150, 150), and suppose for the moment that 3 accepts 2's offer. At this point, with (0, 150, 150) the current threat, country 1 must act, and it has three choices:

1. Counter with the threat (150, 0, 150).
2. Counter with the threat (150, 150, 0).
3. Transfer to 2 so as to make 2 near-predominant and freeze the system.

If 1 proposes (150, 0, 150), 3 rejects 1's offer because it prefers (0, 150, 150) to the continuation value of $\Gamma_{(150,0,150)}$. At this point, given our assumption about continuation values, 3 is choosing between a payoff of 80 ($v_3(\Gamma_{(150,0,150)} = 80$) and a payoff of 150. We indicate this preference by an arrow. Similarly, if 1 approaches 2 by proposing (150, 150, 0), then 2 rejects since $v_2(\Gamma_{(150,150,0)})$ corresponds to a lottery between 150 and 70. Thus, 1 prefers to transfer resources to 2, because each of the other alternatives available to 1 leads to 1's elimination. (We need not consider any other type of transfer: Transfers giving less than 150 are rejected because rejection implements the threat, and a transfer that renders 3 near-predominant is more costly than a transfer to 2.)

Notice, now, that the node corresponding to the point at which 1 must respond to 3's acceptance of 2's counter begins a subgame, $\Gamma_{(0,150,150)}$, with which we initially associated the continuation value $\mathbf{v}(\Gamma_{(0,150,150)}) = (70, 150, 80)$. Since we have shown that 1 transfers to 2, our initial supposition is sustained—to this point at least, subgame perfection and our initial conjectures about continuation values are consistent. Also, country 3 prefers not to play this subgame and instead rejects 2's offer of (0, 150, 150) in favor of implementing the initial threat, because it gives 3 a payoff of 150.

Next consider the lower-right part of Figure 8.9. In this instance, 2 counters the initial threat with (150, 150, 0). Suppose for the moment that 1 accepts. This acceptance marks the beginning a subgame, $\Gamma_{(150,\,150,\,0)}$, in which it is 3's turn to respond. Notice, however, that given the continuation values, country 3 need not consider offering to transfer any portion of its resources since either counterthreat has associated with it a value of eighty for 3. However, this fact implies that 3 is indifferent between countering with (0, 150, 150) or with (150, 0, 150), so let 3 choose between these counterthreats with equal probability. Once again, then, we deduce a value $\boldsymbol{v}(\Gamma(150, 150, 0))$ that is consistent with our initial conjecture. But just as 3 rejects 2's offer of (0, 150, 150), 1 rejects 2's offer of (150, 150, 0), because 1 prefers the certainty of getting 150 units by implementing the original threat to the lottery that $\Gamma_{(150,\,150,\,0)}$ implies.

Thus, both of 2's counterthreats are rejected, leaving 2 with only one choice—transfer enough resources to 1 so as to make one near-predominant. Although we have not drawn in 1's decision as part of Figure 8.9, it is evident that 1 accepts this transfer since it knows that it can never get more than 150 and since it prefers becoming near-predominant by accepting a transfer over implementing a threat. But notice now that the acceptance of the initial threat by 1 or 3 marks the beginning of a subgame, which we have just shown leads

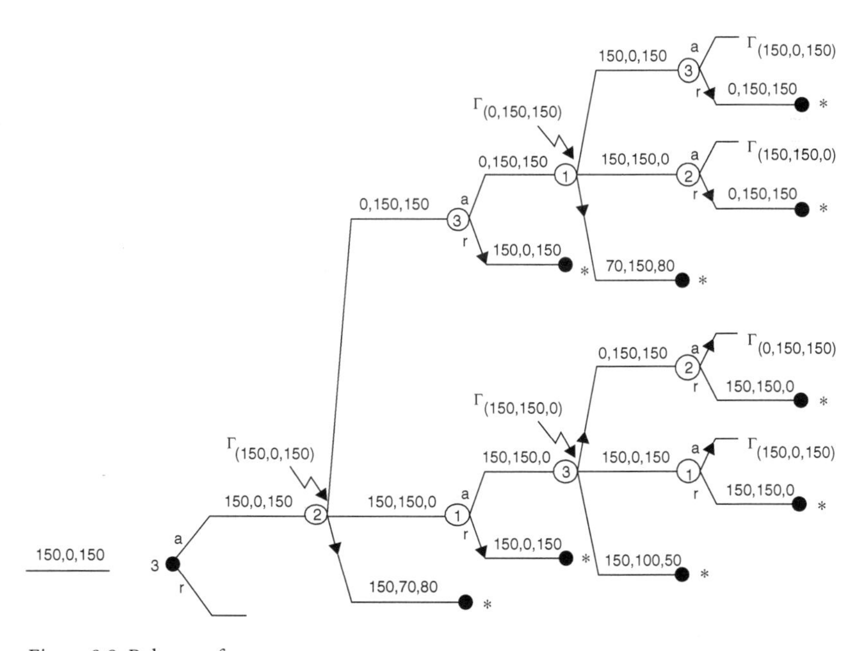

Figure 8.9 Balance of power game

to 2 transferring resources to 1. So as initially conjectured, $\boldsymbol{v}(\Gamma_{(150, 0, 150)}) = (150, 70, 80)$. A similar analysis for initial threats of (150, 150, 0) and (0, 150, 150) confirms the continuation values we posited for them.

What remains in the analysis of this example is a specification of $\boldsymbol{v}(\Gamma)$—the value of the entire game—and an equilibrium identifying the initial choices and responses of countries. We have, after all, merely identified what will happen if certain actions are taken and not what actions will in fact be taken—what *strategies are in equilibrium* in the overall play of the game. And it is here that we return to the debate between proponents of a balance of power view of international affairs versus those who see collective security bolstered by international organizations as the path to stability. Suppose first that $\boldsymbol{v}(\Gamma) = (120, 100, 80)$—in effect, suppose that no initial threat is offered or accepted, and that the status quo prevails. This supposition is sustainable, in fact, if we characterize equilibrium strategies thus: A country does not initiate or agree to a threat unless it gains resources. Since the $\boldsymbol{v}(\Gamma)$'s imply that countries 1 and 2 can each gain from an initial threat, whereas 3 can neither gain nor lose, 3 has no positive incentive to participate in a threat. Thus, the only threat that might be approved is (150, 150, 0). However, 1 and 2 are essentially playing a constant sum game since, as we have already learned, player 3 is never forced to transfer to 1 or 2. So, depending on the probability that 3 chooses one action or another when it is indifferent, neither 1 nor 2 has an incentive to threaten or to approve of (150, 150, 0). Indeed, 3 can counter with a threat that requires the originally threatening country to transfer resources to 3's partner in the counter. Hence, three-country systems can be wholly stable without anyone offering an initial threat.

There is, however, another stationary equilibrium that generates a reallocation of resources characterized by: Accept all initial threats if they promise no loss; otherwise reject. This equilibrium sustains $\boldsymbol{v}(\Gamma) = \alpha(150, 70, 80) + (1 - \alpha)(70, 150, 80) = (70 + 80\alpha, 150 - 80\alpha, 80)$, where α is the probability that 3 coalesces with 1 if it is indifferent between coalescing with 1 and 2. Although country 3 cannot gain resources if it abides by such a strategy (the continuation values for all threats remain as before), 3 has no positive incentive to defect unilaterally to another strategy. However, although 2 transfers to 1 or 1 transfers to 2, no country is eliminated. Thus, our conclusion about this three-country system is that if we see any coalition, it will be either {1, 3} or {2, 3}, with either 1 or 2 becoming near-predominant. And since all three-country systems without a predominant player are equivalent to our example, this fact establishes the possibility of stability in anarchic international systems.

Naturally, we should extend this analysis to larger systems before inferring anything general about balance of power. Rather than consider this generalization, though, which in fact takes us into a morass of algebra, let us consider instead the possibility that there exists a "nicer" equilibrium in which no threats are ever made. Recall that as a partial justification for the Iraq war, President George Bush presented American intervention as a prelude to the formation of a "new world order" in which the confrontational style of the Cold War was to be replaced with a universal agreement among nations to eschew conflict in favor of the mutual benefits of economic cooperation. Of course, this idea met with considerable skepticism, in part because such a scheme had been attempted before in the form of the abortive League of Nations and had clearly failed. One failed experiment, however, does not invalidate a theory, because we can always appeal to the particular circumstances of that experiment to explain the failure (in the case of the League, we have the "circumstance" of the Great Depression). The issue, then, is whether there is any theoretical basis for looking at Bush's proposal with skepticism. To address this issue, we note that this new world order in fact corresponds to the notion of collective security discussed earlier, which is an international system in which all countries initially agree to refrain from making an initial threat and to refrain from accepting one in the event that it is offered. And to enforce such an agreement, countries also agree to punish, whenever possible, those who defect or those who fail to participate in the administration of punishments (regardless of whether the initially threatened states export oil or mushrooms). So the game theoretic question that confronts us is whether states will have an incentive to administer punishments in the event that one or more other states defect from the initial agreement and whether the existence of this incentive is sufficient to preclude any initial threat.

To answer this question, notice first that a collective security equilibrium calls for strategies of a different sort than the type we have thus far considered. Specifically, it calls for an equilibrium of non-stationary strategies in which the target of threats is a function of history (defined in terms of who has defected). However, to make matters precise, assume that if a non-punishing threat is ever made and accepted so that it becomes a standing threat, all countries play stationary strategies thereafter as described previously—that is, the game thereafter reverts to the somewhat nastier balance-of-power mode. Staying now with our three-country example, suppose country 1 defects and, seeking to increase its advantage over 2, proposes (150, 0, 150). If 3 accepts, then, since all countries play as before, we know from out previous analysis that 2 transfers resources to 1 so 3 gains nothing by accepting. Suppose, then, that 3 rejects 1's overture, at which point the question becomes whether punishing 1 is sustainable as an equilibrium. If nature chooses 2 to make the next move and fails to propose to punish 1, then again the game is played as before with 2's continuation value being $\alpha 70 + (1 - \alpha)150$; if instead it proposes to punish 1 by threatening (0, 150, 150), 3 accepts because it has no positive incentive to defect—as the smallest country, we already know that it can never gain resources—in which case 2's continuation value increases to 150. Similarly, if 3 has the first opportunity to punish 1, it is indifferent between defecting and proposing the (0, 150, 150) punishing threat; thus, it has no incentive to defect from the presumed punishment equilibrium. Nor does 2 have any incentive to reject 3's proposal of (0, 150, 150) and to listen to anything that 1 might have to say thereafter, since this guarantees 2 a transfer from 1 and a final payoff of 150. Hence, 3 receives eighty regardless of whether it accepts 1's initial offer or rejects it in favor of administering a punishment, which is to say, as before, that 3 has no incentive to defect from the presumed equilibrium.

At this point, of course, the reader may feel a bit uncomfortable with our assumptions about 3's actions whenever it is indifferent. Indeed, because of such indifference, the presumed collective security equilibrium is weak and vulnerable to the possibility that countries choose "erroneously." However (and we refrain from demonstrating this fact because of the algebra involved), the extension of the analysis to larger systems reveals that in such systems all countries can be targets of "profitable" threats, which thereby gives everyone a positive incentive to avoid the possibility of being the target of a punishment as well as a positive incentive to punish defectors. This model, then, has at least two equilibria that are of opposite types in the sense that one predicts the eventuality of threats and the other predicts that the status quo prevails. Admittedly, this model is far too abstract to allow definitive statements about the viability of foreign policies based on collective security arrangements versus balance-of-power. Among other things, it does not consider uncertainty, the costs of conflict, geography, and the possibility of more complex equilibria (as when, for example, subsets of countries form alliances and agree to play "collective security" among themselves and stationary strategies with respect to everyone else). Nevertheless, this type of analysis does establish the possibility of sustaining cooperation without exogenous enforcement in a world in which "power" is the sole arbiter of disputes while at the same time cautioning that some degree of coordination is required to ensure that players are being coordinated to the same equilibrium.

This model, though abstract and incomplete, offers a final lesson in the value of a game theoretic perspective of politics. In this specific instance, in the absence of such a model the proponents of collective security versus those who see balance of power as the ultimate source of international stability—those who emphasize the importance of international institutions versus those who focus on military capabilities—could, most likely, argue endlessly. Each side of the debate would bring to the table their examples of where collective security or balance of power either worked or failed, while each side laid claim to the state's foreign policy resources (or, in the case of academics, research support). But even if we ignore the specifics of the model just presented, game theory introduces an alternative hypothesis; namely, that collective security and balance of power merely correspond to alternative equilibria, and that the properties of those equilibria will depend on some of the substantive things our model ignores. Any debate over where a state's priorities ought to lie can then be addressed in combination with extensions of this model. Indeed, even if one chooses to forgo such extensions, in lieu of arguing as if one were debating alternative universes, the model focuses that debate on the specific things that might make one equilibrium more attractive than the other. Thus, game theory's contribution to our understanding of politics need not lie in specific abstract models and a blizzard of notation and impenetrable mathematics. Rather, that contribution can lie merely in making us sensitive to the ideas of strategy and equilibrium and to the difference between decision and game theoretic thinking, which of course is where we started this volume.

8.10 Key Ideas and Concepts

coalition
divide-the-dollar game
feasible outcomes
coalitional preference
Schwartz's vote trading theorem
core
dominates
effective
characteristic function
Coase Theorem
internal stability
external stability
V-set
main simple *V*-set
universalism
proposal
viable proposal
competitive solution
minimal winning coalition
satisficing
endogenous satisficing
defensible set
balance of power
collective security

Exercises for Chapter 8

1. Is vote trading in a legislature profitable and most likely to be observed when (a) various minorities find it in their interest to thwart the will of the majority? (b) there is no uniquely stable outcome, and it is possible for every agreement to be upset by something else? (c) congressional rules are sufficiently inflexible that only behind-the-scenes deals will accommodate the majority will? or (d) legislators are constrained by their constituents from voting sophisticatedly?
2. Suppose one of the following four payoff vectors must be chosen by three voters using simple majority rule:

voter 1	voter 2	voter 3
2	−1	−1
−1	−3	3
1	3	−3
−3	1	1

   a. Does this game have a core or a *V*-set?

b. If you assume that utility is transferable, does your answer to part (a) change?

3. Recall our example from section 8.4 of the Valley of the Dump. Suppose, however, that there are only four clans (euphemistically named 1, 2, 3, and 4), each living on its own quarter of the valley, and that every day, the number of bags of garbage produced by each clan equals its name—clan i produces i bags of garbage per day. As before, the legal code of the valley, strictly enforced, requires each clan to produce its allotted bags of garbage, and it allows clan i to dump exactly i bags of garbage in its own backyard or the yards of others. However, the bags must be dealt with as distinct units—no scattering of any bag's contents is permitted. (They have high standards in the valley.) Assume that each clan evaluates the outcome of the garbage problem in the same way as in our original example.

   a. Define and interpret the conditions that must be satisfied if this game is to have a core.
   b. Does a nonempty core exist for this game?
   c. How do your answers change if we suppose that bags of garbage are infinitely divisible?

4. Assume that a five-person committee uses issue-by-issue, majority rule voting, and that its members have the following ideal points on the two-dimensional issue space (let all indifference curves be concentric circles):

   voter 1: (7, 2)
   voter 2: (8, 9)
   voter 3: (3, 4)
   voter 4: (4, 8)
   voter 5: (10, 5)

   a. If communication and coordination are impossible, what is the outcome?
   b. If the committee abandons its restrictive rules in favor of free and open debate, does the game have a core?
   c. More generally, can issue-by-issue voting lead to a different outcome than the core with simple Euclidean preferences? Explain your answer.

5. Consider the following system of representative majority rule among twenty-seven voters: There are three regions, each divided into three districts. Each district has three voters. There are representatives (who are computers, and are not to be considered part of the voting body) at the regional and district levels. District level representatives inherit (over any pair of alternatives) the majority preferences of the voters in their districts, and regional representatives inherit the majority preferences of the representatives in their region. A policy outcome is chosen by a majority vote of the regional representatives.

a. What is the minimum number of voters that can form a decisive coalition (i.e., get their preferred policy positions as the outcome)? Construct an example to support your answer.
b. How does your answer to (a) change if the district level representatives are eliminated?
c. Assuming that all preferences are single-peaked over a single issue, does the final outcome depend on the existence of the district level representatives? Why?

6. Consider a six-person symmetric game where the value of a coalition C, $v(C)$, is given by

$$v(C)=\begin{Bmatrix}0\\1\\2\\3\\3\\4\end{Bmatrix} \text{ if } |C|=\begin{Bmatrix}1\\2\\3\\4\\5\\6\end{Bmatrix}$$

where $|C|$ denotes the number of members of C. Assuming that what a coalition earns is infinitely divisible among the coalition's members, does this game have a core?

7. In economics, a common representation of utility for two commodities is the function $u_i(x_i, y_i) = x_i^{\alpha}y_i^{\beta}$ where both α and β are positive constants. So suppose two persons are bargaining over the allocation of ten units of each of the goods, where both goods are infinitely divisible. Let $\alpha = \beta = 2$ for both persons, and suppose that person 1's initial endowments of the two goods are 2 and 7, respectively (so that 2's initial endowments are 8 and 3, respectively). What is the core to this two-person game?
8. Using the spatial policy positions in problem 4, determine whether the coalition {1, 3, 4} has a stable coalition proposal in the bargaining set. (Hint: Suppose players 1, 3, and 4 tentatively agree to the proposal that is Pareto optimal for them and approximately midway between 1 and 4's ideal. Now let 1 object against 3 with a proposal that is Pareto optimal for {1, 2, 5}, and check whether 3 can counter with any coalition that excludes 1 and with a proposal that gives 3 as much as 3 get from the original proposal with {1, 3, 4}.)
9. Suppose three houses, each with a large front-facing window, are arrayed in a triangle such that a person sitting in any house can see one-fifth of its own garden, all of the garden of the house on the left, and none of the garden belonging to the house on its right. Suppose each homeowner is endowed with one bag of fertilizer and is required by law to use it (suburban values being what they are). A bag of fertilizer, allocated to one garden, improves that garden so that a person having a full view of it enjoys a

benefit of five units of utility, a person having only a one-fifth view enjoys a one unit utility increase, and a person with no view is not benefited at all. Assume that the homeowners can coalesce to allocate their fertilizer, and that people cannot fertilize another's garden without permission. Suppose also that if two bags of fertilizer are allocated to any one garden, then that garden is doubly beautiful. Does the game have a core?

10. Sketch the proof of the assertion that the core of a majority voting game, after being associated with any winning coalition, is a competitive solution. Also, sketch the proof of the assertion that the main-simple *V*-set, in association with the appropriate minimal winning coalitions, is also a competitive solution.
11. Suppose three people, 1, 2, and 3, must use majority rule to choose a rule for dividing $1,000 among themselves, and suppose that only two rules are available: (1) face-to-face bargaining in which persons 1 and 2 have two votes and person 3 has one vote, and (2) a procedure whereby 1 makes a proposal in which he or she receives $500 and in which 2 and 3 divide the remaining $500 in units of $250 (i.e., 2 gets either $500 or $250 or $0). If the proposal is accepted by 2 or 3, it is implemented. If it is rejected by both 2 and 3, then player 3 can make a similar proposal (i.e., $500 for person 2, with persons 1 and 3 receiving the remainder in units of $250). If it is rejected, the outcome (333, 333, 333) is implemented. What is the final outcome?
12. Describe the *V*-set and Competitive Solution to the following vote-trading game (excluding lotteries, letting the defeat of each bill be worth zero to each legislator, and assuming the payoffs across bills are separable):

Legislator	A	B	C	D	E	F
1	3	3	2	−4	−4	2
2	2	−4	−4	2	3	3
3	−4	2	3	3	2	−4

13. Consider the preferences of the following four-person committee over three bills, A, B, and C:

L_1	L_3	L_5	L_7	U_i
A	B	C	A	2
B	C	A	B	1
C	A	B	C	0

Suppose the committee can pass one and only one of the three bills. In order for a bill to pass, it must receive at least two-thirds of the vote. Voting is weighted so that L_1's vote counts once, L_7's vote counts seven times and so on. Legislators are strategic, vote simultaneously, vote for only one bill, and are free to communicate and form cooperative agreements. A legislator receives a payoff of 2 if his most-preferred alternative passes, 1 if his second-most-preferred alternative passes and 0 otherwise. Utility is transferable and infinitely divisible. All cooperative agreements are costlessly enforceable.

a. Which, if any, of the bills pass?
b. If L_1's vote is counted three times instead of once, how does your answer to part (a) change?

14. Bicameralism (a two-chamber legislature) is often assumed to facilitate stability relative to a unicameral legislature. Consider a spatial situation with three legislators in each of two chambers, and assume that all preferences are given by simple Euclidean distance. Indifference contours are circles.

    a. Construct a 2-issue example in which, under simple majority rule, there is no core within each chamber, but there is a core if the two chambers are combined into a single unicameral legislature.
    b. Can there be a circumstance in which the status quo is a core under bicameralism, by which we mean that it cannot be upset by the same alternative in both chambers, but can be upset if the two chambers are combined?

Appendix

Chapter 1 Exercises

Answer Key

1. Portray a utility function for money in which a person is risk averse for small amounts and risk acceptant for large amounts.
   Answer: Risk aversity implies that the utility function is concave such that additional amounts for money have diminishing returns. Risk acceptant preferences, on the other hand, imply convexity. Therefore, the utility function should be concave for smaller amounts and convex for larger amounts. A utility of the form $U(x) = (x - a)^3 + b$ would satisfy these conditions.
3. Consider the following six strict preference orders over three alternatives:

Order 1	Order 2	Order 3	Order 4	Order 5	Order 6
A	A	B	B	C	C
B	C	A	C	A	B
C	B	C	A	B	A

   If you are told that a group of people all simultaneously have single peaked preferences over a single issue, can all six preference orders coexist simultaneously; and if not, what are the various subsets of preference orders that can simultaneously describe individual preferences?
   Answer: The six preference orders cannot coexist simultaneously, given the single peaked preference assumption. The order ABC is not feasible because of ACB and CAB, ACB because of ABC and BAC, BAC because of BCA and CBA, BCA because of BAC and ABC, CAB because of CBA and BCA, and CBA because of CAB and ACB. However, for each possible ordering of the three alternatives on the issue space, one can have a set of four preference orders. For ABC: {ABC, BAC, BCA, CBA}, for ACB: {ACB, BCA, CAB, CBA}, for BAC: {BAC, ABC, ACB, CAB}, for BCA: {BCA, ACB, CAB, CBA}, for CAB: {CAB, ABC, BAC, ACB}, for CBA: {CBA, ABC, BCA, BAC}.
5. Assume 7 people have circular indifference contours preference orders in two dimensions and assume that for the three alternatives, A, B and C, that B is a Condorcet winner. Locate 7 distinct ideal points consistent with this fact.

Answer: The six preference orders given in question #4, with the addition of the following Order # 7: B > A > C, leads to B to be the Condorcet winner.

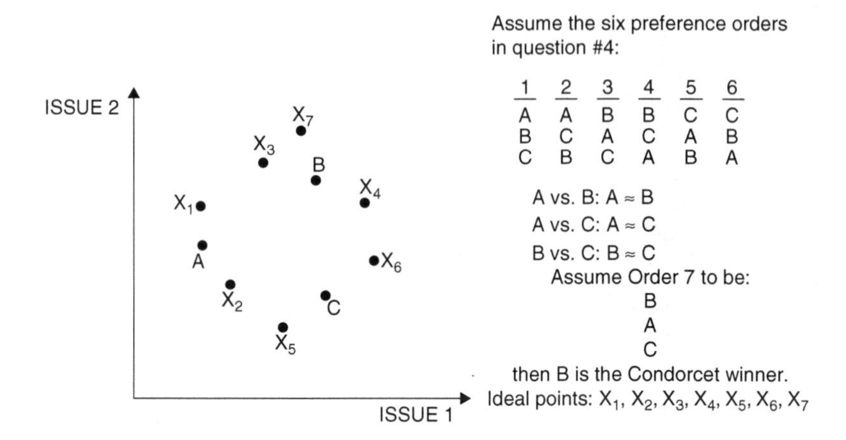

7. Can a person with a single peaked preference be risk acceptant between two alternatives and at the same time risk averse with respect to two other alternatives?
Answer: Yes, the following two graphs are examples of preferences that are single peaked, and that display different attitudes toward risk for different intervals of x.

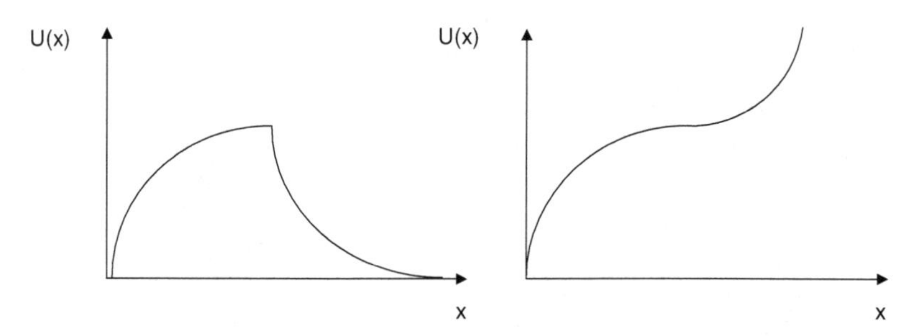

Chapter 2 Exercises

Answer Key

1. Suppose persons 1 and 2 make a sequence of binary decisions, first 1, then 2, then 1, then 2, and so forth, and suppose that all decisions are observed by both persons. Portray an extensive form to represent a situation in which 1 has perfect memory but 2 can only recall the last move of 1.

Answer:

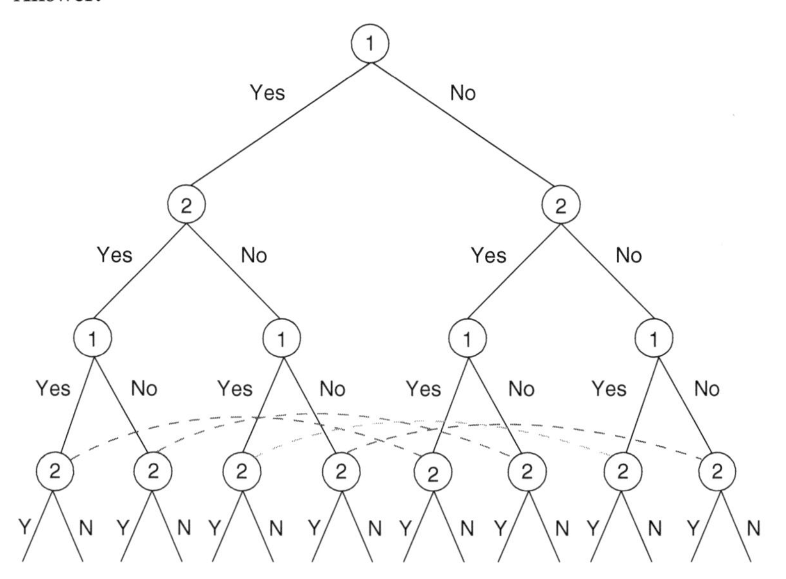

3. Draw the extensive form of the agenda "A versus B, the winner against C" for a three-person legislature in which legislator 3 observes 1's choice, but in which no other legislator observes any other choices.
Answer:

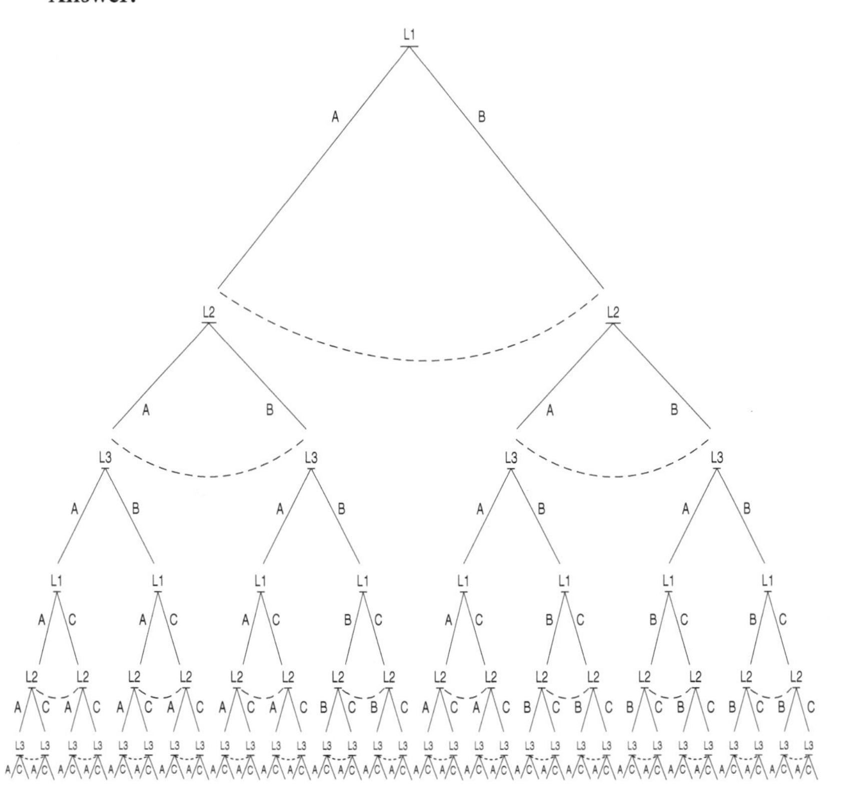

5. Referring to problem 4, suppose you must make your decision before the bureaucrat chooses the program in which to bury the cost overrun and that the bureaucrat can pay you two units to learn beforehand how you intend to assign your auditors. Describe the extensive form for this situation.
Answer:

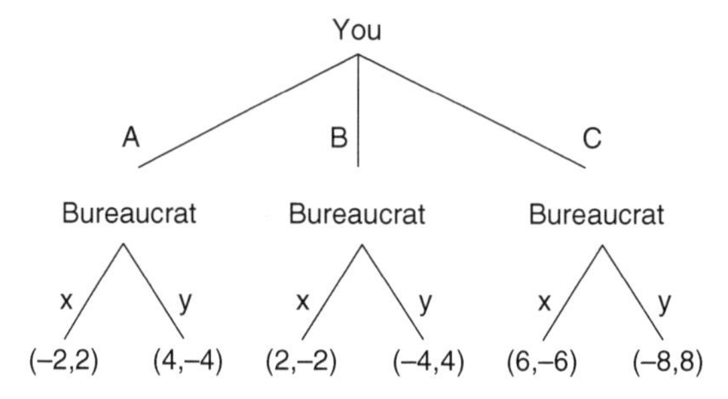

7. If free competition reigns in an industry with two firms, each firm sells 20 million units of that industry's products at a net profit of $1/item. But if they collude to set a higher price, each sells 15 million units at a net profit of $2 each. If one firm defects to the lower competitive price, its sales soar to 35 million units while the other firm sells nothing. Before each firm sets its price (which they do simultaneously) Senator Billie Bob proposes a licensing agreement whereby each pays a tax of $.20/item to produce the product at the fixed cartel price—ostensibly to insure that "destructive competition" does not "leave hard-working Americans unemployed." Construct this situation's extensive form, where each firm must first approve or disapprove of the licensing arrangement, which goes into effect only if both firms agree to it.
Answer:

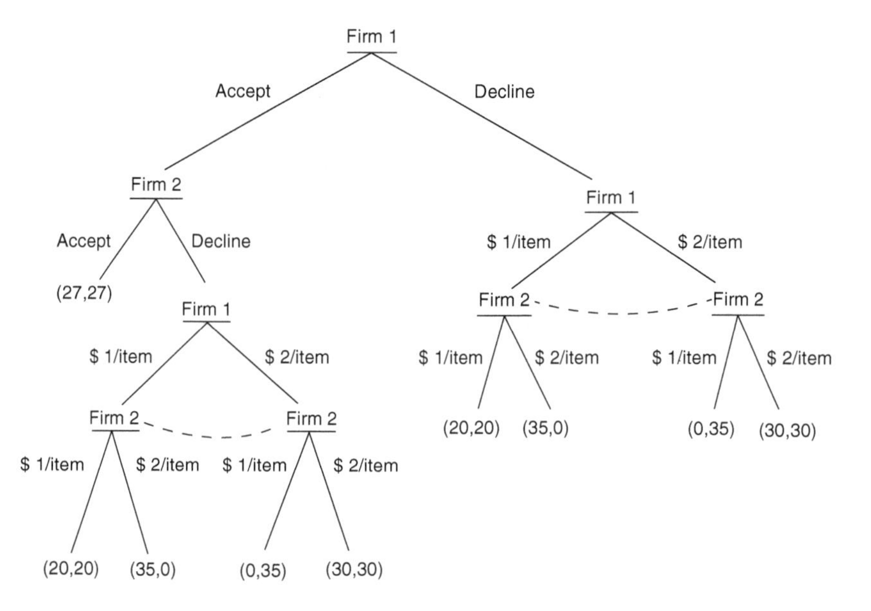

9. Assume the following preferences by a five-member committee, all of whom are sophisticated voters.

1	2	3	4	5
A	B	C	C	B
B	C	A	A	C
C	A	B	B	A

You are voter 1. Each voter receives a payoff of $100 from his or her first choice, $50 from a second choice, and $0 from a last choice. You and voter 2 are competing to set the agenda, but who will do the setting depends on who can "buy" voter 4. You and voter 2 must announce how much of your winnings you will pay, after which voter 4 will choose between you and 2 to be the setter. Because of your sterling character, an indifferent voter 4 will choose you. To simplify matters, suppose that you and 2 must each bid $75, $25, or $0.

a. Assuming that money and utility are equivalent, what is your bid, and what is the final outcome?
b. How does your answer to part (a) change if 4 chooses 2 when indifferent?

Answer: a. Given the preferences of the committee members, a majority of the committee prefers A to B, B to C, and C to A. Since everyone is a sophisticated voter, if you are the agenda setter, you can secure your most preferred outcome, A, by using voting agenda ((A,C),B))—that is, pitting A versus C first, then pitting the winner against B. But if voter 2 secured the right to set the agenda, he would use agenda ((A,B),C) to secure B as the outcome.

Each player can either bid $75, $25, or $0 to win voter 4's favor. The following is the extensive form representation of the game.

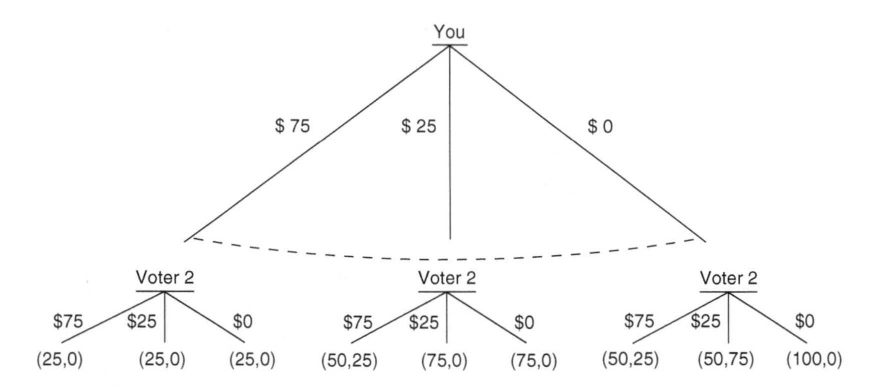

You bidding $25 and voter 2 bidding $75 is the unique equilibrium. B is the final outcome, and your payoff is $50 and voter 2's payoff is $25.

b. If voter 4 chooses voter 2 to set the agenda when indifferent, then the payoffs for the extensive form game are changed to the following:

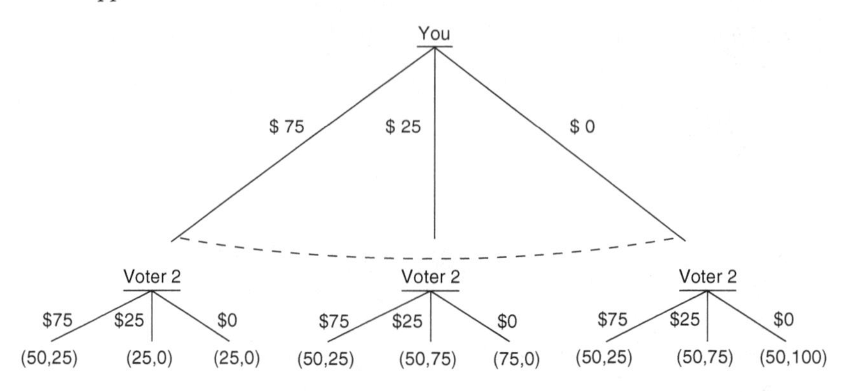

Now each player bidding \$25 becomes the equilibrium. So B is the outcome and the payoffs for voters 1 and 2 are \$50 and \$75, respectively.

11. Portray a congressional voting tree in which, with the alternatives B (bill), A (amended bill) and Q (status quo) already on the agenda, a predesignated legislator must decide before the actual balloting whether to introduce a substitute bill S. Assume that, regardless of whether or not the substitute is introduced, Congress must first decide whether to amend the original bill B.
Answer:

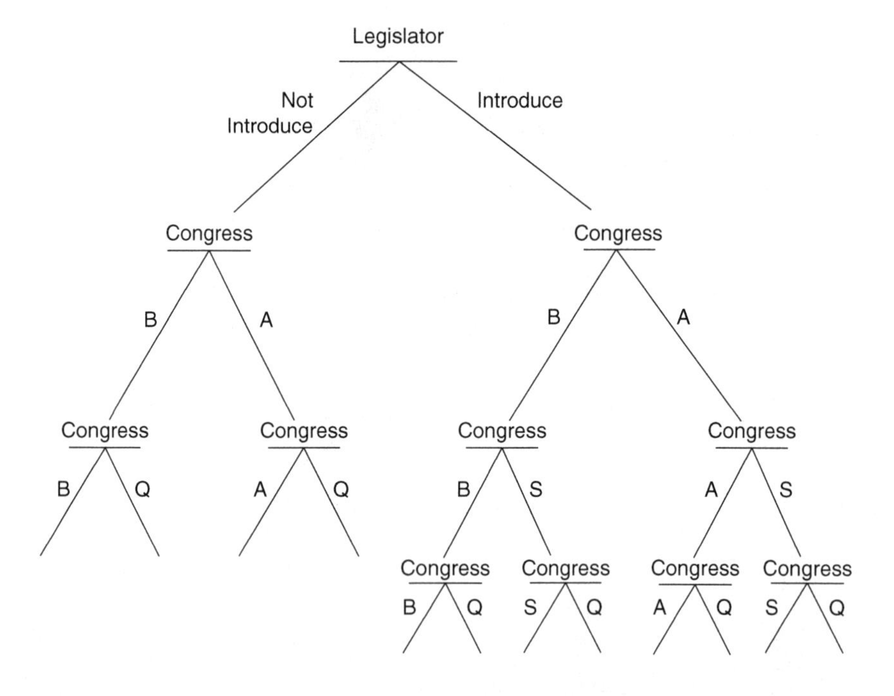

13. Suppose two countries, 1 and 2, each have second-strike nuclear capabilities in that they can retaliate after suffering a first strike or they can launch a second strike after sustaining a retaliation. Nature picks a

country at random, whereupon it, say 1, can launch a preemptive attack (p) against the other's military capabilities. If 1 attacks, 2 can retaliate (r) or capitulate (c). But if 2 retaliates, 1 can choose between launching (l) its second-strike (which 2 no longer has) that moves 2 back to the stone age or it can do nothing (~l). On the other hand, if 1 does not launch a preemptive attack (~p), then the first move is 2's, with 1 and 2's roles reversed. The game ends if neither country attacks, after one country capitulates, or after one country launches its second strike following a retaliation. Assign some "reasonable" payoffs to the outcomes and determine the eventual outcome. What are the preferences that make a preemptive attack inevitable?

Answer: Suppose countries prefer to launch their second strike (over not launching), given that they have done a preemptive strike and there was retaliation. That is, given retaliation, a state does not feel disheartened by the fact that a second strike would annihilate a country and wants to take further action. Under this scenario, retaliation is not a best response—because it leads to further destruction—and countries prefer to capitulate after the initial strike. Given a country's tendency towards capitulation, a preemptive strike becomes inevitable if the country prefers to attack and show dominance through capitulation over the status quo. Thus, a preemptive attack becomes inevitable if the other country's best response is to capitulate instead of retaliate because of fear of a second strike, and the country prefers to the outcome of a preemptive attack and capitulation to status quo because of a need to display an act of power.

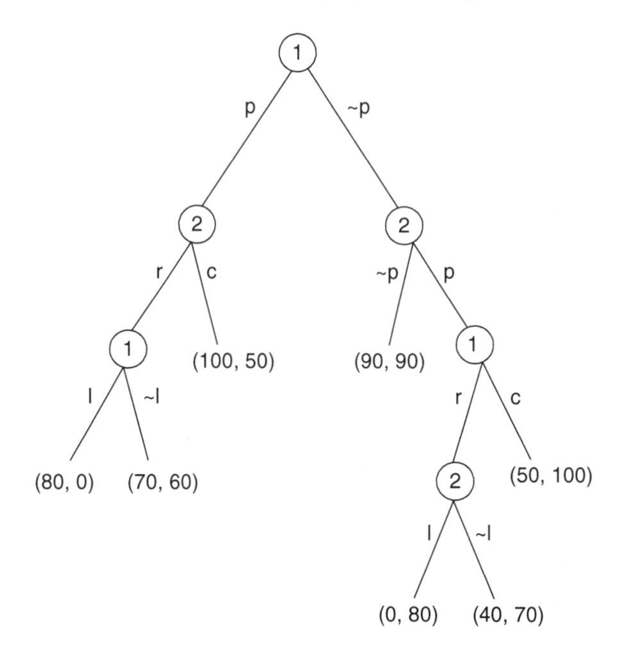

Chapter 3 Exercises

Answer Key

1. Going into the home stretch of a research and development project, you are six months ahead of the competition. To bring the project to completion requires finishing the development stage, and you have two strategies: Risky and Safe. Safe takes two years but is guaranteed to work. Risky takes only a year, but there is a 50% chance it will get you nowhere, in which case you will have to return to the safe strategy and take an additional two years. In six months, your competitor (who cannot observe your decision) will confront an identical choice between Safe and Risky strategies with the same properties as your alternatives, and they too must switch to Safe if their Risky choice fails. Only the first to complete is awarded the patent, and due to limited resources it is not possible to pursue both strategies simultaneously.

   a. Which strategy should you pursue to maximize your chance of winning?

   Answer: The strategic form of the game can be seen below with the relative probabilities of winning and losing the project.

		Opponent	
		Risky	*Safe*
You	*Risky*	$\frac{3}{4}(win)+\frac{1}{4}(lose), \frac{3}{4}(lose)+\frac{1}{4}(win)$	$\frac{1}{2}(win)+\frac{1}{2}(lose), \frac{1}{2}(win)+\frac{1}{2}(lose)$
	Safe	$\frac{1}{2}(win)+\frac{1}{2}(lose), \frac{1}{2}(win)+\frac{1}{2}(lose)$	*win, lose*

If both you and the opponent play *Safe* then you will win the project as you will be done within 2 years whereas your opponent will be done in 2.5. If you play *Safe* and your opponent plays *Risky,* then the outcome depends on your opponent as you will be done in 2 years and there is 50% chance that your opponent will be done in 1.5 (you lose) and 50% chance in 3.5 (you win). Similarly, if you play *Risky* and your opponent plays *Safe*, then for sure he will be done in 2.5 years and there's 50% chance that you'll be done in 1 year (you win) and 50% chance that you'll be done in 3 years (you lose). If both players play *Risky* then you have the advantage in terms of the probability of winning as you have the six-month head start in the project.

There is no pure strategy Nash equilibrium. To find the mixed strategy equilibrium, suppose the opponent plays *Risky* with a probability q, and *Safe* with a probability $(1 - q)$. Additionally, assume for simplicity that when a player loses they get zero and if they win the project they get 1 as utility. The expected return for each strategy is:

$$E_{YOU}(Risky) = E_{YOU}(Safe)$$

$$q\left(\frac{3}{4}\right)+(1-q)\left(\frac{1}{2}\right)=q\left(\frac{1}{2}\right)+(1-q)(1)$$

$$\frac{3}{4}q+\frac{1}{2}=1-\frac{1}{2}q$$

$$q\leq\frac{2}{3}\rightarrow \textit{Safe and } q\geq\frac{2}{3}\rightarrow \textit{Risky}$$

$$E_{OPPONENT}(\textit{Risky})=E_{OPPONENT}(\textit{Safe})$$

$$p\left(\frac{1}{4}\right)+(1-p)\left(\frac{1}{2}\right)=p\left(\frac{1}{2}\right)+(1-p)(0)$$

$$\frac{1}{2}=\left(\frac{3}{4}\right)p$$

$$p\geq\frac{2}{3}\rightarrow \textit{Safe and } p\leq\frac{2}{3}\rightarrow \textit{Risky}$$

Therefore the mixed strategy Nash equilibrium becomes {(2/3, 1/3), (2/3, 1/3)}.

b. Do you want to keep your move hidden?

Answer: Yes. Suppose I let my opponent know that I am playing *Risky*; then he will choose to be *Safe*, and my probability of winning the project is $\left(\frac{1}{2}\right)$. Similarly, suppose I let my opponent know that I'm playing *Safe*; then he will play *Risky*, and again my chance of winning the project is $\left(\frac{1}{2}\right)$. Assuming utility of winning is equal to 1 for simplicity, my expected return from telling my move to my opponent is $\left(\frac{1}{2}\right)$. We need to compare this value to the utility I would get from hiding my move, which is obtained by computing $E_{YOU}(\textit{Risky})=E_{YOU}(\textit{Safe})$ = $1-\frac{1}{2}q=1-\frac{1}{2}\left(\frac{2}{3}\right)=\frac{2}{3}$, which is greater than $\left(\frac{1}{2}\right)$.

3. Consider the following scenario: With one opportunity to bet remaining in the game show, player A has $7,200, player B has $5,000, and player C has $3,601. Assume that there is no benefit to being second versus third, and that the player with the most money wins that amount of money in cash. Each player must decide whether to bet *All* or *Nothing*. Prior to betting (which they must do simultaneously), nature tosses a fair coin to determine which "state of the world" will pertain: in State 1, players A and B win their bets, but C loses; in State 2, players A and B lose their bets, but C wins. Assume that if a player bets "all" and wins, his wealth is doubled. If he loses, his wealth is zero. If a player bets nothing, his wealth does not change. The player with the most money at the end of the game wins. Draw this situation's extensive form and show what each player does in equilibrium.

Answer: Each player has the following strategy set $S_A = S_B = S_C = \{All(A), Nothing(N)\}$. Under State 1, *A* is a strictly dominant strategy for Player A, *A* is a weakly dominant strategy for Player B, and Player C is indifferent between the two strategies. Under State 2, *N* is a weakly dominant strategy for both Player A and B, and *A* is a strictly dominant strategy for Player C. Since State 1 and State 2 occur with equal probability, betting All gives player C a higher expected payoff regardless of Player A and B's strategies. Given that player C will bet All, betting All is the weakly dominant strategy for players A and B. In equilibrium, all three players will bet All. The extensive form of the game is represented in the following.

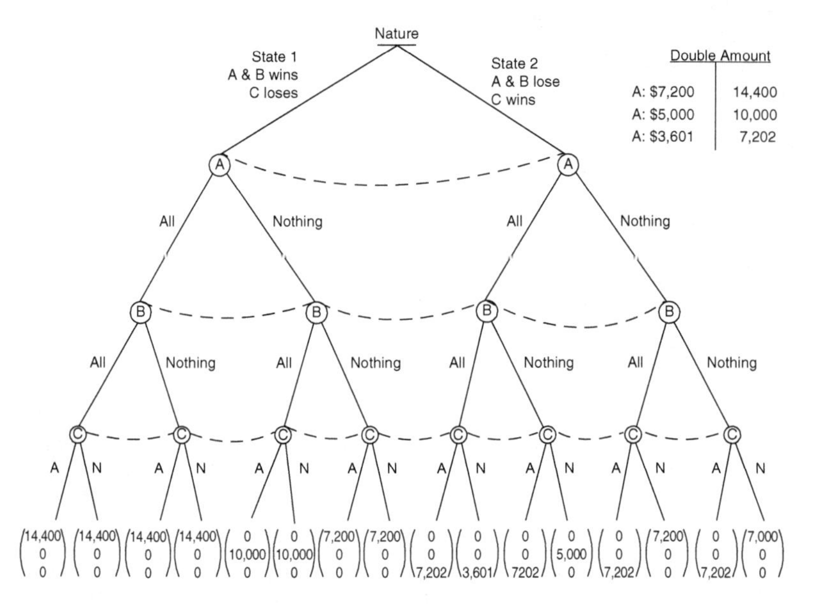

5. Prior to playing a game, a fair coin is tossed to determine which of the following two 2 × 2 games will actually be played (the first number in each cell denotes your payoff—row chooser—whereas the second number denotes your opponent's):

	b_1	b_2
a_1	9, 5	1, −3
a_2	3, 7	4, 6

	b_1	b_2
a_1	3, 0	9, −2
a_2	9, 9	3, 8

a. Assuming that neither person observes the outcome of the coin toss and that both of you must choose simultaneously, portray the situation's extensive form.

Answer:

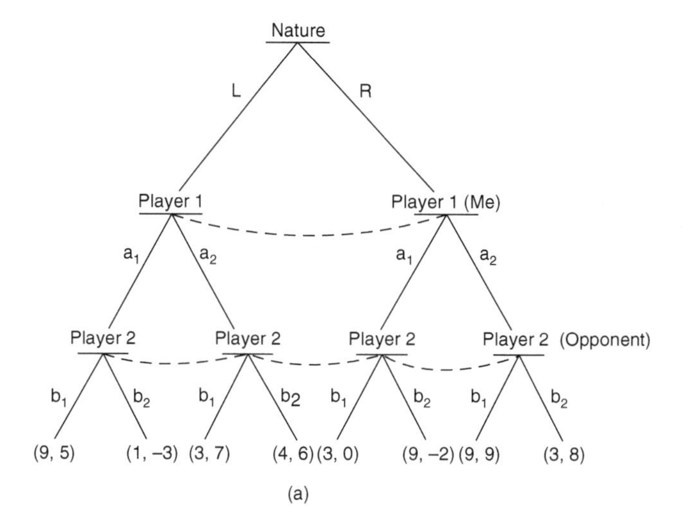

(a)

b. Portray the extensive form, assuming that you can secretly pay \$3 to learn the outcome of the coin toss.

Answer:

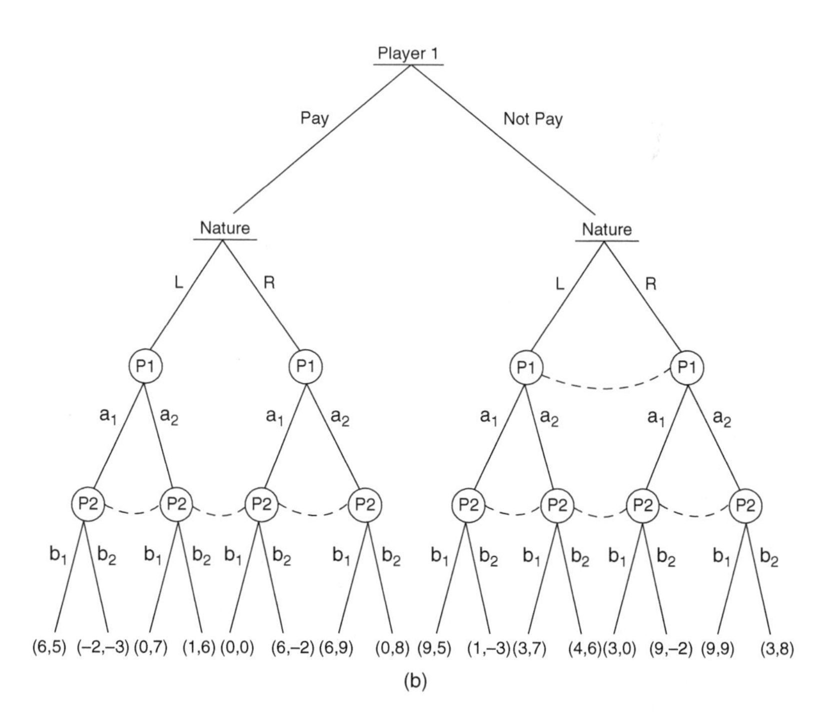

(b)

c. If neither player observes the outcome of the coin toss, how much would you pay to learn the outcome of the toss (assume that all payoffs are in terms of dollars and that utility and money are equivalent) and for what information cost is 2 indifferent between learning and not learning the outcome of the coin toss?

Answer:

For both games, the dominant strategy for Player 2 is to play a_1. Therefore the best response for Player 1 under Game 1 is to play a_1, and under Game 2 is to play a_2. For Player 1, the expected return for playing a_1 is $(1/2)9 + (1/2)3 = 6$ and the expected return for playing a_2 is $(1/2)3 + (1/2)9 = 6$. Therefore, if the coin toss is unknown to Player 1, she expects to receive 6; meanwhile, if she learns the outcome of the toss, she will receive 9 for sure as she will play her best response. Thus, she would be willing to pay 3 maximum to learn the outcome of the toss. Because playing b_1 is the strictly dominant strategy for Player 2, the information cost for her would be zero as she is already playing the optimal strategy available.

7. Let two political candidates each have two strategies, as indicated in the table below, and let the probability that candidate 1 receives specific pluralities from the four possible joint strategy choices also be as indicated. If the objective of both candidates is to maximize their probability of winning the election, what, if any, is the equilibrium to this game? How does this equilibrium change if both candidates seek to maximize their expected plurality? Note that $p(x)$ is the probability of winning a plurality of x votes.

	b1	*b2*
a1	$p(0) = 5/8$ $P(400) = 1/8$ $P(800) = 2/8$	$p(8) = 1/2$ $P(0) = 1/8$ $P(-80) = 3/8$
a2	$P(-400) = 1/2$ $P(400) = 1/2$	$P(0) = 3/4$ $P(-75) = 1/8$ $P(-125) = 1/8$

Answer:

If both candidates maximize the probability of winning, then (a2,b1) is the unique equilibrium.

	b1	b2
a1	3/8, 0	1/2, 3/8
a2	1/2, 1/2	0, 1/4

But if both candidates maximize the expected plurality, then (a2,b2) is the unique equilibrium.

	b1	b2
a1	250, –250	–26, 26
a2	0, 0	–25, 25

9. Recall the game from elementary school called "one, two, three, shoot!" One of the players chooses "even" and the other player gets "odd." On the count of three, each of the two players simultaneously casts out either one or two fingers. If the total number of fingers is even, then the "even" player wins, while if the sum is odd, then the "odd" player wins. Suppose the payoff is 1 for the winner and – 1 for the loser.
   a. Show the extensive form for this game.

   Answer:

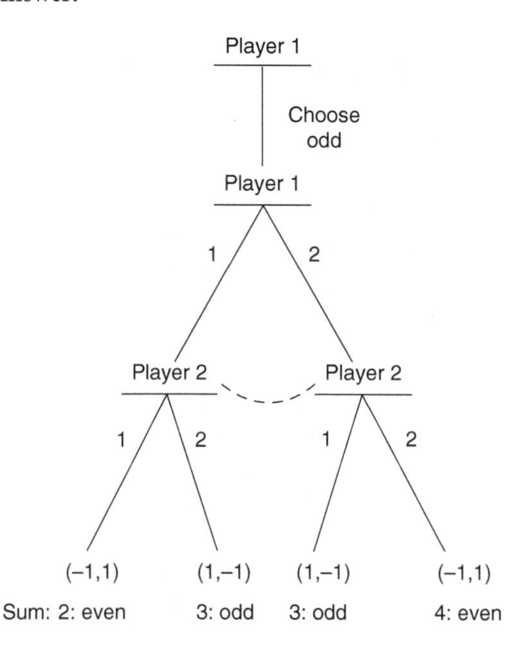

   b. Show that this game has no pure strategy equilibria.

   Answer: The strategic form is as follows and the best responses are highlighted in bold. There is no pure strategy equilibrium.

Player 1 (odd)		**Player 2** (even) 1	2
	1	–1, **1**	**1**, –1
	2	**1**, –1	–1, **1**

   c. Prove that "acting randomly" is the only equilibrium.

   Answer: We need to show that the mixed strategy equilibrium we find will be both players playing each strategy with 0.5 probability, and that they are randomly choosing their strategies.

   Suppose Player 2 plays 1 with a probability of q; then the expected returns for Player 1 are:

$$E_1(1) = E_1(2)$$

$$q(-1)+(1-q)(1)=q(1)+(1-q)(-1)$$

$$q^*=\left(\frac{1}{2}\right)$$

Suppose Player 1 plays 1 with a probability of p; then the expected returns for Player 2 are:

$$E_2(1)=E_2(2)$$

$$p(1)+(1-p)(-1)=p(-1)+(1-p)(1)$$

$$p^*=\left(\frac{1}{2}\right)$$

The mixed strategy Nash equilibrium is {(1/2, 1/2), (1/2, 1/2)}.

d. Suppose the rules of the game are changed so that at the count of three a person can hesitate. If both hesitate, each receives a payoff of 2 whereas if only one hesitates, that player loses 1 and the opponent gets 0. What is the outcome now?

Answer: The new game in its strategic form is:

		Player 2 (even)		
		1	2	hesitate
Player 1 (odd)	1	–1, **1**	**1**, –1	0,–1
	2	**1**, –1	–1, **1**	0,–1
	hesitate	–1,0	–1,0	**2, 2**

There is now one pure strategy Nash equilibrium (hesitate, hesitate) with payoffs (2,2).

11. In a two-player game, the following pairs of actions lead to the following outcomes:

(Player 1's Action, Player 2's Action)	$\rightarrow$ *(1's Utility, 2's Utility)*
(L, L)	*(3, 6)*
(R, R)	*(1, 5)*
(C, C)	*(6, 8)*
(L, R)	*(9, 6)*
(R, L)	*(9, 9)*
(L, C)	*(6, 1)*
(C, L)	*(2, 2)*
(R, C)	*(4, 4)*
(C, R)	*(4, 3)*

a. What is the outcome of this game if 1 goes first and 2 observes 1's action and then chooses his own action?

Answer: Player 2, given L will play either L or R, given C will play C, given R will play L. Player 1 will play R with the highest payoff (9,9). The equilibrium is (R,L).

b. What is the outcome if 2 goes first and 1 observes 2's action and then chooses her own action?

Answer: Player 1, given L will play R, given C will either play L or C, given R will play R. Player 2 will play L and the equilibrium is the same as (a).

c. What is the outcome if each player has to choose an action before he or she knows the other player's action?

Answer:

		Player 2		
		L	C	R
Player 1	L	(3, 6)	(6, 1)	(**9**, **6**)
	C	(2, 2)	(**6**, **8**)	(4, 3)
	R	(**9**, **9**)	(4, 4)	(1, 5)

There are three pure strategy Nash equilibria: (L,R), (C,C), and (R,L).

d. A third player enters the game. If the third player chooses L, he gets the average of the payoff to players 1 and 2 for any given outcome. If the third player chooses R, he gets a payoff equal to whichever of player 1's payoff or player 2's payoff is higher. What is the outcome of the game where 1 goes first, 2 observes 1's action and chooses, and then 3 observes 1 and 2's actions and chooses?

Answer: The Nash equilibria are (R, L, L), (R, L,R), (L, R,R).

e. What is the outcome of the three-player game if all players must choose their actions at the same time?

		Player 3 chooses L			Player 3 chooses R		
		Player 2			Player 2		
		L	C	R	L	C	R
	L	(3, 6, 4.5)	(6, 1, 3.5)	(9, 6, 7.5)	(3, 6, 6)	(6, 1, 6)	(9, 6, 9)
Player 1	C	(2, 2, 2)	(6, 8, 7)	(4, 3, 3.5)	(2, 2, 2)	(6, 8, 8)	(4, 3, 4)
	R	(9, 9, 9)	(4, 4, 4)	(1, 5, 3)	(9, 9, 9)	(4, 4, 4)	(1, 5, 5)

Answer: The Nash equilibria are (R,L,L), (R, L, R), (C, C,R), and (L, R, R).

13. Let four individuals have the following net valuations over the three alternatives, A, B, and C:

person	A	B	C
1	30	0	50
2	45	65	0
3	10	20	45
4	50	35	0

Assume that each person must report a valuation for each alternative and that the alternative chosen is the one with highest summed valuation. Assume also that taxes are collected as described in Section 3.7.

a. How much incremental tax will be paid by each person?

person	A	B	C
1	30	0	50
2	45	65	0
3	10	20	45
4	50	35	0
Total	135	120	95

	Summed net valuations			
Excluded voter	A	B	C	**Incremental tax**
1	105	120	45	15
2	90	55	95	5
3	125	100	50	0
4	85	85	95	10

Answer: The project with the highest sum is A, which would be chosen. If voter 1 is excluded, then project B would be chosen; the difference between the valuation of A and B without voter 1's preferences is equal to 15. If voter 2 is excluded, then project C would be chosen and voter 2 pays the difference in net value between project A and C, without her preferences taken into account: 5. If voter 3 is excluded, it does not change the net value ordering of projects and therefore he does not pay any tax. If voter 4 is excluded, then project C is chosen and voter 4 pays 10.

b. Suppose persons 2 and 3 can hire an agent who will coordinate their responses (reported evaluations), including lies. Should they hire such a person if the fee is not too great?

Answer: No, under the incremental tax scheme telling the true preferences is the strictly dominant strategy. Additionally, for voter 3, project A is the most preferred project to begin with, so she cannot do better by claiming different preferences. For voter 2, even though B is the most preferred outcome, it is still optimal to vote truthfully. Suppose voter 2 claims to enjoy project B with net value of 85 so that the total project value of B increases to 140 and is chosen over A. The incremental tax associated with this decision then would be 40, and the net utility for voter 2 would be 65 (true preference) – 40 = 25, which is less than the truthful outcome where A is chosen and the utility of 45 received.

Chapter 4 Exercises

Answer Key

1. A (row chooser) and B (column chooser) must play the following 2 × 2 game:

8, −8	4, −4
2, −2	6, −6

Beforehand, B can pay four dollars to a third person to learn A's decision.

a. Draw the full extensive form, assuming that A does not know whether or not B purchased the information.

b. Portray the corresponding strategic form.

c. Solve the game for equilibrium strategies.

Answer:

a. The extensive form is as follows:

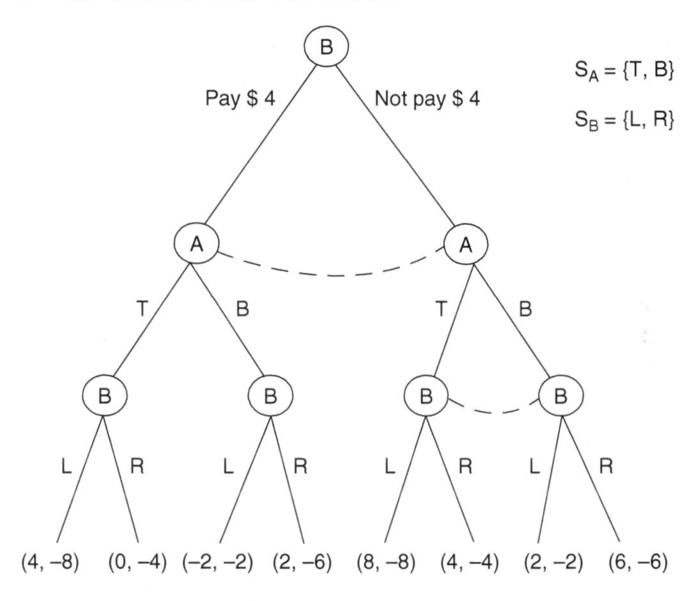

b. Player A plays T or B. Player B pays (Pay, Not Pay) X Strategy where strategy describes either (L,R) X (L,R) where he plays a certain strategy after L and R if he pays, or (L, R) if he does not pay.

	PLL	PLR	PRL	PRR	NR	NL
T	8, −12	8, −12	4, −8	4, −8	8, −8	2, −2
B	6, −10	2, −6	2, −6	6, −10	4, −4	6, −6

c. PLL and PRR are all strictly dominated by NL and PLR is strictly dominated by NR. PRL is strictly dominated by any convex combination of NL and NR. The equilibrium of this game is for player A to play T and B with probability ¼ and ¾ respectively, and for player B to play NL and NR with probability ½.

3. Does the following zero-sum strategic form game possess a pure strategy equilibrium (games within cells are played if that cell is realized as a result of players' initial strategies)?

<table>
<tr><th></th><th colspan="2">b_1</th><th colspan="2">b_2</th><th>b_3</th></tr>
<tr><td rowspan="2">a_1</td><td colspan="2" rowspan="2">10</td><td>1</td><td>3</td><td rowspan="2">6</td></tr>
<tr><td>4</td><td>2</td></tr>
<tr><td rowspan="2">a_2</td><td>7</td><td>10</td><td>10</td><td>20</td><td rowspan="2">6</td></tr>
<tr><td>6</td><td>5</td><td>5</td><td>10</td></tr>
<tr><td>a_3</td><td colspan="2">8</td><td colspan="2">5</td><td>5</td></tr>
</table>

Answer: The pure strategy equilibrium of this game is (a_2, b_3). First, we solve the three embedded games when (a_1, b_2), (a_2, b_1), and (a_2, b_2) are played. If (a_1, b_2) is played, then the column player will play the left and right strategies with probability q and 1 – q to induce row player's indifference, satisfying the equation:

$$-1q - 3(1 - q) = -4q - 2(1 - q)$$
$$q^* = 1/4$$

The expected payoff for the row chooser is 2.5. If (a_2, b_2) is played, then row player always plays top and column player always plays left, and the expected utility of the game is $(10, -10)$. If (a_2, b_1) is played, column player plays left and row player plays top and the expected utility of the game is $(7, -7)$. At which point, examination reveals that (a_2, b_3) is the pure strategy Nash equilibrium.

5. Suppose two presidential election candidates, who maximize their probability of winning, must decide how to allocate three days among six states. Whoever allocates the most time to a state wins that state and whoever wins the most electoral votes wins the election. The states' electoral votes are as follows: 27, 27, 24, 18, 2, and 2. Assume that all ties are broken by a coin toss and that transportation technology renders days non-divisible.
   a. Does the corresponding two-candidate election game have a Nash equilibrium in pure strategies?
   b. Does your answer change if the three largest states have the same electoral weight?
   c. Does your answer to part (a) change if days are divisible?

   Answer:
   a. In terms of probability of winning, allocating one day each to the three largest states, (1, 1, 1, 0, 0, 0), weakly dominates any other strategy.
   b. No, if the 3 states all have the same electoral weight, each player spending 1 day in each is still the equilibrium. A player deviating to spend another day in one of the large states would lead to no change in vote totals, and deviating to one of the smaller states would decrease that player's total.
   c. If days are divisible, then (1, 1, 1, 0, 0, 0) is not a weakly dominant strategy anymore. For example, it loses to (1.5, 1.5, 0, 0, 0, 0). It is fairly straightforward to show that there is no pure strategy equilibrium.

7. In addition to choosing a policy on the real line, suppose the candidates can also try to mask their positions by presenting themselves as lotteries—as probability distributions over the feasible policy space. Suppose that two parameters—the mean and variance—characterize a candidate's strategy (such as when that strategy corresponds to a normal density function), and that voter *i*'s utility for position x is $u_i(x) = -(x_i - x)^2$. Describe the two-candidate election equilibrium if all other provisions of the Median Voter Theorem apply.

   Answer: Given the utility function $u_i(x) = -(x_i - x)^2$, voters are risk averse and will vote for the candidate with minimal variance for a given position. At a given level of variance, a candidate at the median will defeat a candidate not at the median, by the median voter theorem. Therefore, the unique equilibrium is to set the policy at the median with zero variance.

9. Consider a nine-voter electorate in which voter *i*'s ideal point on the issue equals *i*. Suppose a two-candidate election is held and that all of the assumptions needed for the median voter result hold except that candidate I must choose voter 3's ideal point as his stated policy. Candidate II has no such restriction.
   a. Who wins the election?
   b. Below are four possible descriptions of the outcome. Decide which of the following statements are true.
      i. The median voter's ideal point must be the outcome.
      ii. The median voter's ideal point cannot be the outcome, but another voter's ideal point can be the outcome.
      iii. The median voter's ideal point cannot be the outcome and neither can any other voter's ideal point.
      iv. Either the median voter's ideal point or another voter's ideal point can be the outcome.

   Answer:
   a. II will win the election, as candidate I's best response function to all actions by candidate II is mandated to be 3. The best response function to 3 for candidate II is $(3 < x < 7)$ wherein candidate II will win.
   b. IV is the answer because (3, 5) is an equilibrium, but so is (3,4) and any other combination of 3 and a position that defeats 3.

11. Consider the following twenty-nine player game, where two players (I and II) are candidates and the other players, $(1, 2, \ldots, 27)$, are voters. Candidates will compete for office by using one of four possible campaign strategies (A, B, C, D) to recruit voters who will vote for them. The outcome of the election will be determined by which of the two candidates has recruited the most voters. In the first stage of the game, the two candidates simultaneously choose one of the four campaign strategies. For some combinations of campaign strategy choices (BB, CC, DB), the candidates must debate. Candidates choose debate strategies simultaneously, but afterward they both learn the other's campaign strategy. Below is a table that shows the results of different

combinations of campaign strategies. The cell values represent the number of voters recruited by candidate I (row chooser); 27 minus the cell value equals the number of voters recruited by candidate II (column chooser).

<table>
<tr><th></th><th>A</th><th colspan="2">B</th><th colspan="2">C</th><th>D</th></tr>
<tr><td>A</td><td>15</td><td colspan="2">3</td><td colspan="2">21</td><td>7</td></tr>
<tr><td rowspan="2">B</td><td rowspan="2">9</td><td>12</td><td>27</td><td colspan="2" rowspan="2">5</td><td rowspan="2">13</td></tr>
<tr><td>0</td><td>15</td></tr>
<tr><td rowspan="2">C</td><td rowspan="2">17</td><td colspan="2" rowspan="2">19</td><td>24</td><td>17</td><td rowspan="2">23</td></tr>
<tr><td>14</td><td>18</td></tr>
<tr><td rowspan="2">D</td><td rowspan="2">11</td><td>12</td><td>13</td><td colspan="2" rowspan="2">27</td><td rowspan="2">11</td></tr>
<tr><td>11</td><td>14</td></tr>
</table>

After the campaign, the winner of the election is determined by majority rule. Thus, the candidate that recruits a majority of voters, wins the election.

a. Which candidate wins the election?
b. If campaign laws are changed so that strategy D cannot be chosen, which candidate wins the election?

Answer:

a. If both players play B, then in the subgame, candidate I will play T and candidate II will play L and the outcome will be 12. If they both play C, they will mix to maximize their expected plurality and the expected supporters for candidate I will be approximately 17. If the outcome is DB, candidate I will play top and candidate II will play left and the outcome will be 12. At this point, candidate I has an option, C, which guarantees victory in the election.
b. Strategy C still determines victory in the election.

Chapter 5 Exercises

Answer Key

1. For what values of x is the following game a Prisoners' Dilemma?

x, 1	3, –4
1, 3	2, 3

Answer: Assume two strategies, C (cooperate) and D (defect), such that when the row chooser Player 1 and the column chooser Player 2 both play C, the payoffs are (2, 3). Therefore, the payoffs corresponding to strategies (D,C) (C,D) (D,D) become (3,–4), (1,3), and (x, 1), respectively. In order for the game to be a Prisoners' Dilemma, we need (D,D) to be an outcome that is less desirable than (C, C) but that is nevertheless played because of the higher returns from defection: $1 < x < 2$.

3. If free competition reigns in an industry, 20 million units of that industry's products will be sold by each firm at a net profit of $1 per item. But if they collude to set a higher price, each firm will sell 15 million units at a net profit of $2 each. If one firm defects to a lower price, its sales will soar to 35 million units while every other firm will sell nothing, and the creditors will begin to circle overhead. Senator Billie Bob proposes a licensing agreement whereby each member of the industry must pay a tax of $.20/item to produce the product at the fixed price that the cartel prefers—ostensibly to insure that "destructive competition" does not "leave hard-working Americans unemployed." What is the upper limit on how much money he can extract from each firm in the form of campaign contributions to his party?
Answer: Each firm has two strategies, either to sell at the market price, or the cartel price. The game in strategic form is shown below. If every firm sells at the cartel price the profits earned by each firm are $2 * 15 = 30$ million dollars. When sold at market price the profits earned by each firm are $1 * 20 = 20$ million dollars. If any firm defects into selling the goods at the lower market price, then the expected profits equal $1 * 35 = 35$.

	P_{market}	P_{cartel}
P_{market}	20, 20	35, 0
P_{cartel}	0, 35	30, 30

Destructive competition, that is, a firm lowering its price to increase its own goods' sales, can be avoided if the cartel price becomes the law. The senator wants to tax $.20/item at cartel price; therefore, the new amount of profits each firm collects becomes $1.8 * 15 = 27$ million dollars. Thus, the senator can extract, at most, $27 - 20 = 7$ million dollars from each firm as contributions as a substitute for not passing the tax law on cartel pricing.

5. You direct Consolidated Smoke and you must decide whether or not to agree to meet the president of Acme Sludge so that the two of you can fix prices for your similar products, in which case your corporations each earn $220 million. Both of you recognize however that the situation is a Prisoners' dilemma: at the market price, you each earn $90 million, while if only one defects from the agreement, his corporation earns $300 million and the other corporation earns "zip." Being competitive entrepreneurs with your MBAs, neither of you trusts the other to maintain any agreement reached. An additional danger is that federal antitrust investigators (with probability .4) will detect your agreement, negate the price fixing scheme, and impose a fine of $50 million each. Congressman I.M. Crass, however, proposes to offer legislation that will make the cartel legal and enforceable in a court of law, and that will provide the regulatory teeth to maintain it; he demands some assistance in the next election—say, $50 million from each firm. The problem is he wants his money up front, before the

legislature votes, and he can promise only a fifty-fifty chance that the proposed legislation will pass. Assuming that you make the best decision possible in the circumstances, what are your firm's expected profits?

Answer: Let us first look at the game independent of Congressman's proposal:

	P_{market}	P_{cartel}
P_{market}	90, 90	300, 0
P_{cartel}	0, 300	0.4*(90 – 50) + 0.6*(220), 0.4*(90 – 50) + 0.6*(220),

There is a 0.4 chance that price fixing will be detected, cartel pricing revoked and a $50 million fine will be paid if both players choose to stay in the agreement. The expected return is $148 million. The equilibrium is both firms selling at the market price (P_{market}, P_{market}).

The expected return from Congressman's offer is:

$$= 0.5(220) + 0.5(90) - 50 = 105.$$

There is 50% chance that the legislation will pass and each firm earns 220, and 50% chance that it will not pass and the game becomes as shown in strategic form above with an expected payoff of 90 due to both firms' defection. Since $105 > 90$, the optimal strategy is to pay the Congressman and hope that the legislations passes!

7. Two farmers must share an irrigation system, which they use by alternating their access to it day by day. The farmer whose turn it is to extract water on a particular day must choose between taking the allotted share (for a benefit of 0 to himself and a cost of 0 to the other farmer) versus taking more than the allotted share (for a benefit of B to the farmer in question and a cost of C to the other farmer). However, the farmer who must otherwise sit idly by for the day can choose to inspect his opposite number's activities at a cost to himself of K. If an excessive extraction is detected, the farmer is empowered to fine the offender an amount F, which can be kept as compensation for any economic injury.
   a. Assuming that all parameter values exceed zero, and taking a myopic one-day view, for what parameter values is there a pure strategy equilibrium in which the farmer inspects with certainty?

Answer: Let Farmer #1 be the player to decide whether to take the allotted share or more, and Farmer #2 to be the player who inspects. The game in strategic form then becomes:

		Farmer #2: Inspect	Farmer #2: Not Inspect
Farmer #1	Allotted Share	(0, –K)	(0, 0)
	More than Allotted Share	(B–F, –C–K+F)	(B, –C)

Given that $K > 0$, (*Allotted share, Inspect*) cannot be an equilibrium outcome, since *Not Inspect* leads to a higher payoff for Farmer #2. The only possible equilibrium with inspection is (*More than Allotted Share, Inspect*) and it is possible for the following values:

For Farmer #2: $F - K \geq 0 \rightarrow F \geq K$

For Farmer #1: $B - F \geq 0 \rightarrow B \geq F$

Therefore: $B \geq F \geq K$

b. Assuming that there is no pure strategy equilibria, what is the mixed strategy equilibrium?

Answer: Mixed strategy requires that the returns from playing either strategy are equal:

$$\text{Farmer \#2}: \quad E(\textit{Inspect}) = E(\textit{Not Inspect})$$

$$p(-K) + (1-p)(F-K-C) = p(0) + (1-p)(-C)$$

$$F - pF - K = 0$$

$$p \leq 1 - \frac{K}{F} \rightarrow \textit{Inspect}, p \geq 1 - \frac{K}{F} \rightarrow \textit{Not Inspect}$$

$$\text{Farmer \#1}: \quad E(\textit{Alloted Share}) = E(\textit{More than Alloted Share})$$

$$q(0) + (1-q)(0) = q(B-F) + (1-q)(B)$$

$$0 = B - qF$$

$$q \leq \frac{B}{F} \rightarrow \textit{More than Alloted Share}, q \geq \frac{B}{F} \rightarrow \textit{Alloted Share}$$

The mixed strategy equilibrium is $\left\{\left(1 - \frac{K}{F}, \frac{K}{F}\right); \left(\frac{B}{F}, 1 - \frac{B}{F}\right)\right\}$.

c. Suppose that one of the farmers is to be picked at random as the one to use the irrigation system, and suppose the parameters are set such that (take more than allotted share, inspect) is a pure strategy equilibrium. Can we raise the value of F so that the farmers prefer that there not be any pure strategy equilibrium over what they would expect to get from playing the game with the old parameter values?

Answer: Yes, we only need to increase F such that $B < F$. This leads to Farmer #1 choosing *Allotted Share* given Farmer #2 *Inspect*s, and there will be no pure strategy equilibrium.

9. Show that if the "normal" strategy in a society playing the Prisoners' Dilemma is TFT, and if the allowable mutants are STFT, ALLD (always defect), and TF2T, then TF2T can invade if the probability of ALLD is low compared to STFT, whereas the opposite is true if ALLD is more common.

Answer: A strategy will invade the "normal" strategy if the return from playing that strategy exceeds that of the "normal." Consider the following Prisoners' Dilemma game.

	Cooperate	Defect
Cooperate	10, 10	–5, 20
Defect	20, –5	0, 0

The return from playing the normal TFT strategy is:

$$E(TFT) = p_0 U(TFT, TFT) + p_1 U(TFT, STFT) + p_2 U(TFT, ALLD) + p_3 U(TFT, TF2T)$$

$$E(TFT) = p_0\left[\frac{10}{1-\delta}\right] + p_1\left[-5\frac{1}{1-\delta^2} + 20\frac{\delta}{1-\delta^2}\right] + p_2[-5] + p_3\left[\frac{10}{1-\delta}\right]$$

The return from playing TF2T is:

$$E(TF2T) = p_0 U(TF2T, TFT) + p_1 U(TF2T, STFT) + p_2 U(TF2T, ALLD) + p_3 U(TF2T, TF2T)$$

$$E(TF2T) = p_0\left[\frac{10}{1-\delta}\right] + p_1\left[-5 + 10\frac{\delta}{1-\delta}\right] + p_2\left[-5(1+\delta)\right] + p_3\left[\frac{10}{1-\delta}\right]$$

Since the returns with respect to p_0 and p_3 are the same, we need to compare the values that are multiplied with probabilities p_1 and p_2 in order to see if $E(TF2T) > E(TFT)$:

$$p_2[-5] \geq p_2[-5(1+\delta)]$$

Therefore the value associated with p_2 will always be greater for TFT, thus we need the value associated with p_1 to be greater for TF2T in order for $E(TF2T) > E(TFT)$:

$$p_1\left[-5\frac{1}{1-\delta^2} + 20\frac{\delta}{1-\delta^2}\right] \geq p_1\left[-5 + 10\frac{\delta}{1-\delta}\right]$$

With some rearrangement, we can see that the above inequality holds only when $\left[\frac{2}{3} < \delta\right]$. Thus, for values of $\delta < \frac{2}{3}$, TFT will be the surviving strategy against TF2T, for values $\left[\frac{2}{3} < \delta\right]$, TF2T will invade if $p_1 > p_2$, and TFT will survive if $p_1 < p_2$.

Chapter 6 Exercises

Answer Key

1. You are a member of a three-person committee that must choose one outcome from the list (A, B, C, D). Suppose the following preference orders describe the committee (from most to least preferred):

you:	*A*	*B*	*C*	*D*
member 2:	*D*	*C*	*A*	*B*
member 3:	*C*	*B*	*D*	*A*

Which of the following procedures would you prefer to see implemented if you believed that the other two members of the committee were sophisticated: (1) an agenda that first paired B against C, the winner against A, the winner against D; (2) an agenda that first paired C against A, the winner against D, and the winner against B; (3) an agenda that first paired B against D, the winner against C, the winner against A; or (4) you should not care which is chosen.
Answer: 4) Since C is the Condorcet winner, all possible agendas will lead to policy C; therefore, you are indifferent. If agenda 1 is chosen, by backwards induction D will beat A and lose to B and C in the final subgame, then B and C will defeat A in the next subgame, and C will defeat B leading to outcome C. If agenda 2 is chosen again, D defeats A, and C and B defeat D, in the final subgame, B defeats A and C defeats B in the next subgame, and C defeats A at the top of the tree leading to outcome C. If 3 is the agenda in the final subgame, A defeats B, and C and D defeat A. Up the tree, C defeats D and B, guaranteeing the outcome is C.

3. Suppose a majority of the legislature prefers A to B. If you are opposed to A, if A and B must be voted on first regardless of what amendments are introduced, and if everyone is a sophisticated voter, which alternative would you prefer to introduce: C or D? C creates the majority rule cycle "A preferred to B preferred to C preferred to A" while D defeats both A and B. Your preferences are "B preferred to D preferred to C preferred to A."
Answer: Introduce C. In the final subgame, C will defeat A and lose to B. Therefore, at the top of the tree, B will defeat A, leading to your favorite outcome.

5. A legislature (which we assume has three members) can consider four motions, where each motion affects the amount of money going to a legislator's district. Let the amounts (in thousands) to each district from each motion be as follows:

	District 1	District 2	District 3
A (status quo)	300	0	−400
B (committee bill)	500	−600	0
C (possible amend.)	0	800	−900
D (possible amend.)	−900	400	450

Motions A and B are on the floor as proposals and you must decide whether to propose an amended bill. If you propose C, the agenda will be C versus B the winner against A; if you propose D the agenda will be D versus B the winner against A. If D and C are both proposed, the agenda is D versus C, the winner against B, the winner against A.

a. Suppose you are the representative from district 2. Which amendment should you propose: C, which pays your district $800,000 or D, which pays $400,000?
b. Suppose you are chairman (and dictator) of the relevant legislative subcommittee, and that you can report out of your subcommittee either B or C or D as the bill that the legislature must consider. But you are also certain that whatever alternatives you fail to report out will be introduced on the floor as amendments. Thus,

If you report B, the agenda is "C versus D, winner versus B, winner versus A."
If you report C, the agenda is "B versus D, winner versus C, winner versus A."
If you report D, the agenda is "C versus B, winner versus D, winner versus A."

What would you choose as your bill if you were the representative from district 2?

c. With respect to part (b), if majority rule is used by the subcommittee to choose whether to report B, C, or D, what will it report?

Answer:

a. You choose D; if you chose amendment C the outcome will be B and your district loses $600,000. If you choose D, the outcome will be D and your district gains $400,000.
b. You are indifferent. C will never be the final outcome; no matter what order you introduce the bill, D will be the outcome.
c. It will be indifferent for the reasons discussed in (b).

7. A nine-member legislature using majority rule faces a budget that allows them to pass two of three proposed programs (A, B, C). The legislature has the following preference orders (ranked from most to least preferred):

Legislator								
1	*2*	*3*	*4*	*5*	*6*	*7*	*8*	*9*
B	A	B	A	C	C	B	A	C
C	B	A	C	A	B	A	C	B
A	C	C	B	B	A	C	B	A

You chair the legislature and are legislator 1. The voting procedure is as follows:

First, vote whether to keep or veto alternative B. If B is vetoed, the voting ends and alternatives A and C are implemented. If B is kept, then vote whether to keep or veto alternative C. If C is vetoed, the voting ends and alternatives A and B are implemented. If C is kept, then vote whether to keep or veto alternative A. If A is vetoed, the voting ends and alternatives B and C are implemented. If A is kept, one member of the legislature must choose which alternative should be vetoed.

a. As chairman, should you choose yourself to make this veto decision or legislator 4?
b. Assuming that legislator 4 makes the veto decision, design an agenda of the type illustrated in which B and C are nevertheless passed.

Answer:

a) The voting procedure can be illustrated by the following decision tree. If you yourself make the veto decision when none of the programs is vetoed, you will veto your least preferred program, A, and keep B and C. So voters are indifferent between vetoing or keeping A. Moving up the decision tree, program C will be vetoed because A,B are preferred to B,C by a majority of the voters. By the same reasoning, at the beginning of the decision tree, B will be vetoed because (A,C) is preferred to (A,B) by the majority. The final outcome is (A,C). If voter 4 is the veto player, it is straightforward to show that the outcome is again (A,C). So there is no difference between having you or voter 4 to be the veto player.
b) An agenda that produces (B,C) is as follows: keep or veto A first, then B, then C.

Chapter 7 Exercises

Answer Key

1. In late 1941 the British were concerned as to how to respond to an anticipated invasion by Japan of Thailand's (and Myanmar) Kra Isthmus and an eventual invasion of Malaya and the critical British outpost and naval base at Singapore. Thailand was ostensibly neutral and the critical issue that confronted Britain's strategic planners was whether they should attempt to forestall a Japanese occupation of Thailand by invading the country themselves beforehand and thereby block the ports and airfields Japan would most likely use in its move toward Malay and Singapore. As the situation has been described by one historian of the period, "If British forces only entered the Kra Isthmus after a Japanese incursion . . . they would probably set off a war with Japan. If [they moved] . . . before the Japanese entered Thailand . . . Tokyo could use this as a pretext for its own invasion. . . . [The British] had been told, possibly accurately, that the Japanese might even attempt to trick Britain into taking the first step . . . [in which case] Britain might lack for the support of the United States" (Evan Mawdsley, *December 1941*, New Haven: Yale University Press, 2011: 54). More specifically, as the British concern was most clearly stated by Churchill, "We should not

resist or attempt to forestall a Japanese attack on the Kra Isthmus unless we had a satisfactory assurance from the United States that they would join us should our action cause us to be involved in a war with Japan." Of course, this concern became moot in less than a week with Japan's attack on Pearl Harbor. Nevertheless, without assigning values to the possible outcomes, sketch out a description of this situation in extensive form, treating the preferences (responses) of the United States as not known with certainty by either Britain or Japan.

Answer:

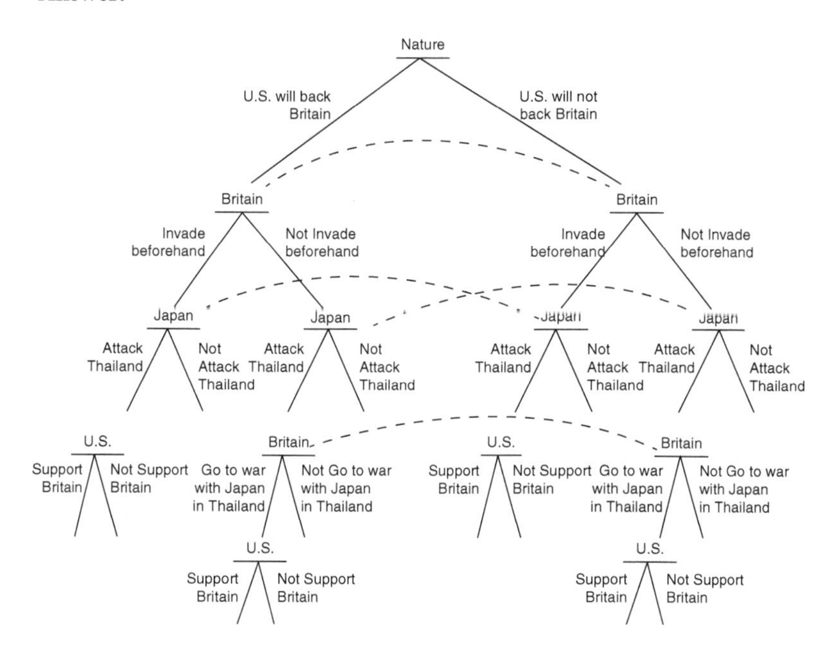

3. In his unsuccessful surgeries (those in which the patient dies) Dr. Ian Competent has only a fifty-fifty chance of not being at fault. With this in mind, the relatives of his latest victim have asked for compensation: $1,000,000. If Competent (who has already had his insurance policy revoked but who as a shareholder in a local savings and loan association is quite wealthy) refuses to settle, the relatives can take the matter to court (or they can forget it). Once in court assume that justice is done. (This is not an exercise that concerns the competence of lawyers.) So if Competent is innocent (and only he knows for sure), he loses nothing and the relatives lose $1,000,000 (attorney's fees being what they are). On the other hand, if he loses, then he loses $3,000,000 and the victim's relatives gain $2,000,000 (again, lawyers take their cut).
   a. Portray this situation's extensive and strategic forms.

Answer:

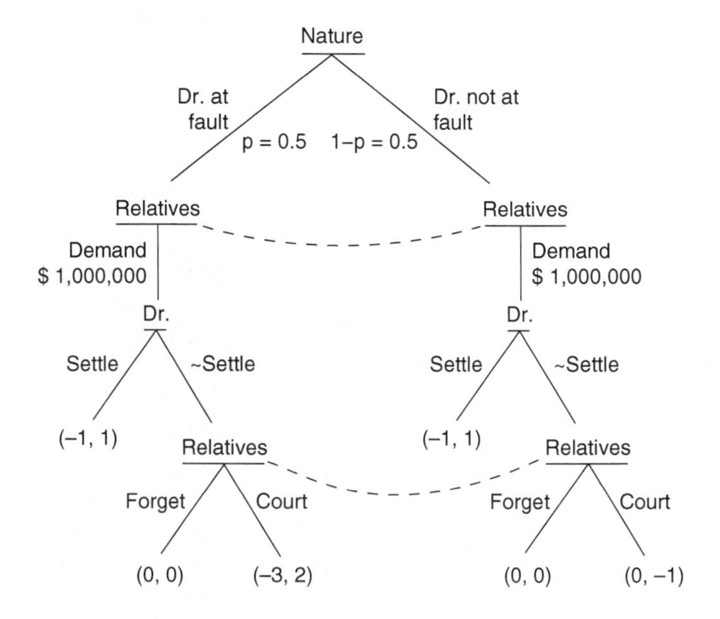

		Relatives	
	At Fault, Not At Fault	Court	Forget
Doctor	(Settle, Settle)	–1, 1	–1, 1
	(Settle, ~ Settle)	–0.5, 0	–0.5, 0.5
	(~ Settle, Settle)	–2, 1.5	–0.5, 0.5
	(~ Settle, ~ Settle)	–1.5, 0.5	0, 0

b. Determine the game's equilibrium.
Answer: For Dr. Competent, (Settle, Settle) and (~ Settle, Settle) are dominated strategies. In the reduced 2×2 game, there is no pure strategy equilibrium. It is straightforward to show that in the mixed strategy equilibrium the doctor will mix between his remaining strategies with equal probabilities, and the relatives will "forget" with two-thirds probability.
c. Interpret this equilibrium.
Answer: The equilibrium suggests that if the doctor is at fault, he will mix between settling and not settling with equal probabilities, and if the doctor is not at fault, he will never settle. On the other hand, the relatives will mix strategies too, going to court one-third of the time and forgetting about the situation two-thirds of the time.

5. Consider the following *sequential elimination* agenda: "Alternatives A and B are first paired. If A wins, it is the outcome; but if B wins, B is paired against C and the winner of this vote is the final outcome." Suppose that only these three preference types are possible (ranked from first preference to last):

   t_1: A B C
   t_2: B A C
   t_3: C B A

   Assuming that a person has type t_i preferences with probability p_i, that the p_i's are common knowledge, and that a person knows his own preferences, show that the selection of a Condorcet winner corresponds to an equilibrium.
   Answer:
   For voters with T_1 preferences, between alternatives A and B, voting for A weakly dominates voting for B because it might help A win, and even if it did not, it would not decrease B's chance of winning against C in the subsequent round. Similarly, for voters with T_3 preferences, voting for A, their least-preferred alternative, is a weakly dominated strategy. For voters with T_2 preferences, if the number of voters with T_3 preferences is greater than the number of voters with T_1 and T_2 preferences, then B defeats A in the first stage of voting regardless of whether they vote for A or B. If the number of voters with T_1 and T_2 preferences is greater than those with T_3 preferences, since B will defeat C in the second stage of the vote, voters with T_2 preferences have no incentives voting for A in the first stage of the vote. In other words, voting for B weakly dominates voting for A for voters with T_2 preferences. Because everyone votes sincerely, the Condorcet winner corresponds to an equilibrium.
7. Refer to section 7.8 and reproduce our analysis of the Centipede Game, assuming that each player's probability of irrationality is .3 rather than .03.

   Answer: This makes the payoffs for our 3 strategies:

	s_1	s_2	s_3
s_1	89.8, 38.2	102.4, 88.6	220, 55
s_2	105.2, 78.8	186.4, 109.6	304, 76
s_3	206, 104	248, 272	640, 160

   Inspection reveals that player 1 will always play s_3. Given this, player 2's best response function is to play s_2.
9. In an upcoming election on insurance rate reform, it is common knowledge that you will cast the decisive vote. You are uncertain about the identity of the reform's sponsor but have (correct) beliefs that there is a 7-in-10 chance that the reform is pro-insurance (INS) and as a consequence will raise your insurance rates (making you poorer) and that there is a 3-in-10 chance that the reform is pro-Consumer (CON), in which case the reform will keep your

insurance rates at their present level. A campaigner, who knows whether the reform is the insurance type or the consumer type, and, in either case, is paid only if the bill passes, must decide whether or not to go to your House to tell you to vote for the bill. (Campaigner chooses "House" or "No.") On election day, you must decide whether to vote "Yes" or "No" on the reform. The payoffs are determined as follows: It costs the campaigner \$5 to go to your house. The campaigner gets paid \$15 if the reform passes and \$0 if the reform fails. If either an insurance-type reform passes or a consumer-type reform fails, your rates go up—you lose \$10. If either a consumer-type reform passes or an insurance-type reform fails, your rates stay the same—you get \$0.

a. Draw the game's extensive form.
b. Specify the pure strategies available to each player.
c. Portray the situation's strategic form.
d. Find all the pure-strategy equilibria.

Answer:

a. The extensive form is as follows:

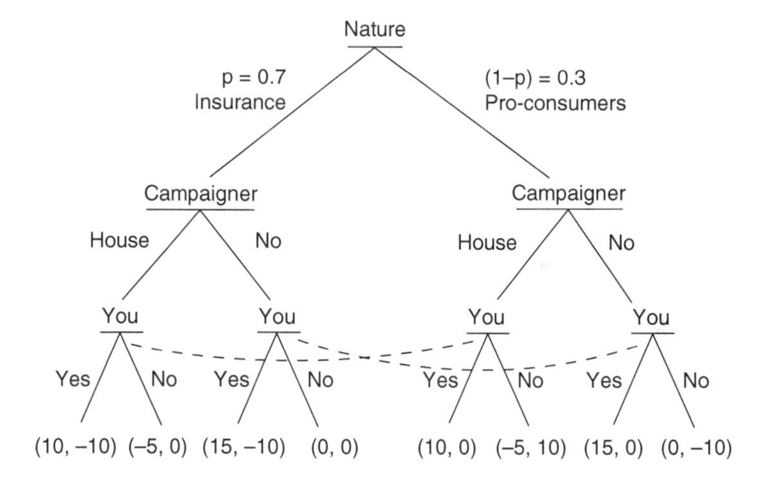

b. The campaigner has four strategies:

a_1: Visit the House regardless of whether the reform is pro-insurance.
a_2: Do not visit the House regardless of whether the reform is pro-insurance.
a_3: Visit the House only if the reform is pro-insurance.
a_4: Visit the House only if the reform is pro-consumer.

You also have four strategies:

b_1: Vote Yes regardless of whether player 1 visits the House.
b_2: Vote No regardless of whether player 1 visits the House.

b_3: Vote Yes if player 1 visits the House, and vote No if player 1 does not visit the House.

b_4: Vote No if player 1 visits the House, and vote Yes if player 1 does not visit the House.

c.

	b_1	b_2	b_3	b_4
a_1	10, −7	−5, −3	5.5, −10	0.5, 0
a_2	15, −7	0, −3	10.5, −10	4.5, 0
a_3	11.5, −7	−3.5, −3	7, −10	1, 0
a_4	13.5, −7	−1.5, −3	10.5, −10	4.5, 0

d. (a2, b4) and (a4, b4) are pure strategy equilibria.

Chapter 8 Exercises

Answer Key

1. Is vote trading in a legislature profitable and most likely to be observed when (a) various minorities find it in their interest to thwart the will of the majority? (b) there is no uniquely stable outcome, and it is possible for every agreement to be upset by something else? (c) congressional rules are sufficiently inflexible that only behind-the-scenes deals will accommodate the majority will? or (d) legislators are constrained by their constituents from voting sophisticatedly?
   Answer: b) There is no uniquely stable outcome, or Condorcet winner, by Schwartz's vote trading theorem.
3. Recall our example, from section 8.4, of the Valley of the Dump. Suppose, however, that there are only four clans (euphemistically named 1, 2, 3, and 4), each living on its own quarter of the valley, and that every day, the number of bags of garbage produced by each clan equals its name—clan *i* produces *i* bags of garbage per day. As before, the legal code of the valley, strictly enforced, requires each clan to produce its allotted bags of garbage, and it allows clan *i* to dump exactly *i* bags of garbage in its own backyard or the yards of others. However, the bags must be dealt with as distinct units—no scattering of any bag's contents is permitted. (They have high standards in the valley.) Assume that each clan evaluates the outcome of the garbage problem in the same way as in our original example.
   a. Define and interpret the conditions that must be satisfied if this game is to have a core.
   b. Does a nonempty core exist for this game?
   c. How do your answers change if we suppose that bags of garbage are infinitely divisible?

Answer:

a) Assume the existence of a distribution $(x1, x2, x3, x4)$ in the core and that the payoff from a bag of garbage on one's lawn is -1. Then it must be that $x1 + x2 + x3 \geq 4$, $x1 + x2 + x4 \geq 3$, $x1 + x3 + x4 \geq 2$, $x2 + x3 + x4 \geq 1$. These conditions state that for $(x1, x2, x3, x4)$ to be in the core, then it must guarantee that no three-clan coalition does any worse than have the garbage of the excluded clan on the coalition's lawns. Similar inequalities can be written for each two-person coalition. Finally, it must be the case that $x1 + x2 + x3 + x4 = -10$, since garbage cannot be exported from the valley.
b) Combining all three-clan constraints we have $3(x1 + x2 + x3 + x4) \geq 10$, and substituting in our total value, we have $-30 \geq 10$, at which point we have a contradiction, so the core is empty.
c) If the bags are infinitely divisible, we still need to satisfy the above constraints, which is impossible, so there is no change.

5. Consider the following system of representative majority rule among twenty-seven voters: There are three regions, each divided into three districts. Each district has three voters. There are representatives (who are computers, and are not to be considered part of the voting body) at the regional and district levels. District level representatives inherit (over any pair of alternatives) the majority preferences of the voters in their districts, and regional representatives inherit the majority preferences of the representatives in their region. A policy outcome is chosen by a majority vote of the regional representatives.
   a. What is the minimum number of voters that can form a decisive coalition (i.e., get their preferred policy positions as the outcome)? Construct an example to support your answer.
   b. How does your answer to (a) change if the district level representatives are eliminated?
   c. Assuming that all preferences are single-peaked over a single issue, does the final outcome depend on the existence of the district level representatives? Why?

Answer:

a) 8 voters are necessary to get majorities (of 2 to 1) in 2 of the 3 districts in 2 of the 3 regions.
b) Eliminating districts means 10 voters are necessary to get majorities in 2 of the 3 regions.
c) The existence of districts changes who the decisive voter is, and thus possibly the final outcome.

7. In economics, a common representation of utility for two commodities is the function $u_i(x_i, y_i) = x_i^{\alpha} y_i^{\beta}$ where both α and β are positive constants. So suppose two persons are bargaining over the allocation of ten units of each of the goods, where both goods are infinitely divisible. Let $\alpha = \beta = 2$ for both persons, and suppose that person 1's initial endowments of the two

goods are 2 and 7, respectively (so that 2's initial endowments are 8 and 3, respectively). What is the core to this two-person game?

Answer: Our answer depends on whether utility is transferable or not. If it is transferable, then $v(1) = 2^2 7^2 = 53$ and $v(2) = 8^2 3^2 = 73$. $v(1,2)$ is then calculated by maximizing the players' summed utility, $x^2y^2 + (10 - x)^2(10 - y)^2$, which occurs when either player is given all of both goods so as to yield a total of 10,000. The core is then the set of utility 2-tuples $\{(u_1, u_2) : u_1 + u_2 = 10{,}000$ and $v(1) \geq 53, v(2) \geq 73\}$. If, on the other hand, utility is not transferable, then given the symmetry between the two goods for both players, only the initial allocation corresponds to the core.

9. Suppose three houses, each with a large front-facing window, are arrayed in a triangle such that a person sitting in any house can see one-fifth of its own garden, all of the garden of the house on the left, and none of the garden belonging to the house on its right. Suppose each homeowner is endowed with one bag of fertilizer and is required by law to use it (suburban values being what they are). A bag of fertilizer, allocated to one garden, improves that garden so that a person having a full view of it enjoys a benefit of five units of utility, a person having only a one-fifth view enjoys a one unit utility increase, and a person with no view is not benefited at all. Assume that the homeowners can coalesce to allocate their fertilizer, and that people cannot fertilize another's garden without permission. Suppose also that if two bags of fertilizer are allocated to any one garden, then that garden is doubly beautiful. Does the game have a core?

Answer: The maximum summed benefit available to the occupants of the three houses here is 18 (6 units of benefit to each household), which occurs if each occupant fertilizes his or her own lawn. Thus, there exists an allocation of the fertilizer that yields a payoff vector of (x_1, x_2, x_3) such that $x_1 + x_2 + x_3 = 18$. Each coalition of two houses, on the other hand, can guarantee themselves a payoff of 17, by putting two bags in the leftmost garden and forbidding the excluded player from fertilizing the rightmost member's lawn (assuming that the excluded member responds rationally by fertilizing his or her own lawn). We thus have:

$$x_1 + x_2 \geq 17$$
$$x_2 + x_3 \geq 17$$
$$x_2 + x_3 \geq 17$$

which, if a core exists (if there is no two-household coalition that can upset any feasible outcome) requires that $2(x_1 + x_2 + x_3) \geq 51$. But substituting $x_1 + x_2 + x_3 = 18$ requires that $36 \geq 51$. Therefore, the core is empty.

11. Suppose three people, 1, 2, and 3, must use majority rule to choose a rule for dividing $1,000 among themselves, and suppose that only two rules are available: (1) face-to-face bargaining in which persons 1 and 2 have two

votes and person 3 has one vote, and (2) a procedure whereby 1 makes a proposal in which he or she receives $500 and in which 2 and 3 divide the remaining $500 in units of $250 (i.e., 2 gets either $500 or $250 or $0). If the proposal is accepted by 2 or 3, it is implemented. If it is rejected by both 2 and 3, then player 3 can make a similar proposal (i.e., $500 for person 2, with persons 1 and 3 receiving the remainder in units of $250). If it is rejected, the outcome (333, 333, 333) is implemented. What is the final outcome?

Answer: If option (1) is chosen, the fact that person 3 has only one vote is irrelevant as to which coalitions are winning—the same coalitions are winning if all three voters have the same voting weight. Hence, the V-set (as well as the Bargaining Set and Competitive Solution) is made up of 3 outcomes (500,500,0), (500,0,500), (0,500,500), and so two of the players will vote to divide all of the money between them. If option (2) is chosen, suppose 1's initial proposal is rejected by both 2 and 3, and that 3 must then make a proposal. If 3 proposes anything but (500, 0, 500) or (0, 500, 500)—if 3 offers (250, 250, 500)—it will be rejected because 1 and 2 both do better at (333, 333, 333). Player 1, then, clearly would not want to make an initial proposal that might be rejected by both players since in that event 1 cannot be certain as to whether it will get 0 or 500. Thus, 1 should propose either (500, 500, 0) or (500, 0, 500), and this proposal will be accepted by either player 2 or 3. In choosing a rule, then, player 1 prefers the second option but players 2 and 3 are indifferent so we cannot say definitively what the final outcome will be.

13. Consider the preferences of the following four-person committee over three bills, A, B, and C:

L_1	L_3	L_5	L_7	U_i
A	B	C	A	2
B	C	A	B	1
C	A	B	C	0

Suppose the committee can pass one and only one of the three bills. In order for a bill to pass, it must receive at least two-thirds of the vote. Voting is weighted so that L_1's vote counts once, L_7's vote counts seven times and so on. Legislators are strategic, vote simultaneously, vote for only one bill, and are free to communicate and form cooperative agreements. A legislator receives a payoff of 2 if his most-preferred alternative passes, 1 if his second-most-preferred alternative passes and 0 otherwise. Utility is transferable and infinitely divisible. All cooperative agreements are costlessly enforceable.

a. Which, if any, of the bills pass?
b. If L_1's vote is counted three times instead of once, how does your answer to part (a) change?

Answer:

a) A winning coalition needs 11 votes to pass a bill, therefore no bill passes without the support of L_7. Passing bill A yields a total group payoff of 5 (i.e., 2 + 0 + 1 + 2), B yields 4, C yields 3. Therefore all transfers of utility in the core are those that follow passing A and give legislator L_7 at least 2.

b) A majority now requires 12 votes, so L_7 remains essential to any winning coalition. Thus, the argument of part (a) still holds.

Index